The Time Museum

Catalogue of the Collection

Catalogue of the Collection

General Editor: Bruce Chandler

The Time

Museum

Volume 1

Time Measuring Instruments

Part 1

Astrolabes

Astrolabe Related Instruments

by A.J. Turner

Rockford 1985

THE TIME MUSEUM
Director: Seth G. Atwood
Curator: William Andrewes

Coordinated by
William Andrewes

Photographs by
Stephen Pitkin

Designed by
Lucy Sisman
Richard Weigand

Drawings by
David Penney

Typeset by
Talbot Typographics, Inc.

Printed by
Toppan Printing Co.

ISBN 0-912947-02-0

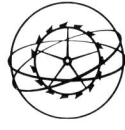

Introduction

What is time? Many are driven by the quest for an answer to this question. Although we never fully receive an answer, we accept the challenge and exciting adventure of trying to gain a better understanding. The thrill of momentary "revelations" that bring order to chaotic and miscellaneous thought comes to all of us from time to time.

Such a revelation struck me at the age of seventeen when I was a freshman at Carleton College in Northfield, Minnesota. During that year, 1934, I was especially intrigued by astronomy and philosophy. One evening in May, about 10:00 p.m., I wandered alone toward the arboretum and stretched out on a grassy hillside to enjoy a balmy, clear night with its sounds of field crickets and cicadas. During that night on the hillside, I experienced one of those choice moments of grasping a partial understanding of the universe. What was the universe all about? The temporary answer was very exciting and satisfying: In a pseudo-scientific way, I perceived that everything was somehow interrelated to everything else, and that with patience one could find this interrelationship by means of overlapping sine curves generated by plotting dominant characteristics. The universe was, then, a cohesive whole, unified through all of its interrelated parts. But what about the beginning or the end? I received no answers to this question, but my mind suddenly darted to another engaging question: What is time? Surely someone knows, I mused. Time has certainly been central to human history and involved with every phase of human existence. What did time have to do with space? Clearly, the concept of time was also involved with an understanding of space. What about those learned articles referring to relative time, slowing biological aging, and time dilation? The only thing I had learned from Einstein, at that juncture, was that nothing could go faster than the speed of light, and, therefore, I couldn't stop to see myself coming.

After this youthful questioning and groping for answers, I did not think about time seriously again until 1965, when I discovered that others who shared my query and quandary had failed to define time. St. Augustine's memorable statement made about four-hundred years after the birth of Christ stands today: "What then is time? If no one asks me, I know. If someone asks me to explain, I know not."

My readings on time naturally brought me into touch with the use and development of timekeeping devices. Pictures of these artifacts stirred early childhood memories of the wonders and mysteries of clocks and watches. Wouldn't it be fun, I thought, during my travels with my wife, to find and purchase a few of these pieces so that we could have a historical record? Nothing to it, I thought. How deceptive and naive! Assembling the collection turned into an adventure for us much like following a path with numerous branches and no apparent end.

Fortunately, before we bought anything, I established a single, clear objective: the purchase of only those pieces significant to the history of timekeeping devices. In this connection, even at the outset I bought only pieces of museum quality so that the collection would not have to be upgraded. Secondly, I bought nothing before I had studied horology for about two years. I purchased my first piece at a Sotheby auction in May, 1968 – a pear-cased Tompion quarter-repeater watch.

I am happy that someone didn't bother to tell me I was crazy to try to build this collection. I thought that less than one-hundred pieces would be adequate. The Time Museum now has over three-thousand pieces and an horological library with hundreds of books. Of course the collection is not complete, but the fact that it isn't continually stimulates and challenges my interest.

The joys of collecting are many. Horology is a particularly mind-stretching subject, for it encompasses history, science, and technology. The development of timekeeping devices has been strongly influenced by the demands of religion, astronomy, physics, and navigation; and timekeepers have affected the course of history. Timekeeping devices are the products of peoples, nations, cultures, and of economic and political systems. In gathering a collection of significant time-measuring instruments, I have increased my knowledge of diverse cultures, art, architecture, styles, and manufacturing methods, and have gained a richer understanding of the techniques of historical research.

Collectors add purpose to their travel and see and do things no other travelers have the privilege of experiencing. Furthermore, a serious collector not only expands the horizons of his knowledge, but enriches his life with new and lasting friendships. And surely the greatest reward was my wife's enthusiasm in every aspect of this horological adventure. She was completely responsible for the museum's decor and the way in which the artifacts are displayed. All of our children and in-laws embrace the museum as a part of their lives and a heritage to be improved and conserved.

In each volume, I plan to write a short introduction that tells the story of collecting one or more of the pieces described in that publication. This volume covers astrolabes – elegant, complex devices developed almost two-thousand years ago and used for centuries.

Few people know about this instrument or have seen one. Fewer still understand how to use it. From a collector's point of view, further complications exist: (a) study and examination of astrolabes is difficult since only a few survive, (b) there is not much written about them, (c) most astrolabes are engraved in Arabic, and (d) they require an intimate knowledge of astronomy.

All of these factors add up to "buyer beware," because fakes abound. Knowledgeable collectors and dealers know about and have seen most of the important artifacts of a given type. One day, Alain Brieux, in Paris, noticed what he thought was a similarity between an instrument he had and one he had recently examined. Brieux noticed a characteristic engraving touch; for example, the numeral five had a distinctive serif. Further research convinced him that a number of widely diversified sundials and astrolabes – ostensibly the works of famous instrument makers, most of whom worked in the 16th century – had been made by the same talented hand. This led to his suspicion that the instruments were forgeries.

In short, after painstaking worldwide investigation, the expert forger was identified, by Scotland Yard. At his trial, this unusual man readily admitted the work he had done, and he served a term in prison. Many museums and discriminating collectors were surprised and I daresay dismayed to learn they had been duped.

Some years earlier, I had felt very fortunate indeed to find a rare "de Rojas" astrolabe that had been viewed without suspicion by a number of the world authorities on scientific instruments. Apparently, less than a handful of astrolabes designed with a de Rojas projection of the universe and the solar system on a flat plate exist. Yet this one turned out to be questionable. I could have returned it to the innocent, reputable London dealer for full reimbursement, but this de Rojas astrolabe remains in the museum as a teaching example. It proves the advisability of acquiring knowledge and of being inquisitive before making a definitive judgment.

There is another side to this coin, which I freely admit. I have, indeed, purchased pieces without full knowledge or understanding simply because they emotionally excited me and I felt they were "right." (Frankly, when I have turned down pieces that interested me it has usually turned out to be a bad decision.) Here is a story of an occasion when just this happened.

One day in Alain Brieux's handsome, private study, he showed me an astrolabe with two *retes*, an unusual construction. Brieux pointed out to me that he was in the process of investigating the piece, but agreed to sell it to me before he had probed its mysteries. It turned out to be one of the most important astrolabes in my collection, because it is a very rare instrument that can be used in both northern and southern hemispheres. Curiosity and emotion led to this purchase, not knowledge. Because of our friendship, Brieux sold it to me before its value was ascertained. Collecting involves personal interaction much more than mere examination of artifacts!

SETH G. ATWOOD
Rockford, April 1984

Foreword

This volume of the catalogue of The Time Museum, Rockford, Illinois, is the first of those which describe all the time-finding and time-measuring instruments in the collection which are not clocks or watches in the accepted sense of the word – that is, not geared movements driven by weights, springs, or electricity. The parts which together make up Volume I of the whole catalogue include:

1. Astrolabes
   Horizontal Instruments
   Astrolabe quadrants
2. Sundials
   Nocturnals
3. Water-clocks
   Sand-glasses
   Fire-clocks
4. Calendars
   Astronomical and Other Instruments

Each volume, although part of the set, is complete in itself with its own bibliography and index. A narrative history of the development of each class of instrument included in the collection has been provided since no general synthesis of the history of time-measuring instruments seems to exist in English. These sections, therefore, may be read continuously as an introduction to the subject. Descriptions of individual items in the collection are arranged as nearly as possible in chronological order. Where there are several items of the same date, they are arranged in alphabetical order of the maker's name (if known) or that of the region of origin. Technical details are then given in the following order:

1. Object.
2. Date. Where an object is dated in a calendar system other than that of the Christian era, the date as marked on the instrument is given first, and its European equivalent, indicated by the initials A.D., is shown in brackets immediately after it. Dates in the Islamic Hijra era are indicated by the letters A.H.[1]
3. Materials.
4. Overall key dimensions, in inches and millimetres.

1. These have been converted following Grenville. For the background, see Taqizadeh.

5. Signature transcribed or transliterated with contractions expanded and translated where necessary. Transliterations of Arabic names follow the forms adopted by L.A. Mayer[2] and the forthcoming revision of this work.[3] An enlarged photograph of the signature or inscription follows each transcription or translation.

6. Museum inventory number (which should be cited in any communication with the museum about the item).

7. General description of object, method of use, and commentary.

8. Provenance, when known.

9. Biographical notes on the maker, where possible.

In compiling this catalogue of a collection which covers many centuries and widely dispersed civilizations, I have inevitably been heavily dependent on the published works of others and the kindnesses of colleagues. Bibliographical references in the footnotes, and the list of acknowledgments below, can in some measure show the extent of my indebtedness and express my gratitude; they do not excuse faults which remain. For his constant kindness, enthusiasm, and patience, I must first thank Seth G. Atwood, and for their cheerful co-operation, the staff of The Time Museum particularly William Andrewes, Curator, and his assistants—successively Monique Rouzet and Jan Dolman. From the General Editor, Bruce Chandler, flowed much valuable comment, criticism, and hospitality, which have greatly enhanced the work, while the vigilant eye and judicious comments of Laura Tennen have saved this, and other volumes in the series, from solecisms and other errors. The explanatory diagrams were drawn especially for the work by David Penney, and the indexes were prepared by Jennifer Drake-Brockman. To Francis Maddison I must acknowledge a special debt, not only for help with the Islamic instruments here described, but for advice, criticism, and information freely given over many years, without which this volume would be immeasurably poorer. For much general help, advice, and acts of kindness, whether scholarly or administrative, I wish to thank Josette Alexandre, Paris; Margarida Archinard, Geneva; Silvio A. Bedini, Washington, D.C.; Michèle Bachelat, Paris; Alain Brieux, Paris; Giuseppe Brusa, Milan; Maria Celeste Cantù, Florence; Catherine Cardinal, Paris; Richard Good, London; Beresford Hutchinson, Greenwich; John Leopold, Groningen; Anne-Marie de Narbonne, Paris; Emmanuel Poulle, Paris; Wade Provo, Rockford; the late Maria Luisa Righini-Bonelli, Florence; Philip Rogers, London; Francesca Saba, Florence; Alain Segonds, Paris; Jean-Pierre Verdet, Paris; Clare Vincent, New York; Roderick and Marjorie Webster, Chicago; Johann Willers, Nuremberg.

2. Mayer (I), Mayer (II).

3. Brieux & Maddison. I am indebted to Alain Brieux and Francis Maddison for providing much information from this work in advance of publication and for allowing me to use their translations of inscriptions on Islamic instruments.

The source of illustrations is given at the end of each caption, and grateful acknowledgment is here made to all the institutions that have supplied photographs. Uncredited illustrations are from the collections of The Time Museum.

A.J.T.
Le Mesnil-le-Roi, France
April 1983

Contents

INTRODUCTION

FOREWORD

1 ASTROLABES

1 Planispheric Astrolabes

1 Description and Use

10 Origins, Development, and Diffusion to *circa* 1300

20 Islamic Astrolabes

21 Persia & Eastern Islam to *circa* A.D. 1500

23 Spain & The Maghrib

23 Persia & Indo-Persia after *circa* A.D. 1500

29 European Astrolabes

59 CATALOGUE

151 Universal Astrolabes

167 CATALOGUE

184 Linear Astrolabes

185 Spherical Astrolabes

189 ASTROLABE RELATED INSTRUMENTS

191 Horizontal Instruments

197 CATALOGUE

202 Astrolabe-quadrants

211 CATALOGUE

229 Astrological Astrolabes

231 CATALOGUE

239 GLOSSARY

243 BIBLIOGRAPHY

263 INDEX

Map of principal places mentioned in text.

CIALIST REPUBLICS
Volga
MONGOLIA
Aral Sea
Syr Darya
Caspian Sea
Amu
casus
RMENIA
TURKESTAN
CHINA
Nisibis
SOPOTAMIA
Rayy
Tigris
PERSIA
Baghdad
Isfahan
Lahore
TIBET
AFGHANISTAN
Himalayas
Persian Gulf
Indus
Ganges
ARABIA
Gujrat
INDIA
Bengal
OMAN
Jaipur
Arabian Sea

Astrolabes

Astrolabes

This word Astrolabe is as much to say as the handle or instrument of the Starres, by helpe whereof the manifolde motions and apparences of the heauens and of the Starres therein contained are known. . .

Thomas Blundeville[1]

The astrolabe[2] is an astronomical model, usually portable, which represents a part of the celestial sphere in relation to the earth and, by simulating the apparent movement of the stars around the celestial pole, makes possible the mechanical solution of various astrological and astronomical problems. The instrument is thus a form of analogue computer. The addition of a simple sighting device allowed observations of celestial bodies to be made in order to obtain information needed to set the instrument for the solution of a variety of problems. Astrolabes were not, however, instruments of serious observational astronomy, being too small to supply accurate data. There are three main types of astrolabe: planispheric, linear, and spherical. Planispheric astrolabes are designed either for use in a single latitude or for general use anywhere (in which case they are called universal).

PLANISPHERIC ASTROLABES FOR USE IN SINGLE LATITUDES

Description and Use

The planispheric astrolabe for single latitudes, or the 'particular' astrolabe as it was called by 16th-century writers, is historically the most important of the different types of astrolabe. Most surviving astrolabes are of this type. A two-dimensional representation of the surface of a sphere on a plane, the astrolabe is drawn by means of stereographic projection, which has the important property of projecting circles on the sphere as circles on the plane and retaining in projection the true values of angles between arcs on the sphere. In the astrolabe (fig. 2), a projected representation of the celestial sphere (fig. 1) with a selection of stars lies over a projection of the terrestrial horizon-zenith co-ordinate system for a particular latitude (fig. 3).

1. Blundeville, 281.
2. The 'Mariner's Astrolabe' and other instruments which are called by the name 'astrolabe' but which are used exclusively for altitude measurement are here excluded from consideration. For a detailed discussion of the 'Mariner's Astrolabe,' its relations (if any) with the stereographic astrolabe, its origins, and use, see the volume of this catalogue devoted to *Navigation and the Chronometer.*

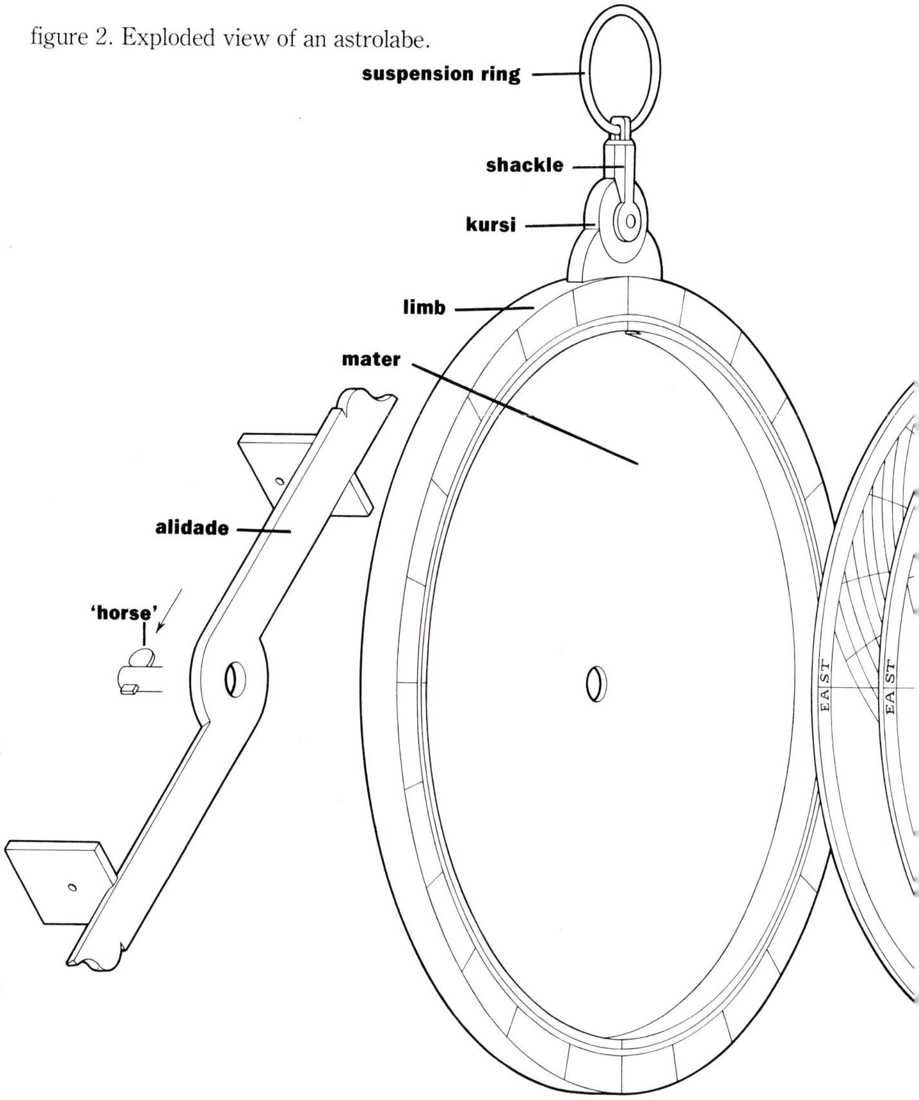

figure 2. Exploded view of an astrolabe.

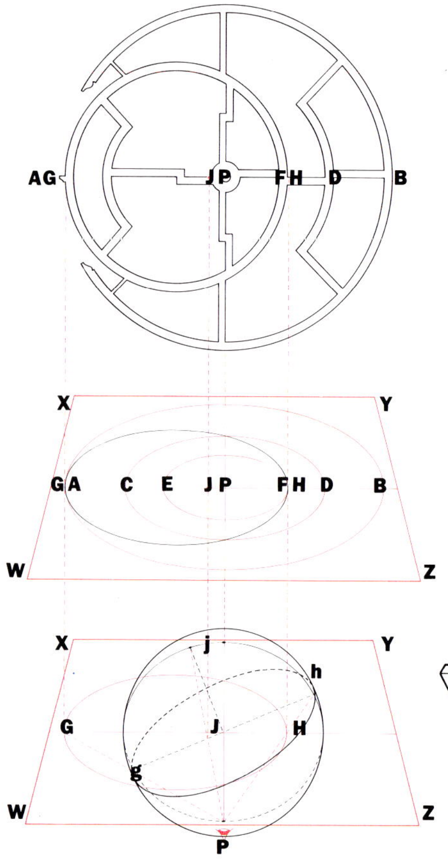

figure 1. The principle of stereographic projection: *rete*. The observer's eye is imagined to be placed on the surface of the celestial sphere at one of the poles, P. From this point, visual rays (represented by dotted lines) pass from the eye to the circles on the sphere. These rays cut the equinoctial plane WXYZ on which they locate the points required for the projection. From the pole P, the visual ray is projected through the two points g and h on the sphere where the ecliptic touches the tropics of Cancer and Capricorn to points G and H. These latter points, therefore, coincide with points A and F in fig. 3. J is the projection of the pole (j on the sphere) of the ecliptic. The central diagram shows the configuration obtained when the ecliptic is added to the circles projected in fig. 3, and the actual arrangement, as adapted for use on astrolabe *retes*, is shown in the upper diagram.

These two stereographic projections are made from one pole onto the plane of the equator (fig. 1). For astrolabes to be used in the northern hemisphere, the centre of the projection is the south pole; the north pole is used as centre for astrolabes to be used in the southern hemisphere. The resulting star map is a mirror image (since it is viewed from outside the celestial sphere) of the heavens as seen from the earth (inside the celestial sphere).

The instrument, which has varied little in form over the thousand years of its recorded history, was usually made of brass or was drawn or printed on paper or vellum sheets glued to wood or paste-boards. Very occasionally examples were made in gold or silver, but only rare examples in the latter material have survived (catalogue no. 14). The astrolabe consists of a backplate (*mater,* Arabic *umm*[3])

3. The English terminology of the astrolabe is a curious mixture of terms derived from Greek, Arabic, and Latin, a mixture which accurately reflects its transmission. An early attempt to explain the names of the parts was made by Michael Scot (see Thorndike (III), 16). *Cf.* al-Bīrūnī's accurate etymology and ascription of the instrument to the Greeks (Kennedy (II), 111). The term most generally accepted in English is here given first, followed by its Arabic equivalent. For recent studies of astrolabe terminology, see Kunitzsch (III & IV), *passim*.

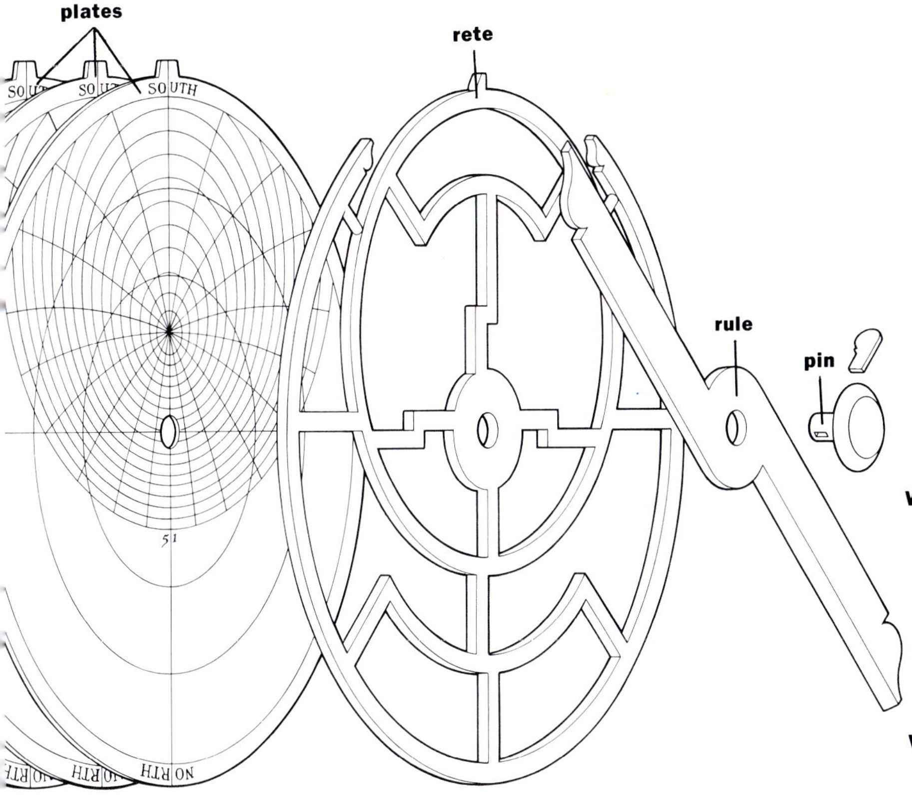

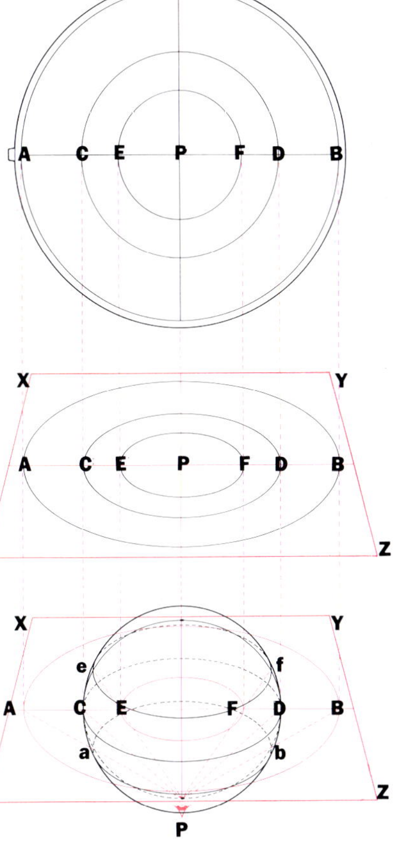

with a raised circumference (limb, *ḥajra*) on which is engraved a scale of degrees and, usually on Western instruments, an hour scale (fig. 111). At the top there is a triangular or ogee-shaped projection (throne, *kursī*[4]) which carries the shackle (*'urwa*) through which passes the suspension ring (*ḥalqa*). *Mater*, limb, and *kursī* may be cast together in one piece (most Islamic astrolabes are made in this way) or may be made up from pieces separately formed and rivetted or soldered together. Inside the *mater*, which is sometimes engraved with scales, one or more plates are placed one on top of the other with some device, such as a projecting lug, to prevent them from turning. Each plate is engraved with a stereographic projection of the equator and the tropics of Cancer and Capricorn and, for a particular terrestrial latitude, with a similarly projected horizon-zenith co-ordinate system made up of azimuths and almucantars (figs. 3 &

4. Since there seems to be little warrant for the term 'throne' in early European literature, *kursī* is used throughout. Indeed, European astrolabes seldom have 'thrones' in the Islamic sense but rather a composite suspension piece made up of mount, shackle, swivel joint, and suspension ring.

figure 3. The principle of stereographic projection: plate. The observer's eye is imagined to be placed on the surface of the celestial sphere at one of the poles, P. From this point, visual rays (represented by dotted lines) pass from the eye to the circles on the sphere. These rays cut the equinoctial plane WXYZ on which they locate the points required for the projection. A and B are the projections on the plane of the points a and b of the tropic of Capricorn; C and D of the equator; and E and F of the tropic of Cancer. The projection thus formed results in the diagram shown in the central figure, whereas the circles as they appear on an astrolabe plate are shown in the upper diagram. The horizon-zenith co-ordinate system for a given latitude on earth is projected in a similar fashion onto the equinoctial plane and appears on the astrolabe plate as seen in fig. 4.

figure 4. Scales on an astrolabe plate.

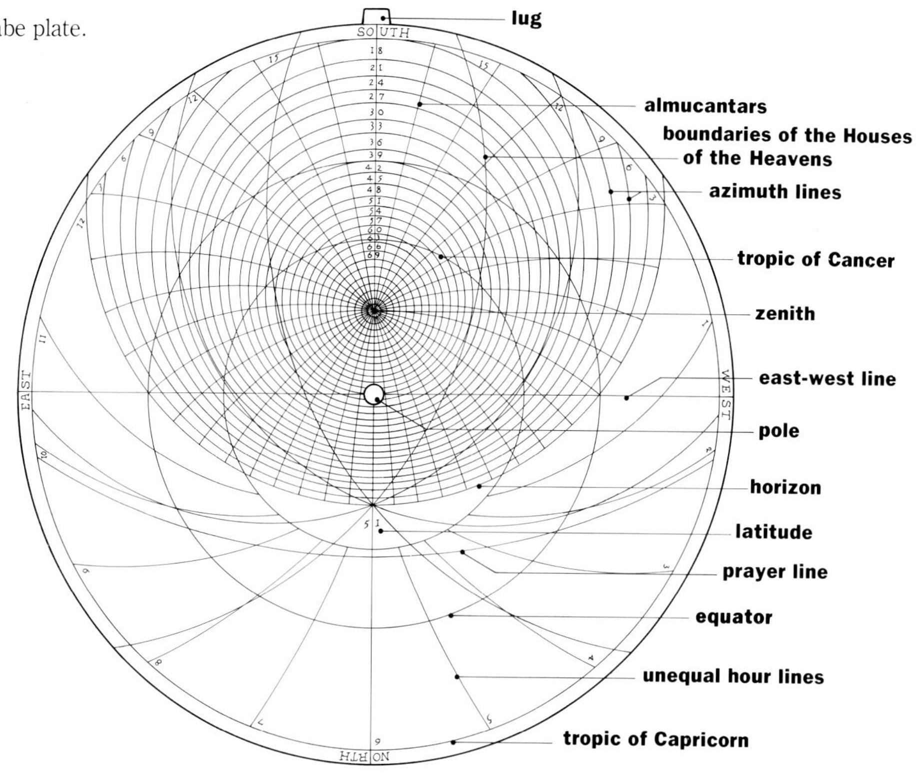

figure 5. The scales normally found on the back of an astrolabe.

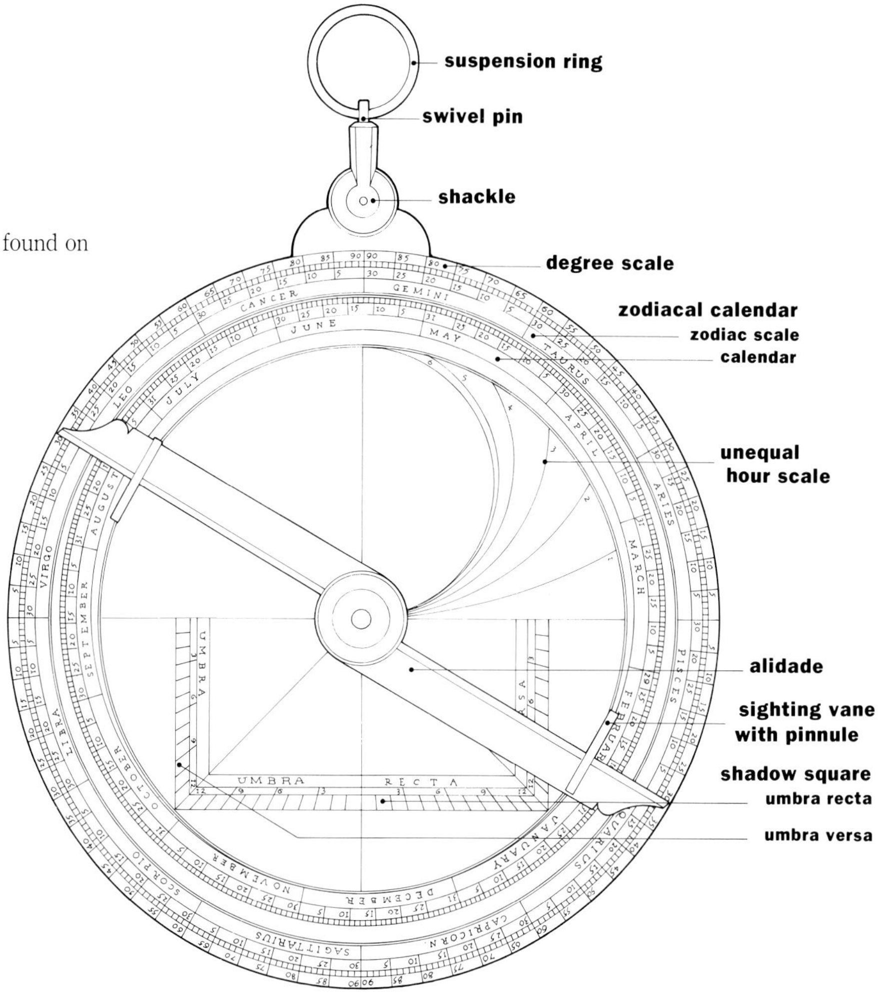

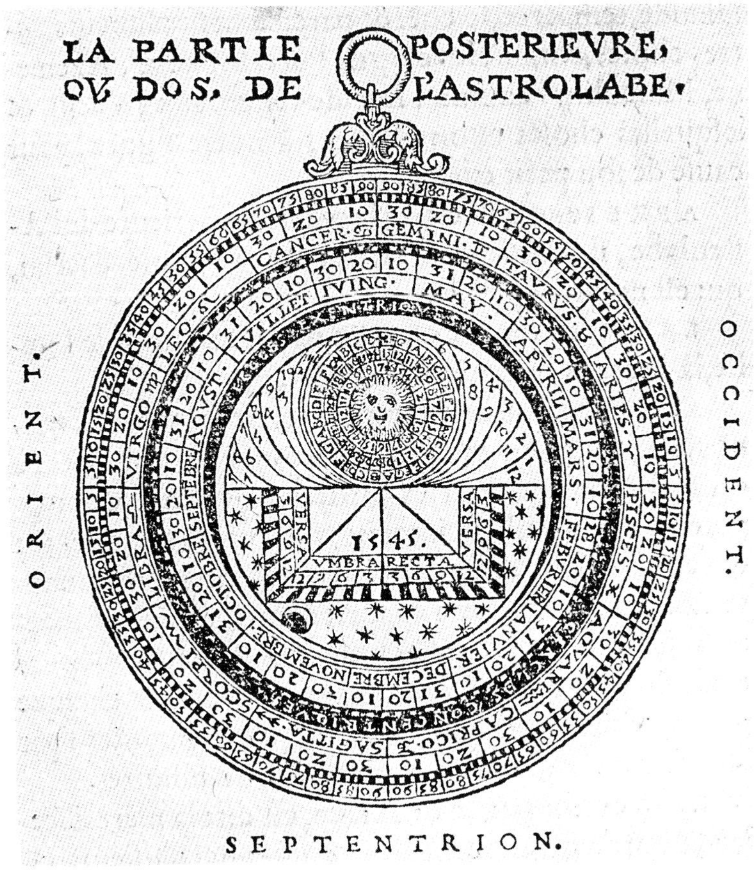

figure 6. Back of an astrolabe. From Jacques Focard, *Paraphrase de l'astrolabe* (rev. by Jacques Bassentin), Lyons, 1555, 19. The date 1545 on the astrolabe is the date of the first edition of Focard's book.

4). Whether the almucantars are drawn for each degree or for groups of two, three, five, or six depends upon the size of the instrument. Lines marking the hour of twilight, unequal hours, and the astrological 'houses of the heavens' may also be engraved upon the plates (figs. 4 & 8).

Above the plates and level with the limb is the *rete* (*'ankabūt*), an open-work skeletal plate showing stereographically the ecliptic, and the positions of a number of stars (fig. 7). The plates and the *rete* are held together in the *mater* by a bolt-shaped pin which passes through a hole in the centre of the *mater*, all the plates, and the *rete* and is secured by a cross-piece – the 'horse' (*faras*) in the form of which animal it is often shaped (figs. 2 & 9). The *rete* in this arrangement is free to be rotated around the pin over the plates. On the back (*ẓahr*) of the astrolabe (figs. 5 & 6) a considerable variety of scales may be engraved. These are described below as they occur on individual instruments in the collection and are briefly defined in the Glossary. Almost always, however, there is a degree scale engraved on the outer edge which is used in conjunction with the alidade (*al-'idāda*), a two-vaned sighting device (figs. 10 & 11) for altitude measurements, and in the lower centre there is a shadow square used with the help of the alidade for determining heights and distances by means of proportions between similar triangles. On some European astrolabes a rule (mounted on the pin) is added on the face to facilitate readings (figs. 2 & 9).

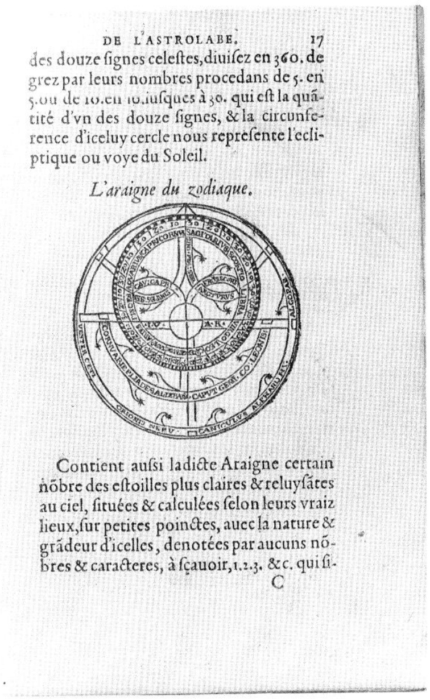
DE L'ASTROLABE. 17

des douze fignes celeftes, diuifez en 360. degrez par leurs nombres procedans de 5. en 5. ou de 10. en 10. iufques à 30. qui eft la quãtité d'vn des douze fignes, & la circunference d'iceluy cercle nous reprefente l'ecliptique ou voye du Soleil.

L'araigne du zodiaque.

Contient aufsi ladicte Araigne certain nõbre des eftoilles plus claires & reluyfãtes au ciel, fituées & calculées felon leurs vraiz lieux, fur petites poinctes, auec la nature & grãdeur d'icelles, denotées par aucuns nõbres & caracteres, à fçauoir, 1.2.3. &c. qui fi-

C

figure 7. *Rete* of an astrolabe. From Dominicque Jacquinot, *L'Usage de l'astrolabe*... (2nd edit., rev. by Jacques Bassentin), Paris, 1559, 17v.

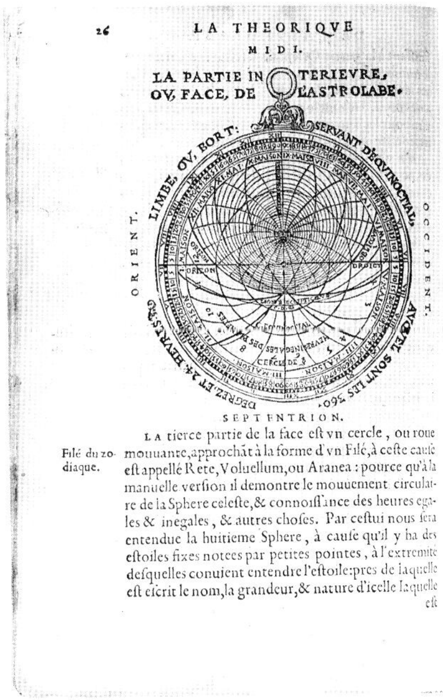
26 LA THEORIQVE

MIDI.

LA PARTIE INTERIEVRE, OV FACE, DE L'ASTROLABE.

Filé du zodiaque.

LA tierce partie de la face eft vn cercle, ou roue mouuante, approchãt à la forme d'vn Filé, à cefte caufe eft appellé Rete, Voluellum, ou Aranea : pource qu'à la manuelle verfion il demontre le mouuement circulaire de la Sphere celefte, & connoiffance des heures egales & inegales, & autres chofes. Par ceftui nous fera entendue la huitieme Sphere, à caufe qu'il y ha des eftoiles fixes notees par petites pointes, à l'extremité defquelles conuient entendre l'eftoile: pres de laquelle eft efcrit le nom, la grandeur, & nature d'icelle laquelle eft

figure 8. The scales for a latitude plate, shown here drawn on the *mater* of the instrument. From Jacques Focard, *Paraphrase de l'astrolabe* (rev. by Jacques Bassentin), Lyons, 1555, 26.

figure 9. The rule, alidade and horse of an astrolabe from a 14th-century copy of the treatise on the astrolabe of Pseudo-Māshā'llāh. London, British Library, Ms Egerton 844 fol. 61r. (Photograph: British Library)

figure 10. Alidades.

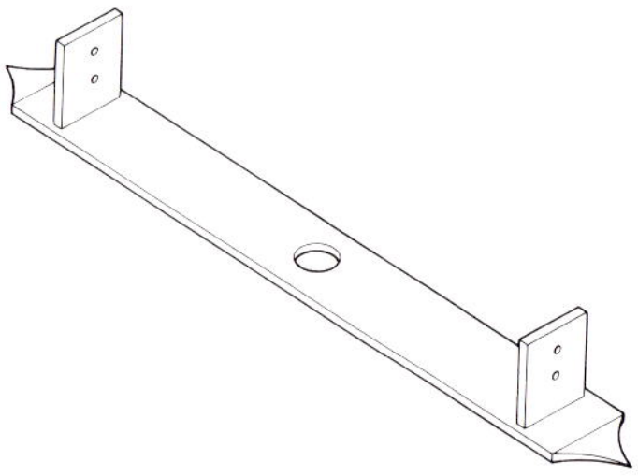

figure 10a. Conjectural representation of a possible early form of alidade.

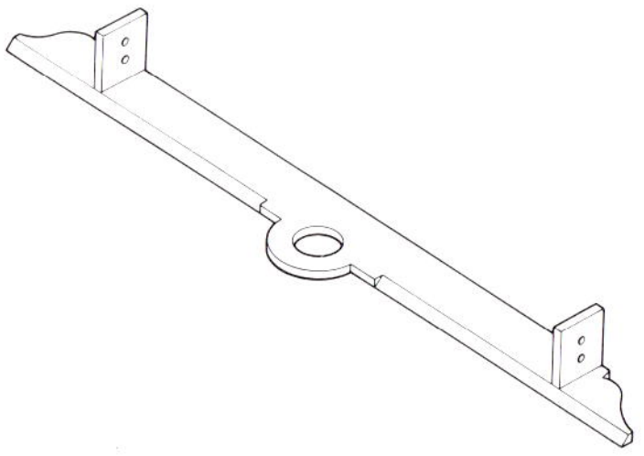

figure 10b. Straight-bar alidade as found on many Islamic astrolabes.

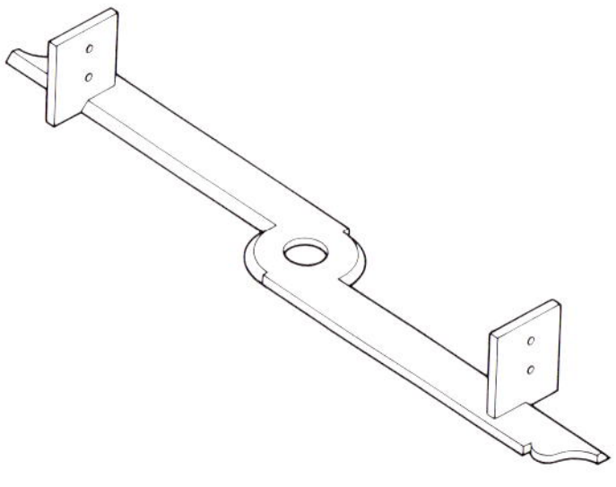

figure 10c. Counter-changed alidade used most commonly on Western astrolabes.

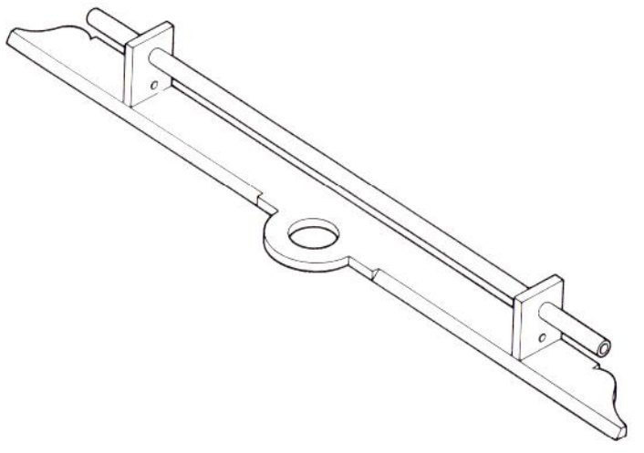

figure 10d. Straight-bar alidade with sighting-tube commonly found on Indian astrolabes.

The astrolabe has a multitude of uses which it is not possible to describe here in detail. During the many centuries of its use, a vast body of literature grew up,[5] and later authors strained their ingenuity to invent new problems which could be resolved by means of the astrolabe.[6] Primarily, it was the adaptability and versatility of the instrument that gave it importance. With it, time could be found by day or by night, latitude could be found, elementary surveying operations could be performed by angular measurement, and the position of the stars for any moment of time could be determined or demonstrated. In Islam, the astrolabe was valuable for finding the times of daily prayer, which were astronomically determined; although planets are not shown on astrolabes, by using a set of planetary tables together with the astrolabe, astrologers could solve problems concerning the position of a planet in the heavens and its relationship with stars or astrological houses. In order to give some idea of how the astrolabe is set and used, the procedure for finding the hour is here described.

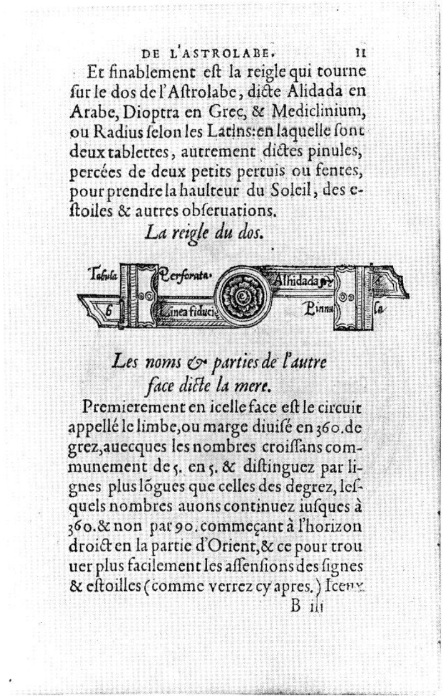
DE L'ASTROLABE. 11

Et finablement eſt la reigle qui tourne ſur le dos de l'Aſtrolabe, dicte Alidada en Arabe, Dioptra en Grec, & Mediclinium, ou Radius ſelon les Latins: en laquelle ſont deux tablettes, autrement dictes pinules, percées de deux petits pertuis ou fentes, pour prendre la haulteur du Soleil, des eſtoiles & autres obſeruations.

La reigle du dos.

Les noms & parties de l'autre face dicte la mere.

Premierement en icelle face eſt le circuit appellé le limbe, ou marge diuiſé en 360. degrez, auecques les nombres croiſſans communement de 5. en 5. & diſtinguez par lignes plus lõgues que celles des degrez, leſquels nombres auons continuez iuſques à 360. & non par 90. commeçant à l'horizon droict en la partie d'Orient, & ce pour trouuer plus facilement les aſſenſions des ſignes & eſtoilles (comme verrez cy apres.) Iceux

B iii

figure 11. Alidade. From Dominicque Jacquinot, *L'Usage de l'astrolabe*... (2nd edit., rev. by Jacques Bassentin), Paris, 1559, 11r. Note that the alidade has vanes which may be folded flat for transportation in a case (*cf.* the alidade of the astrolabe by Georg Hartmann, cat. no. 17) and that each has two pinnules, one slightly larger than the other. Which was used depended on whether a star or the sun was being sighted (fig. 12 a & b).

Firstly, the observer places beneath the *rete* the plate appropriate to the latitude of his position. Using the alidade and the degree scale on the back of the instrument, he then measures the altitude of the sun (fig. 12a) or, at night, one of the stars (fig. 12b) marked on the *rete* (42° in fig. 12c). From the zodiacal calendar that is usually engraved on the back of Western instruments, or from a set of tables, the sun's position in the zodiac for the date of the observation is found (24 April = 13° Taurus, in fig. 12d), and, turning the instrument over, the *rete* is turned until that point of the ecliptic lies above the corresponding almucantar marked on the plate beneath (fig. 12e). If a star has been observed, the *rete* is turned until the pointer for that star lies above the appropriate almucantar for the observed altitude (fig. 12f). The astrolabe is thus set and represents the heavens for the moment of observation. To find the time, it is only necessary to take the point diametrically opposite that of the sun's position in the ecliptic and read off the unequal hour from the scale engraved on the plate beneath the *rete* (figs. 12g, middle right). If an equal-hour scale is engraved on the limb, then it is only necessary to extend a straight line through the axis and the point of the sun's declination on the ecliptic circle out to the limb to read off the equal hour (fig. 12g, bottom right). The rule fitted to many European astrolabes facilitates this reading.

5. The classic account of the astrolabe in English is that by Geoffrey Chaucer. Written in 1391, probably but not certainly, for the benefit of his son Lewis, Chaucer's work is largely an English translation of a Latin version of the Pseudo-Māshā'allāh treatise (for which see Kunitzsch (II)). Versions in modern English of both texts are given in Gunther (III), but the most reliable text of Chaucer is that included in Robinson, 543–63. Of the large number of Latin works on the subject, that by Stöffler is comprehensive and was widely used in early modern Europe.

6. The astrolabe and all the main problems for which it was used are described in detail in several modern works. The classic treatment is that by Michel (III). Shorter but equally valuable are Hartner (I), Morley (I), and García, who combines descriptions of some 24 instruments with a general treatise on the subject. An introduction in Danish is given by Møller. Three shorter accounts are accompanied by usable examples of astrolabes made from modern materials (plastics or laminated paper). See Saunders, Webster, and NMM. There is also an elegant brief exposition by North (II) and an attractive description by Archinard.

figure 12. Reading the time from an astrolabe.

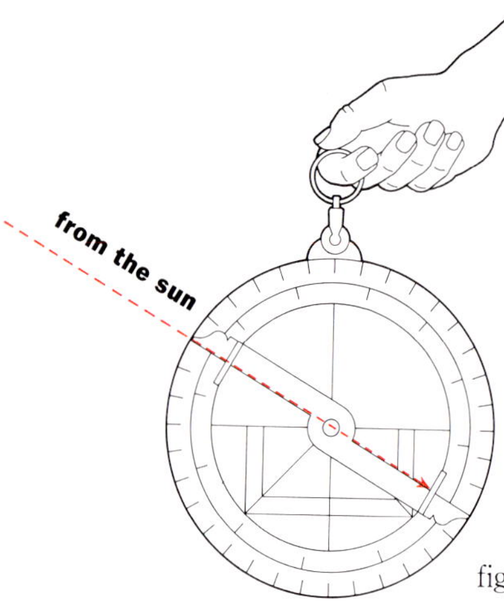

figure 12a. When an observation of the sun is being made to set the astrolabe, in order not to blind himself by looking through the pinnules directly at the sun, the observer holds the instrument at a convenient height and turns the alidade until a spot of light passing through one of the pinnules falls directly on the corresponding pinnule on the opposite vane.

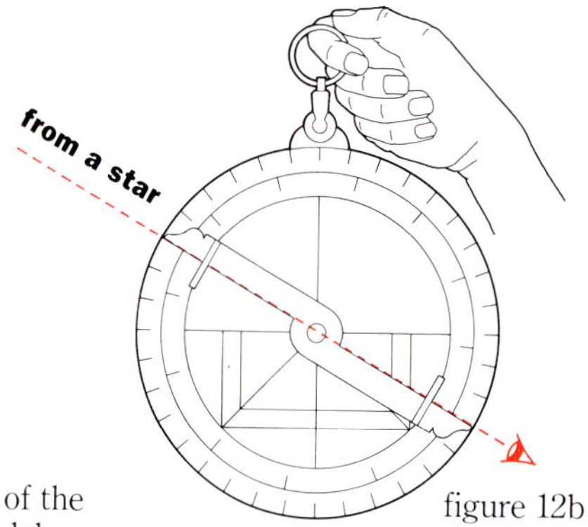

figure 12b. When a star is observed, the observer looks directly at it through the pinnules.

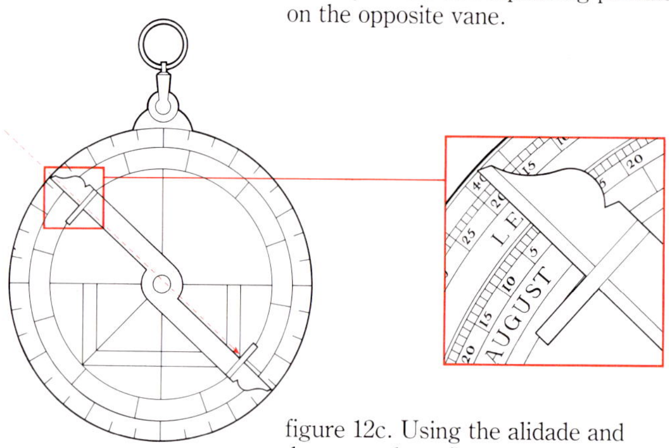

figure 12c. Using the alidade and degree scale to find the altitude of the sun or a star (42° in the illustration).

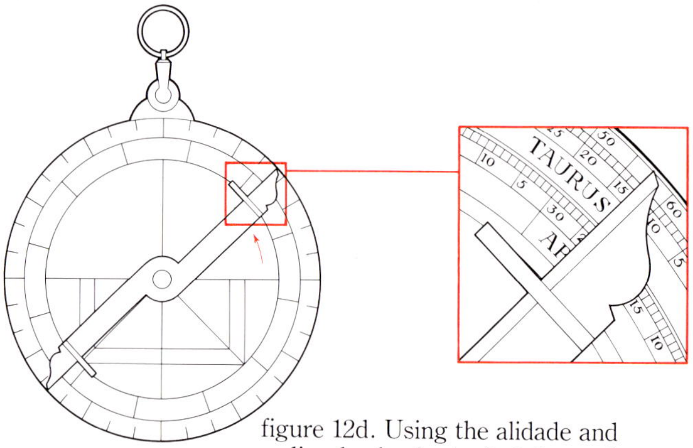

figure 12d. Using the alidade and zodiacal calendar to find the sun's position in the ecliptic for the date of the observation (24 April, 13° Taurus in the illustration).

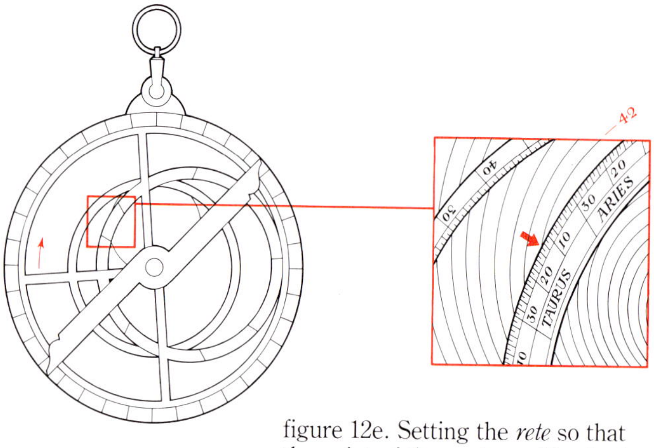

figure 12e. Setting the *rete* so that the point of the sun's position in the ecliptic (13° Taurus) lies over the corresponding almucantar (42°).

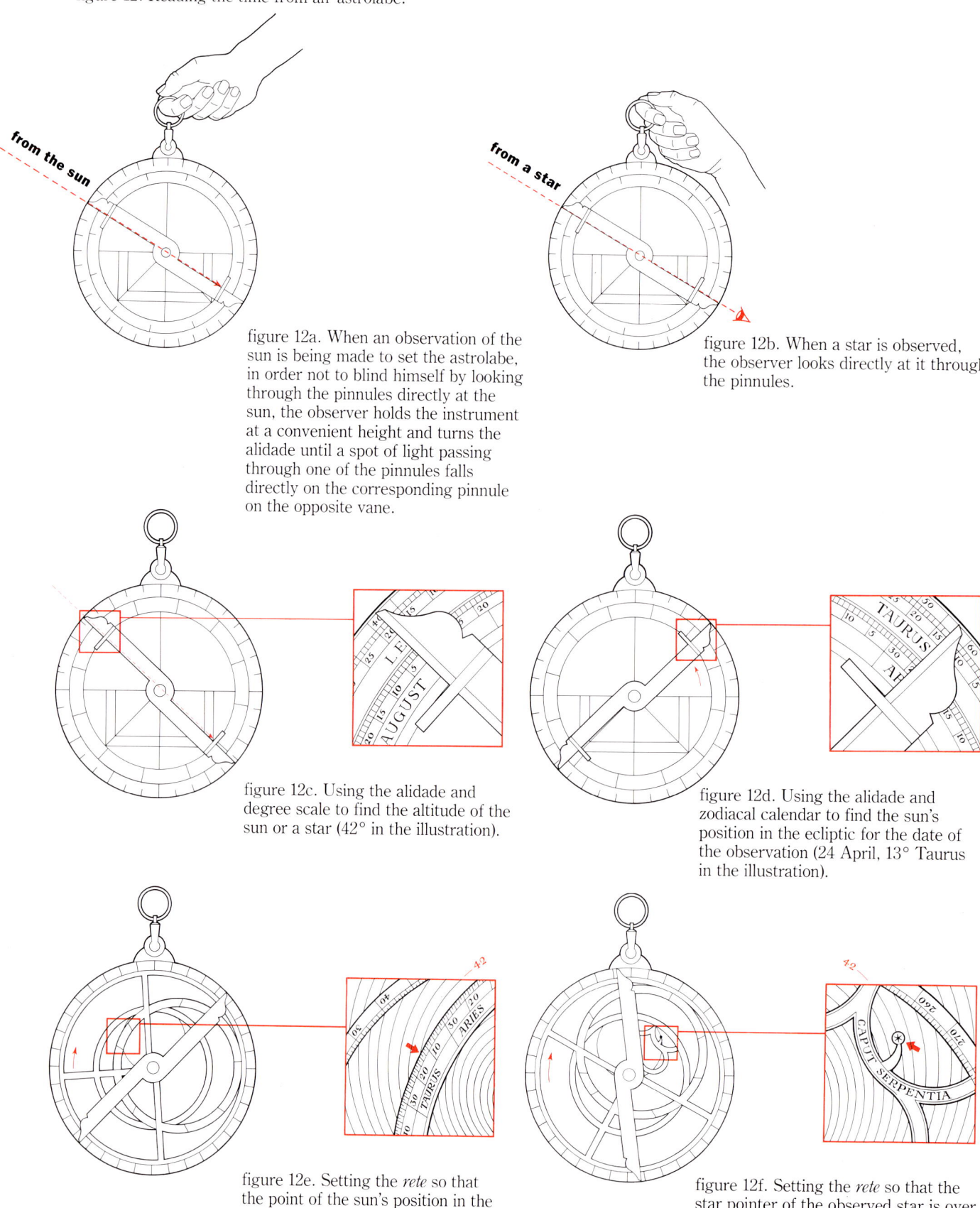

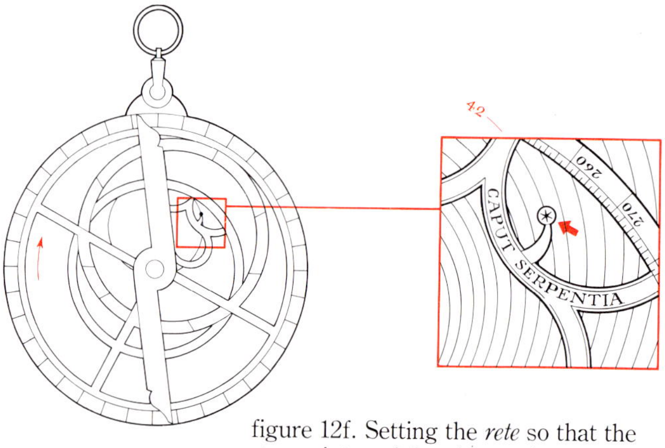

figure 12f. Setting the *rete* so that the star pointer of the observed star is over the corresponding almucantar (42°).

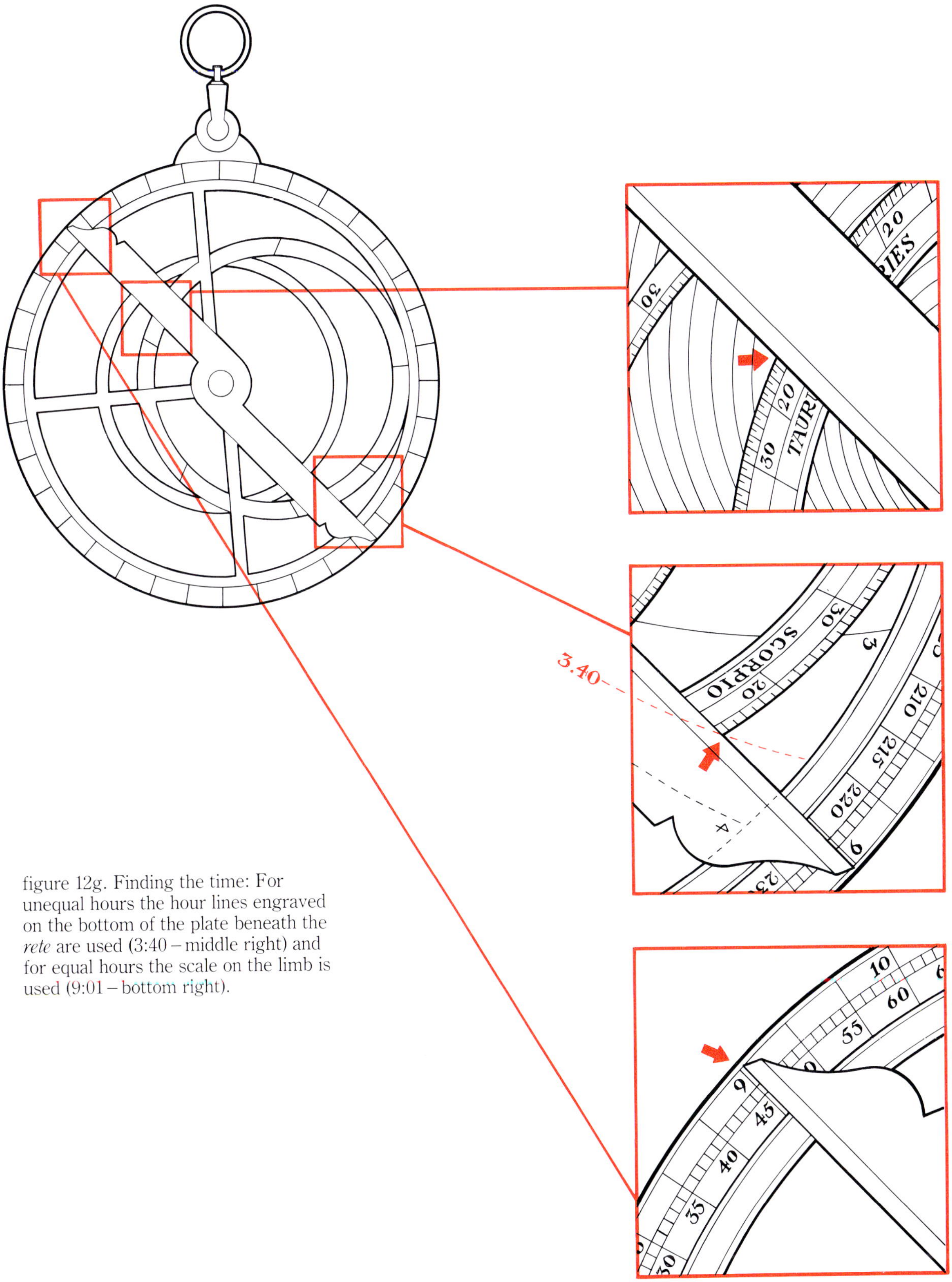

figure 12g. Finding the time: For unequal hours the hour lines engraved on the bottom of the plate beneath the *rete* are used (3:40 – middle right) and for equal hours the scale on the limb is used (9:01 – bottom right).

Origins, Development, and Diffusion to *c.* 1300

Without stereographic projection the astrolabe is inconceivable. It cannot, therefore, predate Hipparchus of Nicaea (*c.* 150 B.C.), the earliest astronomer for whom a knowledge of the projection has been claimed.[7] That it was well-known by the end of the 1st century B.C. seems certain, since the anaphoric clock described by Vitruvius[8] (d. *post* A.D. 27) employs stereographic projection in an arrangement of its dials which is the reverse of that found in an astrolabe. That is, the star map normally found on the *rete* of an astrolabe is engraved on the solid plate, and the projection of the local horizon and the hour lines replaces the star map on the *rete*, which is made to rotate over the constellation map once a day by means of an inflow clepsydra.[9] In the following two centuries, knowledge of stereographic projection spread and was given a firm theoretical basis by Ptolemy in his treatise *Planisphaerium*.[10] It was also practically applied in instruments. Fragments of the dials of two anaphoric clocks both dating from the period 1st to 3rd centuries A.D. have survived,[11] and stereographic projection of the tropics of Cancer and Capricorn, the equator, and the unequal-hour lines is found inside the lid of a small portable sundial now in the Kunsthistorisches Museum, Vienna[12] (fig. 13). Since the lid itself is formed from a medallion of Antoninus Pius which may be dated to A.D. 143–4 by its inscription, and since this particular series is known to have circulated for only a short period, the dial may have been constructed in the late 2nd century and even, perhaps, within a few years of the writing of Ptolemy's *Planisphaerium*, which is conventionally dated to *c.* A.D. 160.

For the history of the astrolabe, however, the Vienna sundial may have more significance than the evidence it supplies for the early use of stereographic projection. The complete instrument consists of a small box 39 mm in diameter and approximately 14 mm thick. The top and bottom of the box are made from the medallion of An-

7. Synesius: Letter to Paeonius. The Greek text may be found in Migne, columns 1583–4. Partial English versions are given by Hartner (I) 289, n. 1; Dicks; and Neugebauer (II), ii, 874–6. The full text of the letter is translated in Fitzgerald, 258–66. French translation of paragraphs 4 and 5 in Segonds (II), 56–60. That Hipparchus knew stereographic projection has, however, been disputed by Dicks, Appendix. For Synesius' life, see Lacombrade, viii–xliv.

8. *de Architectura*, XI, 8. The best edition is that of Soubiran.

9. Soubiran 290ff; Neugebauer (II), ii, 869–70; Drachmann *(passim)*.

10. απλωσις επιφανειας [? = εξαπλωαις] σφαιρας. Neugebauer (II), ii, 870–1. The Greek text of this treatise is lost. There was a translation into Arabic by Maslama b. Aḥmad al-Majrīṭī of Cordoba (d. c. A.H. 397 [*c.* A.D. 1007]) from which a Latin translation was made by Hermannus Dalmatus. See Heiberg, ii, 227–59.

11. Best described in brief by Neugebauer (II), ii, 870 n. 5 & 6. For the disc from Grand. *cf.* Maitzner in Tardy, i, 17–20, but it should be noted that some uncertainty still exists as to the interpretation of this dial.

12. Inv. VI 4098. Price (V), no. 6; Buchner (I & II) following recent cleaning, which revealed greater detail. Buchner's findings have been summarized by Segonds (I). A second such sundial medallion is in private ownership (Buchner (II)), while a third has recently been identified at Munich.

figure 13. Portable sundial, ?2nd century A.D., with latitude plates and using stereographic projection. Vienna, Kunsthistorisches Museum no. VI 4098. Lid (upper left), inside of lid with stereographic projection (upper right), base (middle left), inside of base with pin (middle right), both sides of one of the sundial plates (lower left and right). (Photograph: Kunsthistorisches Museum)

toninus Pius; the stereographic projection is on the inside of the lid. The box contains four plates, each engraved on both sides with sundial scales and each scale for use in a different latitude. Fixed to the base of the box is a vertical pin which passes through a hole in the plates, thus securing them in position. The resemblance of four features of this instrument to those of an astrolabe is striking: the portable box, containing a series of plates; each plate being drawn for a different latitude; the plates being secured by a pin; the use of stereographic projection.[13]

13. Indeed, it is probable, for Ptolemy states that the *αστρολαβων ωροσκοπιων* is the only instrument which can give the hour of birth in minutes. *Tetrabiblos* I: iii:2, Robbins, 229; Hartner (I), 2532 and (II), 289, n. 1 thinks this refers to the planispheric astrolabe. Neugebauer (II), ii, 871, however, thinks the reference is to the observing armillary sphere elsewhere described by Ptolemy in Bk V:1 of the *Almagest*. An independent examination of the question by Segonds (II), 21–2, concludes that it was the planispheric astrolabe that Ptolemy had in mind. This position is strengthened by the fact that the term 'horoscopion' seems normally to have been used in the 1st–3rd centuries A.D. as a generic term for a timepiece (see discussion in Volume I, part 2, ch. 1 of this catalogue). The observing armillary sphere hardly enters into this category.

At some point in the later 4th century, Theon of Alexandria (b. *c.* A.D. 335) wrote a treatise on the astrolabe, which, although it has not survived, can be shown to have described an instrument similar in all essential respects to the instrument known from surviving examples of the 10th century and later.[14] These essential elements of the astrolabe are:

1. A stereographically projected map of the heavens in which the projection of the stars and ecliptic is marked on the skeletal *rete* which rotates over the local co-ordinate plate that usually carries hour lines
2. A set of plates, each projected for a different latitude
3. A sighting device
4. A suspension apparatus

Several of these elements appear in instruments, all based on stereographic projection, developed between the 2nd century B.C. and the 4th century A.D. Alongside the instrument of Hipparchus referred to by Synesius, the anaphoric clock, and the Antonine sundials mentioned earlier, there was the 'horoscopic instrument' mentioned at the end of Ptolemy's *Planisphaerium*. This was designed for use in one latitude, employed stereographic projection, and showed all the risings and settings visible during the year at Alexandria.[15] This last characteristic suggests that the device was arranged with the horizon and equatorial co-ordinates fixed and the stars and ecliptic rotating over them. Another instrument for which some details have survived is that sent by Synesius of Cyrene (*c.* A.D. 370–413) to Paeonius at Constantinople.[16] Although the description is imprecise, it seems clear that the object, perhaps primarily intended to convert ecliptic to equatorial co-ordinates, employed stereographic projection and was, perhaps, fitted with a *rete* of the anaphoric form.[17] That these five objects exhaust the number of devices based on stereographic projection we may reasonably doubt. It was surely from just such a context where the possibilities of stereographic projection were being explored that the astrolabe is likely to have emerged.

And indeed the astrolabe has resemblances with all the instruments listed above. With all of them, it shares stereographic projection. With the Antoninus sundials, it shares portability and a set of plates designed for different latitudes contained in a box. With anaphoric clocks, with Synesius' instrument, and with the 'horoscopic instrument' of Ptolemy, it shares the stereographic star map and the rotatable *rete*–with Ptolemy's instrument this might have been in the same relationship of stars and zodiac turning over the local co-ordinates on the fixed plate. Only two elements of the

14. Neugebauer (I); Neugebauer (II), ii, 877–9; Segonds (II), 30–8.
15. Neugebauer (II), ii, 871.
16. Above, n. 7.
17. For discussion of Synesius' instrument in great detail, see Vogt & Schramm. *Cf.* Neugebauer (II), ii, 874–6. Segonds (II), 60–3.

instrument described by Theon are lacking: the sighting device and a suspension apparatus. The origin of the suspension piece remains unclear, but for the sighting device we may find a source in the dioptra, a sighting instrument used on its own, but also incorporated in the observing armillary sphere ('αστρολάβον) of Ptolemy,[18] which, when added to the portable star map with multiple plates, led to the transfer of the old name 'astrolabon' to the new composite instrument. For the astrolabe was, indeed, unique and new. It combined three functions – time-finding, representation of the stars, and observation – with portability. No other instrument of the period did so much nor was so convenient.

In succeeding centuries we gradually learn more about the astrolabe. The first-known treatises which describe it are those in Greek by Theon Alexandrinus (*c.* A.D. 375), Ammonius (*fl. c.* A.D. 390), and John Philoponus (A.D. 530), and one in Syriac by Severus Sebokt, Bishop of Nisibis (pre *c.* A.D. 660).[19] By this time the astrolabe had achieved a familiar form, although it was not as elaborate either in decoration or graduation as it would later become. The treatise of Sebokt is an example of those Syriac books based on Greek sources, of which some were translated into Arabic in the mid- to late 8th century A.D. A treatise on the astrolabe, *Kitāb al-'amal bi'l-asṭurlāb al-musaṭṭaḥ* (Book of the Use of the Plane Astrolabe), was ascribed by the bibliographer Ibn an-Nadīm to al-Fazārī (*fl.* late 3rd/8th century), whom he says was the first Muslim to make an astrolabe.[20] At the end of the 8th century we have a glimpse of the astrolabe in use when, describing a horoscope he had calculated, Stephanus the Philosopher mentions using it. Stephanus was an astrologer who migrated from Persia to Constantinople, and such displacements may represent one way in which knowledge of the instrument was diffused in non-Islamic regions.[21]

Early manufacture of astrolabes centred on the pagan Ṣābian city of Ḥarrān, which was also a centre for translation from Greek and

18. The term 'dioptra' subsumes several different forms of sighting device, including the alidade of an astrolabe (so Philopon, e.g., in the heading of ch. 2, Segonds (II), 143). The dioptra, mentioned by Archimedes and Ptolemy as an instrument for measuring the apparent diameter of the sun, as described by Pappus, had one fixed and one movable sight. In other instruments both sights were fixed, being furnished with sighting holes or slits. Since the usefulness of sighting tubes for aiding observation at a distance was already known by the time of Aristotle (see *de Generatione Animalium*, V 1; 780b & 781a), it may also have been used in sighting devices (*cf.* Polybius X. 46). Indeed we may wonder if the sighting tube typically used on the alidade of Indian astrolabes may not be a survival from an early Greek tradition. For a brief general discussion of the dioptra, see Martin 488–9.

19. For the relationship of these treatises to each other, see the remarkable paper by Neugebauer (I), and the introduction to Segonds (II), 30–8, his French translation of Philopon. Complete English translations of the latter two treatises may be found in Gunther (V), i, 61–81 and 82–103, although these should be treated with caution. There is an earlier French translation of Sebokt with introduction and notes by Nau, who anticipated Drachmann in recognizing a relationship between the astrolabe and the anaphoric clock but arrived at quite different (and largely erroneous) conclusions. See also Kunitzsch (III).

20. Pingree (I), 103–4. Dodge, ii, 649. Elsewhere, however, (Dodge, ii, 644, 670) Ibn an-Nadīm states that Abīyūn al-Baṭrīq was the first to make an astrolabe and he also is said to have written a treatise.

21. Neugebauer & Hoesen, 158, 160, 190. It is interesting, however, that Stephanus' modern editors suggest that errors in the horoscope may have been caused by this use of the astrolabe.

Syriac into Arabic. 'Later they were distributed, becoming common and increasing in number.'[22] Jāḥiẓ (d. A.H. 255 [A.D. 868/9]) mentioned astrolabes as time-finders for daytime use,[23] but in *c.* A.H. 243 (A.D. 857/8) al-Farghānī (d. post A.H. 247 [A.D. 861/2]) noted doubts about the theoretical principles of astrolabe construction. These doubts had arisen from the poor presentation of the theoretical basis in existing manuals. He intended his own work, therefore, to resolve these problems and included extensive tables in it for constructing the circles on the latitude plates.[24] The earliest Islamic astrolabe to survive dates from late in the 9th century,[25] and the earliest dated example is by Basṭūlus, A.H. 315 (A.D. 927/8), whose name suggests that he was of Greek extraction but of whom little else is known.[26] A little later in the same century we catch a glimpse of how knowledge of the astrolabe spread when, in the dedicatory introduction to his treatise on the armillary sphere, Dunas ibn Tamīm (first half of 4th/10th century) recounts how he had described the different kinds of astrolabes to his patron, the Shaykh Abu'l-Ḥasan Muḥammad b. al-Ḥusayn, who evinced his interest, and complained that most people followed instruction books for using astrolabes without understanding the theory of the projection on which astrolabes are based.[27]

The eleven known astrolabes which may be dated before the year 1000[28] come from different parts of eastern Islam and thus attest the very early transmission from the Syro-Egyptian region through Ḥarrān to Iraq and Persia. Very soon, however, transmission occurred to the West – to North Africa (the Maghrib) and to Muslim Spain (Andalusia). Alongside this, however, there may also have been some transmission from a different cultural area, similar to what seems to have occurred with water-clocks – that is, from the Byzantine Empire directly to the Latin West.[29] One Byzantine astrolabe dated 6570 (= A.D. 1026) survives, although this was made for

22. Ibn an-Nadīm, cited from Dodge ii, 670.
23. Siddiqi, 246; more generally Maddison (I), 16–7; Maddison & Turner, 88.
24. Sergeyeva & Karpova, A.I. Sabra in *D.S.B.*, iv, 543, who comments that al-Farghānī's work 'ought to be counted among the more respectable treatises devoted to this subject in Arabic.' See also King (II), 53–5.
25. Maddison (I), 16, who gives reasons for this date.
26. Maddison & Brieux; King (III); Sezgin, 288.
27. Stern, 375.
28. These are by:

a) Khafīf, an apprentice of 'Alī b. 'Isà. Museum of the History of Science, Oxford.
b) Aḥmad b. Khalaf. Bibliothèque Nationale, Paris.
c) Basṭūlus, A.H. 315 (A.D. 927–8), Private Collection.
d) Basṭūlus, Museum of Islamic Art, Cairo.
e) Ḥāmid b. 'Alī al-Wāsiṭī, A.H. 348 (A.D. 959–60), Museo Nazionale, Palermo.
f) Ḥāmid b. 'Alī. . . . Museum of Islamic Art, Cairo.
g) Muḥammad b. Shaddād. Location not known.
h) Ḥāmid b. al-Khiḍr al-Khujandī, Private Collection.
i) Aḥmad and Muḥammad, the sons of Ibrahīm, Museum of the History of Science, Oxford.

To these may be added:

j) an unsigned *rete* now in the Museum of the History of Science, Oxford.
k) an unsigned astrolabe in the Museo di Storia della Scienza, Florence.

For the signed examples, see Brieux & Maddison under the maker's name. For brief descriptions of all except nos. d & e, see Maddison & Turner, 96–103 and references there given.
29. Turner (V), *s.v.* 'Water-clocks.'

a Persian and is inspired by Islamic models.[30] Ryan has suggested that the Western term *astrolabium* is more likely to be a Latinized form of Greek ’αστρολᾶβιον than a reconstruction from the Arabic *asṭurlāb*.[31] Most likely such transmission as occurred from Byzantium was more one of skills and men, of tools and artifacts, than of scholars and texts. It is perhaps for this reason that we know little about it.[32]

figure 14. Astrolabe by Muḥammad b. aṣ-Ṣaffār, Cordoba, A.H. 417 (A.D. 1026/7). Front (left) and back (right). Edinburgh, Royal Scottish Museum No. 1959-62. (Photograph: Royal Scottish Museum)

The first treatise on the astrolabe known to survive from Muslim Spain is by Ibn aṣ-Ṣaffār. He was the brother of the maker of the earliest dated astrolabe to survive from this region (fig. 14), Muḥammad b. aṣ-Ṣaffār (A.H. 417 [A.D. 1026/7]).[33] Knowledge of

30. It is now in the Museo Civico, Brescia; see Dalton. It is also discussed by Gunther (V), i, 104–8, and Destombes (I), 9–10. A photograph of a plate with Greek inscriptions, perhaps from an astrolabe and said to have been found in or near Lake Stryon, Salonica, soon after 1918, was shown to M. Henri Michel by Commandant Vivielle between 1933 and 1936. All traces of both object and photograph have now disappeared. Price (V), 262. That there was some late Greek/Byzantine instrument-making is attested by surviving objects.
31. Ryan, 155–6, but *cf.* the critique by Segonds (II), 23–4 who concludes that the form ’αστρολᾶβιον never existed in Greek.
32. See Price (V); Brieux, (IV), 110–6. For the astrolabe in medieval Byzantium, see the excellent survey of treatises in Segonds (II), ch. 2, *passim.*
33. Maddison & Turner, 104. The instrument is now in the Royal Scottish Museum, Edinburgh. See also Plenderleith. The treatise by Ibn aṣ-Ṣaffār was translated into Latin before *c.* A.D. 1145 by Plato of Tivoli. Millás (VI), 151. For the life of Ibn aṣ-Ṣaffār and the Arabic text of his treatise, see Millás (V), 41, 47–8.

the instrument, however, was probably transmitted earlier. Destombes[34] has pointed out the role of Maslama b. Aḥmad al-Majrīṭī (d. *c.* A.H. 398 [A.D. 1007/8]), the teacher of Ibn aṣ-Ṣaffār, in transmitting star tables and astrolabe treatises, notably Ibn aṣ-Ṣufī's enormously comprehensive treatise with 386 chapters, to Spain at the end of the 10th century. Already in the same period there is some evidence of the beginning of a transmission of Arabic learning from Spain to Christian Europe. The route this transmission followed seems to have been from Cordoba (capital of the Ummayad Emirs) to Barcelona, and from there across the Pyrenees, via Vich and Ripoll, to Lorraine, Southern France, and Germany. A crucial document which supplies evidence for the early transmission of Arabic science to Christian Europe is a miscellaneous collection of scientific treatises translated from Arabic and assembled, it would seem, in the middle decades of the 10th or early decades of the 11th century in the *scriptorium* of the monastery of Santa Maria de Ripoll, a Benedictine monastery at the foot of the Pyrenées, for the use of scholarly brothers within the house itself. This collection, Ms Ripoll 225 (fig. 15), now in the Royal Archives of Aragon, includes at least eleven sections dealing with aspects of the astrolabe. Several of these are direct translations from Arabic.[35] From Catalonia, knowledge of the astrolabe spread, from the end of the 10th century onwards, into Lorraine, to Liège,[36] to Gorz, to Cologne, and so to the heartlands of Europe and beyond. An important agent in popularizing Arabic knowledge at this stage was the great scholar Gerbert, later Pope Sylvester II, who in 967 had been sent to Vich to study with Hatton. He spent three years in Spain studying music and mathematics, and was later renowned throughout Europe for his astonishing learning. Gerbert remained in close contact with Spain. Some of his letters requesting books and translations have survived, and it was possibly through him that early translations into Latin of texts on the astrolabe made by Llobet of Barcelona reached Latin Europe.[37] Possibly some examples of the instrument were also diffused at this time. One astrolabe has been claimed as deriving from the Ripoll-Barcelona region at this period. Neither signed nor dated, it carries letter numerals in Latin script (a script not without some Byzantine characteristics) and has a general resemblance in its *rete* design to slightly later examples from Cordoba by Muhammad aṣ-Ṣaffār, with the exception of the shape of the star pointers.[38] Knowledge of the instrument spread to Northern Europe

34. Destombes (II), sect. vi. See also Millás (V), 38–40. Lindberg, 59–62. For a later possible influence of Maslama on the astrolabe treatise in the *Libros del Saber*, see Samsó.
35. Millás (VI), 139–42. See also Destombes (II). For Ripoll, see Riché 157–61, and for its close relations with Fleury, 145.
36. For Lorraine, see Welborn, Thompson. For Liège and other urban centres of learning, Riché, 165 ff.
37. Destombes (II), 420–44; Millás (VI), 142–3; Poulle (I). 83–4. For Llobet or Lupitus, see Lattin (I). Lattin (II), 214–5, however, argues strongly that Gerbert did not use an astrolabe in his teaching even if he was acquainted with it. For the wider context of European learning in which all this should be set, see Riché, 131–3.
38. Destombes (I), 2–7, and *cf.* Beaujouan, 658–61. For Muḥammad b. aṣ-Ṣaffār, see Mayer (I), 75; Brieux & Maddison; Plenderleith.

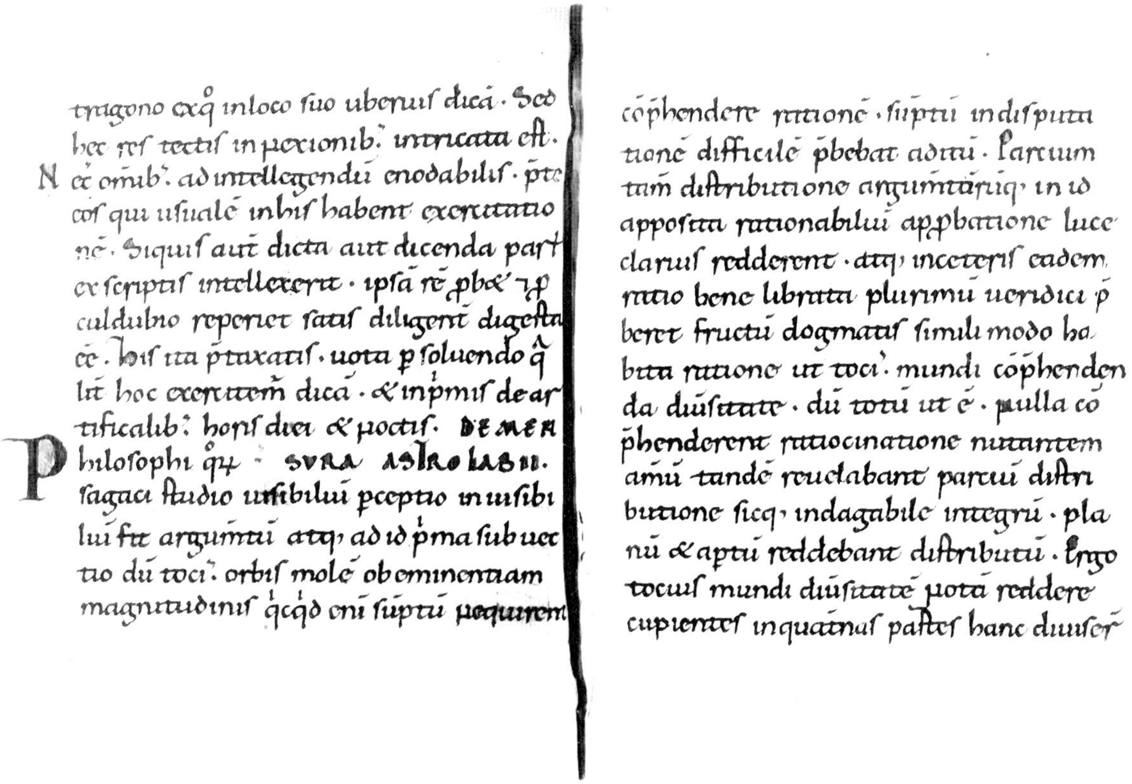
trigono exq̄ in loco suo uberius dicā. Sed
hec res tectis in plexionib: intricata est.
N ec omib: ad intellegendū enodabilis. ptē
eos qui usualē in his habent exercitatio
nē. Siquis aut̄ dicta aut dicenda part
ex scriptis intellexerit. ipsā rē pbē ꝛ p
culdubio reperiet satis diligent̄ digesta
ēe. his ita ptaxatis. uota p soluendo q̄
lit hoc exercitiū dicā. & in pmis de ar
tificalib: horis dici & noctis. DE MEN
P hilosophi q̄ꝝ · SVRA ASTROLABII.
sagaci studio uisibiliū pceptio inuisibi
liū fit argumtū atq; ad id pma sub uec
tio dū toci: orbis mole obeminentiam
magnitudinis qcq̄d oni suptū pequirent
cōphendere ratione. suptū indisputa
tione difficile pbebat adiri. Partium
tam distributione argumtāriq; in id
apposita rationabiliū appbatione luce
clarius redderent. atq; in ceteris eadem
ratio bene librata plurimū ueridici p
beret fructū dogmatis simili modo ha
bita ratione ut toci: mundi cōphenden
da diūsitate. dū totū ut ē. pulla cō
phenderent ratiocinatione nutantem
amū tandē reuelabant parciū distri
butione sicq; indagabile integrū. pla
nū & aptū reddebant distributū. Ergo
tocius mundi diūsitate pota reddere
cupientes in quatnas partes hanc diuser

during the 11th century. The great teacher, Fulbert of Chartres (*c.* 970 – 1028), who was possibly a pupil of Gerbert, included among his poems a mnemonic jingle for remembering the names of the ecliptic stars on the astrolabe and a list of Arabic names for parts of the astrolabe with their Latin equivalents.[39] In 1025, Rudolph of Liège, who had been a pupil of Fulbert, writing to a monastic friend, Ragimbold, says that he has an astrolabe: 'I would willingly send it to you in order that you might examine it, but it is my model. If you wish to know what it is, come to the festival of St. Lambert, you will not be sorry for it. It would be useless for you just to *see* an astrolabe.'[40] Some measure of the speed of transmission is suggested by the fact that at the monastery school of Reichenau in Carinthia, Hermann Contractus (1013–54), despite his distance from the centres of Arabic learning, wrote treatises on the astrolabe which are largely adaptations of the translations of Llobet of Barcelona and thus closely related to Ms Ripoll 225.[41] The historical importance of this development is enormous, although the treatises are, to later eyes, somewhat defective. A few decades later Walcher of Lorraine had an astrolabe in England which he used during an eclipse of the moon on 18 October 1092.[42]

figure 15. A page from Ms Ripoll 225 ff. 1v-2r. Archives of the Crown of Aragon.

There were, however, still very few Latin treatises. Those that existed, moreoever, were poorly arranged and erroneous in the

39. Behrends, 260–1 and *cf.* xxvii–viii; Behrends & McVaugh, *passim*.
40. Cited from Welborn, 191. *Cf.* Beaujouan 661.
41. Millás (VI), 146; Poulle (I), 84; Riché, 273–4. For Hermann's life, see Kren.
42. For this episode, see Haskins, 113–5; Southern, 166–7. Its possible significance is underlined by Pedersen (II), 69, but for a quite different interpretation, see Poulle (XIII).

graduation of the ecliptic and the treatment of right ascension.[43] Even in the 11th century, however, some improvements had taken place in the presentation of the treatises, and as knowledge of their contents spread, new treatises were produced either by way of new translations from Arabic or by the combination of matter from existing works. In the course of the 12th and 13th centuries, a corpus of writings on the astrolabe was established. In 1143, Ptolemy's *Planisphaerium*, which provides a general treatment of stereographic projection, was translated by Hermann the Dalmatian (Herman of Carinthia), who moved from France to northern Spain and then to Languedoc in order to study the new philosophical works there available. In 1143, Hermann was living at Bezières, where he was in correspondence and close contact with the Archdeacon of Pamplona, the Englishman Robert of Ketton (Robert of Chester). By 1147, Robert had returned to London, where he wrote a treatise on the astrolabe calculated for the latitude of that city.[44] In making his translation of *Planisphaerium*, Hermann was assisted by his pupil Rudolph of Bruges, who himself wrote a treatise on the construction of the astrolabe, adding to it a few notes on its use. This treatise Rudolph dedicated to Johannes David, one of the earliest of the important group of translators working at Toledo.[45] Shortly afterwards Johannes Hispalensis, who may perhaps be identified as John of Seville,[46] translated the treatise on the use of the astrolabe by Ibn aṣ-Ṣaffār, later to be falsely attributed to Māshā'allāh or (but less frequently) to Maslama.[47]

In all this work of translation and transmission an important role was played by Jewish scholars working both independently and in collaboration with Mozarab and Christian writers. As examples of such men in the early 12th century, we mention Mosé Sefardi of Huesca, who converted to Christianity under the name of Pedro Alfonso;[48] Abraham bar Hiyya of Barcelona, who collaborated with Plato Tiburtinus (of Tivoli) to produce Latin versions of Arabic texts, among which was Ibn aṣ-Ṣaffār's treatise on the astrolabe; and Abraham ibn'Ezra of Toledo, who wrote a Hebrew treatise on the astrolabe which itself had received a Latin version by 1160.[49] The tradition of translation into Hebrew and Latin thus established continued to flourish in Spain throughout the 12th and 13th centuries whence it spread to southern France, particularly to Provence and Languedoc, where perhaps the most notable exponents

43. Poulle (I), 84; Vyver, who describes the 10th-century treatises as 'anonymous, brief and badly composed.' Poulle (XI), 33, suggests that the deficiences of description and terminology in the early treatises resulted from the fact that, lacking Western instruments, scholars were forced to use imported Arabic instruments which they could only imperfectly comprehend.
44. For Hermann, see Haskins, ch. III, esp. 54–6; for Robert's treatise, *Ibid.*, 122, & Thorndike (II), 467, n. 1.
45. See Kunitzsch (II), Appendix, who argues that Rudolph's treatise was not a translation.
46. The identity of Johannes Hispalensis is a matter of much debate. For an introduction to the problems, see d'Alverny.
47. Kunitzsch (II), 49.
48. He was the teacher of Walcher of Lorraine mentioned above.
49. Millás (VI), 147–53, and Levy, who raises a doubt about the ascription. For some mss of Ibn Ezra's work, see Gandz, 471–2.

of this phase of transmission were the Qimh and Ibn Tibbon families. Their work continued over many decades,[50] but alongside them there were dozens of other equally able, if less prolific, and long-enduring translators. To give a single example: Rudolph of Bruges, in the preparation of his work on the astrolabe, is said to have been assisted by Abraham Savasorda (d. 1136). The involvement of Jewish scholars in the work of translation had two results for the history of the astrolabe. First, more texts were made available from the Arabic more quickly than would otherwise have been the case; second, the astrolabe itself became familiar to Hebrew scholars who had previously known nothing of it. In the 13th and 14th centuries, indeed, the astrolabe became sufficiently well-known to play a part in biblical exegesis and criticism,[51] although it never became familiar in Hebrew culture and never enjoyed, in Jewish society, the role that it held in the Latin West.[52]

By the 13th century, the astrolabe was absorbed into the learning of Christian Europe and had an established place in its culture. In the late 12th century a number of works were produced, such as the treatises by Robert of Chester, Arialdus, and Raymond of Marseilles, which were not translations.[53] Two theoretical treatments also appeared: *de Plana Spera* by Jordanus de Nemore (*fl.* late 12th or early 13th century), on the stereographic projection,[54] and the *Demonstrationes pro astrolapsu* ascribed, on weak evidence, to Jordanus and, on no evidence at all, to Campanus de Novara (d. 1296).[55] At the same time, two important developments took place which illustrate how well-assimilated in Christian learning the astrolabe then was. First, one or several scholars in the course of the 13th century developed a composite 'Making and Use of the Astrolabe' treatise based in part on the treatise of Ibn aṣ-Ṣaffār, in part on other Arabic texts, and in part on the Latin version of Ibn aṣ-Ṣaffār by Johannes Hispalensis, revised and augmented.[56] This compilation was widely used – almost 200 manuscripts of it have survived[57] – and was falsely ascribed to Māshā'allāh, who seems in fact not to have written on the astrolabe. Nonetheless it is of great importance, for it provided a sort of canon, a standard textbook for teaching and a basis for new developments.[58] Second, in the latter half of the 13th century, Pierre

50. Millás (VI), 158–9, 164–5. *Cf.* below on the work of Profeit Tibbon and the 'new quadrant.' We should also note here the importance of the peripatetic life of many Hebrew scholars in spreading scientific knowledge.
51. Gandz, 480–2.
52. A very few astrolabes inscribed in Hebrew survive which are of late-medieval origin.
53. Poulle (I), 87–8; Poulle (VII), 867, who describes Raymond's treatise as the first written in the Latin West which was neither a translation nor a servile imitation of an Arabic text, but a true original work.
54. Edited by Thomson, who indicates (52) that it is more purely theoretical and more formal and complete than Ptolemy's *Planisphaerium*.
55. *Ibid*, 55–8, and Appendix 2.
56. Kunitzsch (II), *passim*.
57. *Ibid*, 42.
58. Poulle (I), 94. 'All the treatises resemble each other, and it is not impossible that the text of Pseudo-Messahalla, in its corrected and widespread thirteenth century version, had played a polarizing role offering a "canon" that each writer adapted to his personal taste while retaining the general framework.'

de Maricourt (Petrus Peregrinus, *fl.* 1269)[59] and Henry Bate of Malines (1246– *c.* 1310)[60] both wrote treatises in which they describe their attempts to develop the astrolabe in original ways.

Islamic Astrolabes

By the 11th century, the astrolabe was familiar to learned men throughout Islam. Sources of metal were abundant in the Islamic world, and in both the Syro-Egyptian region and in Persia there were long-established traditions of fine metal-working which extended to North Africa and to Islamic Spain. With these earlier traditions of metal-working it is clear that the astrolabe- and globe-makers had close technical and familial connections,[61] but it is also clear that most astrolabists were specialists. This was perhaps inevitable, for the maker of an astrolabe had to be more than a metal-worker. He had to have the geometric skill to lay out the instrument and the astronomical skill to determine what he wished to place on it and to carry out the necessary calculations. Many of the early astrolabists, in fact, were primarily astronomers[62] working in metal only in the service of their astronomical studies; it is also true that even from a relatively early period some division of labour occurred between the designer and the constructor of an instrument. In later periods, especially among the astrolabe-makers of Safavid Persia and Mughal India, division of labour became more common, leading on occasion to three men working on an instrument – a designer, an engraver, and a calligrapher/decorator.[63]

Such, however, was an extreme case. More usually the astrolabist worked independently. He might pass on his knowledge orally to pupils and apprentices; occasionally he might embody it in a treatise.[64] Certainly he took a pride in his work,[65] justified not only by the beauty and utility of the objects he produced, but also by the prestige and value accorded to the astrolabe in Islamic society: 'For

59. Poulle (I), 91–3, who says of Peregrinus' treatise, 'par sa clarté, sa precision, son souci d'être complet sans prolixité, il est une des meilleures productions de ce genre au Moyen Age.' (By its clarity, its precision, its concern to be complete without prolixity, it is one of the best medieval works of this kind.)
60. *Ibid*, 93–4. Bate's astrolabe was a purely astrological instrument. For the Latin text, see Gunther (V), 368–76. *Cf.* Levy.
61. Mayer (I), 21.
62. *Ibid*, 13–4. The most striking example is perhaps that of Ḥāmid b. al-Khiḍr al-Khujandī (d. A.H. 390 [A.D. 999/1000]), an important astronomer and mathematician of ar-Rayy from whom one signed astrolabe survives. See Maddison & Turner, 102. Some 700 years later we may note a comparable attitude. Astrolabes made by mere instrument-makers, Sir John Chardin noted in Persia, were not esteemed as highly as those made by astronomer/mathematicians. But 'Il faut ajoûter à cela qu'un *Astronome* n'est point mis au rang des *Savans,* s'il ne sait faire tous les *Instruments* lui-même, et s'il n'y travaile mieux qu'un habille artisan.' (One must add that an astronomer is not considered a *savant* if he does not know how to make all the instruments himself, and if he does not work them better than a skilled craftsman.) Chardin. ii. 121.
63. E.g. the superb astrolabe made for Shāh 'Abbās II of Persia in A.H. 1057 (A.D. 1647/8) now in the Museum of History of Science, Oxford. This is signed 'Made by Muḥammad Muqīm al-Yazdī under the supervision of Muḥammad Shāfī, the astronomer of Janābād' and separately 'Engraved by Faḍl Allāh as-Sabawarī.'
64. E.g. 'Alī b. 'Isā al-Asṭurlābī, astronomer and astrolabe-maker.
65. Mayer (I), 15.

a piece of copper worth 5 dirhams, when wrought into an astrolabe is worth 100 dirhams, and this price is not for the matter but for the form that has been impressed upon it.'[66]

By the 10th century A.D., from which the earliest astrolabes survive, the instrument had already attained its basic form which was to remain unaltered during the succeeding nine centuries. This basic form, however, had only recently been established. Al-Bīrūnī in his *Ifrād al-maqāl fī amr aẓ-Ẓilāl* (The Exhaustive Treatise on Shadows) reports that invention of the shadow square was ascribed in some treatises to al-Khwārizmī (d. *post* A.H. 233 [A.D. 847/8]), whereas his remark upon the alidade, 'that it can be complete, according to the old custom, or it can be halved and edged sword wise, as is the modern custom,' may indicate the moment when the alidade was changing in form. This change was perhaps from a broad strip carrying at each end sight vanes with a pointer in the centre, to the familiar straight bar alidade with the diametral edge passing through the centre of the astrolabe chamfered like a sword blade[67] (fig. 10a & b).

In later periods, however, there was very little development of the planispheric astrolabe. More scales were added to the instrument, giving it a greater versatility, and decoration was elaborated. The fundamental principles, however, remained the same, and although the astrolabes produced in the different regions of Islam developed characteristic stylistic features which serve to differentiate them one from another, there was no radical innovation which would have made it impossible for an astronomer of the 10th century to use one made in the 19th century.

PERSIA AND EASTERN ISLAM TO *c.* A.D. 1500

From the Syro-Egyptian region, the making of astrolabes spread rapidly to Eastern Islam and Persia. It is from these regions that most of the earliest surviving astrolabes derive. They are characterized by a spare, undecorated appearance and a triangular *kursī* pierced by two as-yet-unexplained holes, one on either side of the shackle, and with short, straight, dagger-shaped star pointers springing from squarish bases, with subsidiary points on either side (catalogue no. 1). Generally, rather few stars – perhaps less than twenty – are marked. The back of the instrument is usually rather bare, with few scales, and the inscriptions are marked in a plain, elegant *kufic* script.[68]

66. Ikhwān aṣ-Ṣafā' (Brethren of Purity), cited from Lewis (I).
67. Kennedy (II) i, 115, 116; ii, 55, 57. *Cf.* the remark at i, 236.
68. For more detailed discussion of the early astrolabes, and of regional variation, see Maddison & Turner, 96–123; Michel (III), 149–57. *Kufic* was the earliest formal hand developed by the Arabs to become generally used throughout Islam. It takes its name from the Iraqi town of Kufah and was already perfected by the 2nd/8th century. *Nashkhī* is a cursive hand which dates back to the 2nd/8th century. Reformed and perfected by Ibn Muqlah and Ibn al-Bawwāb during the 4th/10th century, it gradually displayed *kufic* in eastern Islam during the 9th–10th/15th–16th centuries. In western Islam, however, *kufic*, or scripts derived from it, remained in use until modern times. For a detailed discussion of Islamic scripts, see Safadi, especially 10–3, 19, 42–51, 62–7.

figure 16. Astrolabe by Ḥāmid b. al-Khiḍr al-Khujandī A.H. 374 (A.D. 984/5). Face (left) and back (right). Private Collection. (Photograph: Alain Brieux)

Even among the earliest astrolabes, however, the two Persian examples that survive – by al-Khujandī (fig. 16) and the brothers Aḥmad and Muḥammad b. Ibrahīm – are a little more ornate than the non-Persian examples. They have a higher and more decorative *kursī*, more scales on the back of the instrument, and rather more fluid star pointers, some of which are treated zoomorphically. They also mark more stars on the *rete*, although this is otherwise close in design to those of the non-Persian instruments. The elaboration of the Persian astrolabes seems to have begun at an early date, for al-Bīrūnī in his *Kitāb at-Tafhim li awā'-il ṣinā'at at-Tanjīm* (Book of Instruction in the Elements of the Art of Astrology), A.H. 420 or 421 (A.D. 1029/30), can already list most of the scales characteristic of the later elaborate instruments. 'In the larger astrolabes the back is not only used for measuring altitude and shadow, but affords space for a great deal of astrological information. The signs are there with their faces, terms and their lords, the Mansions of the Moon, the parallelogram of two shadows with a table of triplicities in the interior; sines, cosines in the upper-left quadrant and, in the right, the parallels of the signs, the meridian altitude of the sun at various latitudes... and its altitude at various places, when it crosses the azimuth of Mecca...'[69] These features remained characteristic of Persian astrolabes although in about the middle of the 9th/15th century a more fundamental change occurred with the displacement of the *kufic* script by the *naskhī* script.

69. Wright, 195. It seems clear from al-Bīrūnī's remark that only large astrolabes were treated in this way. This may account for the relatively plain backs of the surviving small astrolabes.

SPAIN AND THE MAGHRIB

In the same way that knowledge of the construction and use of the astrolabe was quickly diffused throughout Eastern Islam, so in the 3rd/9th and 4th/10th centuries it penetrated westwards into Muslim Spain and the Maghrib. The first Spanish treatise was written by Abū-l-Qāsim Aḥmad b. 'Abdallāh b. 'Umar, known as Ibn aṣ-Ṣaffār (d. A.H. 426 [A.D. 1034/5]). His brother, Muḥammad b. aṣ-Ṣaffār, was a famous astrolabist and the maker of the earliest astrolabe from this region to survive.[70] (fig. 14). Already, despite a family resemblance to the instruments of Eastern Islam, this astrolabe shows some features characteristic of Western instruments. The *kursī* is lower and generally undecorated. On later instruments it changed and sometimes became high, with a European-type shackle and swivel-ring suspension. The star pointers are already a little curved in form and would develop into fine wavy pointers often studded with brass or silver knobs. The name of the place for which it was intended to be used, as well as the latitude were marked on each plate, and the back received the major addition of a zodiacal calendar. The shadow square tended to be divided only into six or twelve parts, and not to use the seven-part division found on Eastern astrolabes.

From the time of Muḥammad aṣ-Ṣaffār onward, there is a steady tradition of astrolabe-making in Spain, which came to an end only with the *Reconquista*, (i.e. the prolonged reconquest of Islamic Spain by the Christian princes of the north), the fall of Granada in 1492, and the expulsion of the Moors. Even then the tradition did not end. For not only had it already influenced the design of the earliest Latin astrolabes, but it also now moved to North Africa. Here a strong line of astrolabe-making developed which was conservative in style, although a few changes were introduced (viz. those to the suspension apparatus mentioned above). This tradition flourished throughout the 11th/17th and 12th/18th centuries, and survived, indeed, almost to the present day. The manufacture seems to have been particularly well-established in Morocco, where the prolific maker Muḥammad b. Aḥmad al-Baṭṭuṭī was working in the early 12th/18th century,[71] and where one of the few surviving Islamic astrolabists, Muḥammad Laḥbābī, in 1974 was *muwaqqit* (i.e. calculator of the hours for prayer) of the Qarawiyyīn Mosque, Fez.[72]

PERSIA AND INDO-PERSIA AFTER *c.* A.D. 1500

It is a testimony to the conservative nature of later Islamic culture that at just the epoch when the astrolabe was obsolescent in the Christian West (the late 16th/17th centuries A.D.), it should have been the subject of a remarkable efflorescence of manufacture and

70. It is now in the Royal Scottish Museum, IC/CCA 3650. See Plenderleith; Brieux & Maddison. *Cf.* n. 33 above.
71. About twenty astrolabes by him are known. See Brieux & Maddison; Mayer (I), 60.
72. Brieux (II).

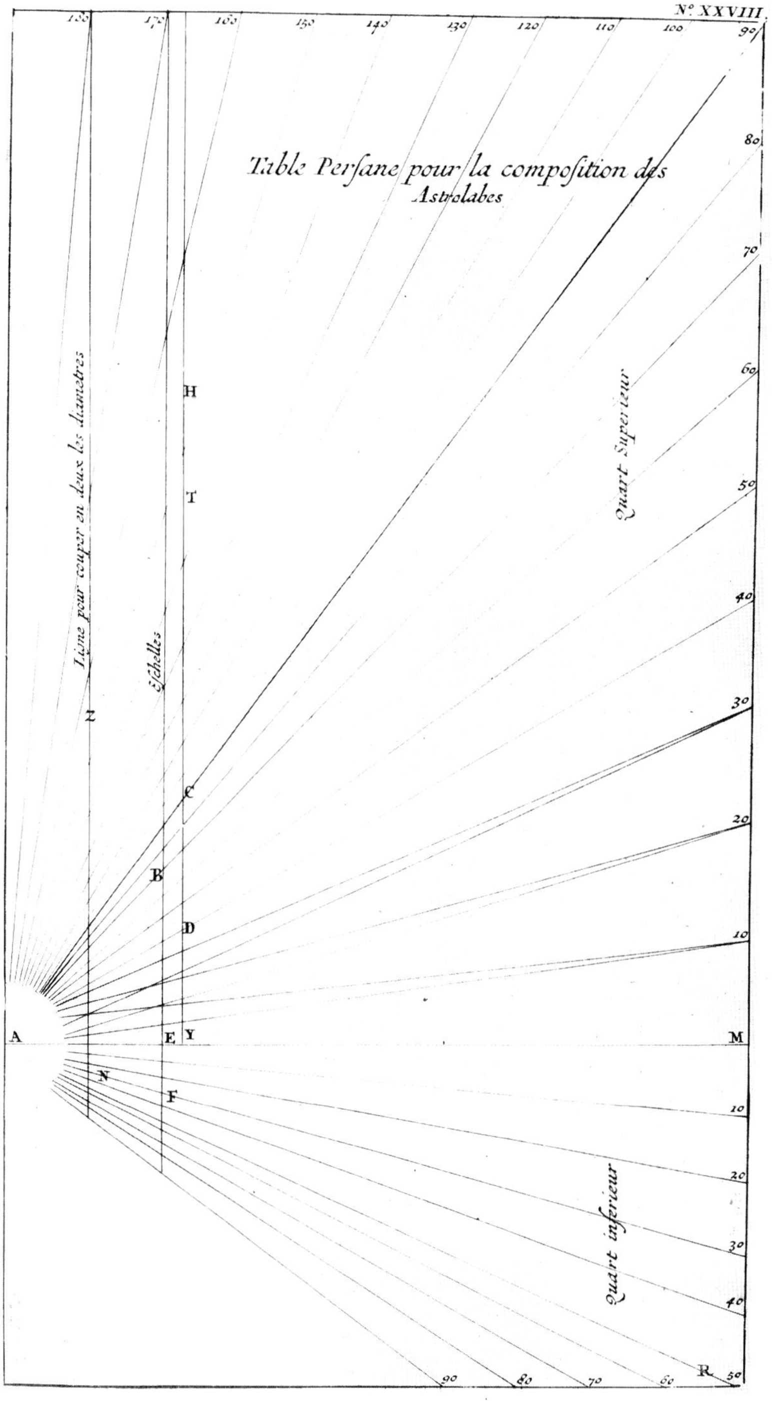

figure 17. Illustration of a *Dastūr*. From Sir John Chardin, *Voyage en Perse*... (new edit.), Amsterdam, 1735, iii, pl. XXXVIII facing 171. Chardin's description of the instrument translates freely as follows: 'But the main instrument that they have for the precise and exact construction of their astrolabes (and which I believe they alone use, to the exclusion of Europeans) is a board which they call *dastūr*, or 'rule,' which among them is a name common to all ways of working. This plate is of brass, of the thickness of an *ecu*, of the length of a foot, and of the width of half a foot, well-polished and untarnished...

'At a quarter of the plate – that is to say, a height of 3 inches – they take the centre A, where they draw a semi-circle, of which the semi-diameter is cut by a line at right angles to its diameter, that is... AEM, by which the figure is divided into two quarters of 90, the one 9 inches large, that is the upper, and the other small, which is called here the lower quarter and has only 3 inches. The upper quarter is divided into 180 equal parts, or degrees, of which the lines, drawn from the centre to the circumference, end at the edge of the plate, leaving only some space to mark the number by tens starting from the semi-diameter AEM. The lower quarter is also like the upper quarter divisible in 180 equal parts, but they only mark half the lines or degrees as you see, and leave the part from the other 90 degrees empty, without drawing anything there as it does not serve them for anything....' (Photograph: British Library, London)

decorative achievement in Persia and in Mughal India. There were no fundamental innovations, but the astrolabes produced in Safavid Persia and Mughal India developed their own characteristic styles which distinguish them from the earlier astrolabes.

Surviving instruments from 11th- and 12th-century (17th- and 18th-century) Persia are relatively numerous (catalogue nos. 6–10). They are distinguished by a highly ornate, but balanced, style with delicate *retes* designed as a mass of flowing stalks and tendrils, the star pointers in the form of leaves. The *kursī* is high and engraved with patterns or with inscriptions. Religious or dedicatory inscriptions or Persian verses may appear on any blank space which the instrument offers, including the edge of the limb (catalogue no. 6). The script is *naskhī* or *nasta'līq*, never *kufic*. The decorator or engraver of the instrument was sometimes a different person from the astrolabist who calculated it, and quantities of scales of purely astrological use were usually added to the back of even quite small instruments. The astrolabes were very highly prized. In his remarkable eye-witness description of the making of astrolabes at Isfahān, Sir John Chardin remarked how carefully they were preserved and how they were regarded, even by the common people, as precious as jewels.[73] Such, indeed, they almost were, so extensive was the decoration added to them. Although magnificent, they are self-conscious and somewhat mannered pieces.

Interest in the astrolabe, and perhaps knowledge of its construction, appears to have passed to India in the wake of its Mughal con-

figure 18. Observers at work in the observatory of Taqī ad-Dīn at Istanbul *c.* A.H. 985 (A.D. 1577/8). The two observers at the right behind the table are working with an astrolabe, and a *saphea* lies on the table slightly left of centre. Full-page miniature from the *Shāhinshāh-nama* (History of the King of Kings), an epic poem by 'Alā ad-Dīn Mansūr- Shīrazī, written in honour of Sulṭān Murād III (reigned A.H. 982-1003 [A.D. 1574/5–94/5]). Istanbul Üniversıtesi Kutüphanesi, Ms Yıldız 2652/ 260 (=F–1404) Fol. 57r. (Photograph: Robert Harding, London)

73. 'Car les *Persans* le tiennent toûjours dans les étuis & les sacs, quoi que l'air de *Perse* n'enrouille, ni ne salisse & ne ronge pas les corps comme il fait dans nos Païs Septentrionaux: parmi le commun peuple même chacun garde son astrolabe comme un bijou.' (Because the Persians always keep [astrolabes] in cases or bags so that the air of Persia does not rust, dirty, or eat away the body, as it does in our Western countries; even among the common people each keeps his astrolabe like a jewel.) Chardin, iii, 121. The passage concerning astrolabe manufacture has been partly reprinted with commentary by Michel (II).

querors. Muḥammad Humāyūn (reigned A.H. 937–63 [A.D. 1530–56]), the son of the founder of the dynasty, Zāhir ad-Dīn Bābūr (reigned 932–7 [A.D. 1526–30]), was immersed in astrology, which he himself taught while in temporary exile and by which he regulated his own affairs and those of the court.[74] It was for this reason that astronomy and the astrolabe were of interest and that a manufacture of globes and astrolabes was developed in Mughal India. The name of one of Humāyūn's instrument-makers, Maulānā Maqṣūd Hirawī, is mentioned in the chronicle of his son's reign (Jalāl ad-Dīn Akbar, reigned A.H. 963–1014 [A.D. 1556–1605]), the *'Aīn-i-Akbarī*, but otherwise only one group of globe- and astrolabe-makers is known. It was established at Lahore and its members belonged to the same family in different generations. Most of the several unsigned surviving Indo-Persian pieces can be ascribed to their workshops on stylistic grounds.[75] The earliest member of this dynasty was apparently the Ustādh Shaikh Allāh-dād Āsṭurlābī Humāyūnī Lāhūrī. Two of his astrolabes, one dated A.H. 975 (A.D. 1567/8), survive[76] and he had six known successors[77] who span the 11th/17th century. That few other Indo-Persian astrolabists have been identified[78] suggests that astrolabes were far from being widely spread in Mughal society, and perhaps remained the exclusive preserve of a court coterie.

The astrolabes produced by the Lahore school are recognizably influenced by Persian instruments of the same period, but tend to have more complicated *retes* because of the greater number of stars marked (catalogue no. 4). By contrast, there is far less decorative engraving. The *kursī* is high and decorative, as on the Persian instruments, but it is pierced rather than chased.[79] The back of the instrument carries even more scales than the Persian instruments, and the graph of the meridian altitude of the sun is plotted on equidistant rather than stereographically projected arcs of the signs of the zodiac (fig. 187). It is only on Indo-Persian instruments that the more unusual *rete* designs and arrangements described in the earlier treatises are found executed,[80] a fact which might suggest a degree of conscious antiquarianism about the development.

INDIA

In his account of his travels in India, al-Bīrūnī remarked that while there he taught the (Hindu) pandits how to use the astrolabe.[81] Several centuries later, in Śaka 1292 (A.D. 1370), the court astrologer of Bhringapura, Mahendra Sūri, a pupil of Madana Sūri, compiled a Hindu treatise (*Yantrarāja*) on the astrolabe from earlier (pre-

74. Nadvi (I), 625.
75. Nadvi (I & II) *passim*; Maddison & Turner, 117–9.
76. Maddison (I), 20, no. 159.
77. For a family tree, see Maddison & Turner, 119.
78. For two examples, see *Ibid*, 26, no. 164; Brieux & Maddison, *s.v.v.* Muḥammad Ṣalīḥ and Balhumal.
79. Maddison (I), 21.
80. See, for example, two specimens of the 'ship' *rete* in Gunther (V), i, 218–9.
81. Maddison & Turner, 120.

figure 19. An astrologer (? al-Harīth) (top right), probably using an astrolabe to time the moment of birth to cast a horoscope. Miniature, illustrating *Maqāma* 39 of the *Maqāmat* of al-Harīrī painted by Yahyā b. Maḥmud al-Wāsitī, A.H. 634 (A.D. 1236/7). Paris, Bibliothèque Nationale, Ms Arabe 5847 fol. 122v. (Photograph: Bibliothèque Nationale)

sumably Arabic and Persian) works. This treatise was intended to be a summary. 'Many *Yavanas* [Muslims] have composed scientific works on the astrolabe in their own language and according to their own particular understanding . . . Having found them like oceans I now compose this work, like nectar as the essence of them all.'[82] If al-Bīrūnī's teaching and Mahendra Sūri's writing stimulated interest in the subject, we as yet know little of it except a commentary on Mahendra Sūri's treatise written by his pupil Malayendu Sūri in about 1382.[83] No surviving Indian astrolabe predates the later 16th century,[84] but a number of treatises on the instrument were composed in Gujarat and Rajasthan between the 15th and 18th cen-

figure 20. An astrologer observes with an astrolabe while his assistant consults tables in a book. Detail from a Persian or Indo-Persian engraved metal disc, perhaps once part of a monumental astronomical clock. ?17th century A.D. London, Victoria & Albert Museum, 1577–1904. (Photograph: Victoria & Albert Museum)

82. Kaye (I), 3–4. Traces of Islamic influence on Indian astronomical concepts have also been noted in works of the 10th to 12th centuries A.D. Pingree (II), 625–6. Pingree (III) *passim*. The sine table and list of astrolabe stars from Mahendra Sūri are printed in Pingree (II), 626, 628.
83. Pingree (III), 318, and n. by Pingree in Kennedy (II), ii, 53.
84. At least one early instrument, signed in *kufic* by Maḥmūd b. 'Alī b. Yusha' al-[. . .]ri and dated A.H. 669 (A.D. 1270/1), however, arrived in India and was there modified since it had added to the *umm* a gazetteer of 12 Indian places between latitudes 33°45′ and 16°20′ inscribed in *devenāgari* script. There is also a double plate for latitudes 24° & 30°/27° & 32° also marked in *devenāgari* characters (Morley [I], 32–4; Mayer [I]. 57). an instrument by 'Alī b. Ibrāhīm of Taza (Morocco) dated either A.H. 724 (A.D. 1323/4) or A.H. 709 (A.D. 1309/10) now in the Musée d'Histoire des Sciences. Genève (Inv. 1051, see Archinard) has been more drastically modified, the *rete* and all the plates having been replaced by those of local Indian manufacture. Clearly such instruments could have influenced the Indian astrolabe-making tradition, but in the absence of knowledge of the dates of their arrival in India, it would be unwise to base any speculations upon them.

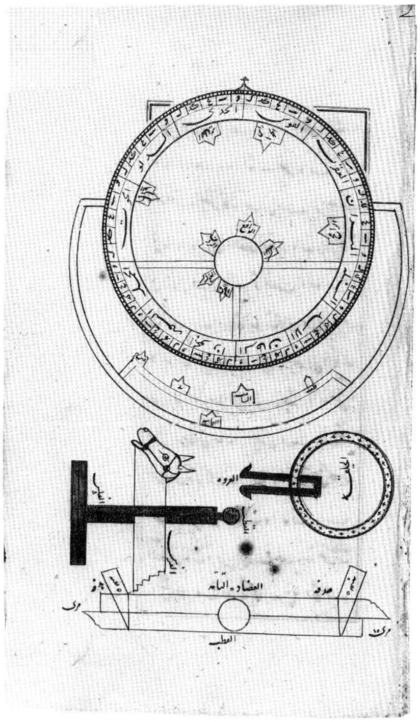

figure 21. *Rete*, alidade, pin with its horse, and suspension ring for an astrolabe. From a copy dated A.H. 822 (A.D. 1419/20) of al-Bīrūnī's *Kitāb fī istī'āb al-wujūh al-mumkina fī ṣanāt al-asṭurlāb* (A Book of Detailed Treatment of All Possible Ways of Making an Astrolabe). Oxford, Bodleian Library Ms Marsh 701 fol. 227r. (Photograph: Bodleian Library)

turies.[85] Among the instruments that survive, we may discern two traditions. The first of these stemmed from the manufactory established at Lahore already described. The second is quite different, being exemplified by a group of instruments inscribed in the *devenāgari* script of northern India and characterized by a marked resemblance in their design, both on the *rete* and on the back, to the earliest astrolabes known from Syro-Egypt and Persia (catalogue nos. 12–14). This circumstance suggests[86] that their design was influenced by illustrations in copies of early treatises on the astrolabe such as those by as-Sijzī, al-Bīrūnī (fig. 21), or aṣ-Ṣūfī. Since most of the surviving astrolabes appear to be no earlier than the 18th century, however, one may conjecture that they were produced following the stimulus given to astronomy by the activities of the Mahārāja Jai Singh II (1686–1743), who between 1728 and 1743 built five astronomical observatories for his master, Muḥammad Shāh,[87] and to whom 'a conventional Sanskrit treatise on the astrolabe is attributed.'[88]

The typical Hindu astrolabe (*Yantra rāja:* 'king of instruments') consists of a single sheet of metal over which rotates the *rete* (catalogue nos. 12 & 14). There is no *mater*, limb, or set of latitude plates, the instruments being made for use in one latitude only–that for which the co-ordinates are engraved on the side of the plate to which the *rete* is affixed. On the back, a limited number of scales, usually only those customarily found on the backs of the earliest astrolabes, are marked. Unlike the early Syro-Egyptian and Persian astrolabes, however, which the *rete* design resembles closely, the *kursī* is generally high, plain, and solid with several ogival lobes. It thus rather resembles the *kursī* of the later Persian and Indo-Persian instruments in form, although not in decoration, suggesting influence from Lahore, where single-plate instruments were also produced.[89] That the single-plate instrument became characteristic of Hindu astrolabes suggests that Hindu makers were working from two-dimensional illustrations rather than from direct knowledge of the instrument. Alternatively, a simpler explanation may be found in the supposition that each astrolabe was made specifically for a particular place, for use in a private observatory, but not as an instrument with which to travel. Other characteristics of the Hindu instrument are in accord with this supposition: the general clarity of the engraving, the large size (few are less than eight inches in diameter), and the use of a sighting tube mounted on vertical struts on the alidade (which is usually straight [fig. 10d]) rather than pinnule sights. Two of the instruments at Jaipur are seven feet in

85. Pingree (II), 628.
86. Maddison (I), 30.
87. They were at Jaipur 27°, Delhi 28°40′, Benares 25°20′, Ujayyin 23°30′, and Mathwā 27°30′. See Kaye (I) *passim*: Sayili 359–61.
88. Pingree (III), 329
89. In this context it is interesting to note that the illustration of an astrolabe in an A.D. 19th-century copy of al-Bīrūnī's *Tafhīm* (British Library Add Ms 23,566) shows exactly this kind of high, lobed but plain *kursī*, combined with the severe *rete* of the earliest astrolabes.

diameter,[90] and the Hindu astrolabes give the impression of having been made for a limited number of purposes. They are often, in addition, rather crudely made. They are nearly always without decoration, date, or signature, and they eschew astrology.

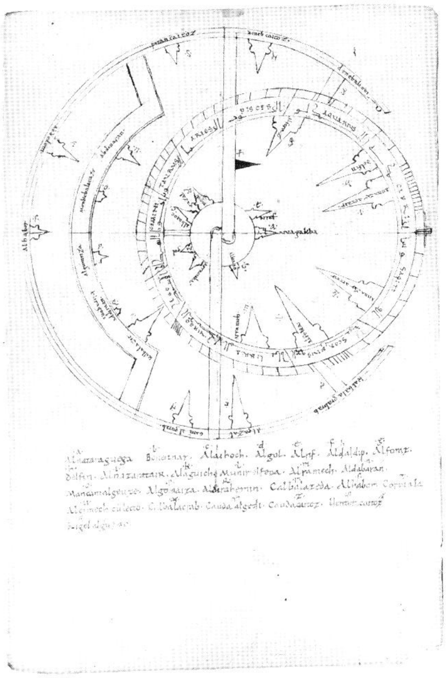

figure 22. *Rete* of an Hispano-Moorish astrolabe, 11th century A.D. Paris, Bibliothèque Nationale Ms Lat. 7412 fol. 9v. (Photograph: Bibliothèque Nationale)

European Astrolabes

By the mid-13th century the astrolabe was established in Europe. Its theory now comprehended and its uses known to scholars, it became a valuable teaching instrument and a useful aid in calculation and observation.[91] Along with texts transmitted from Islam we may assume that astrolabes themselves were imported from Spain and the Maghrib. Several examples survive of Islamic astrolabes which were later engraved with the Latin names of the zodiac signs, the months, and the stars, alongside the original Arabic.[92] That this practice was widespread and that imported instruments supplied the majority of those in use during the 12th and early 13th centuries is suggested by the fact that the illustrations to an 11th-century treatise on the astrolabe also show an imported Islamic instrument (figs. 22 & 23). The *rete* is similar in design to those of instruments by Muḥammad aṣ-Ṣaffār although with larger, straighter star pointers. The Latin names of the zodiac signs and the months are written alongside the Arabic, and the star names are Latinized without being translated.[93] The practice continued throughout the Middle Ages. The astronomer and teacher Martin Bylica of Olkusz (1433–93), for example, used an astrolabe made in Cordoba in A.H. 446 (A.D. 1054/5) until 1486, when he was able to commission a new one from the instrument-maker Hans Dorn (*c.* 1430–1509).[94] Clearly, there could have been few purely Latin astrolabes available although some local manufacture, probably limited to individual scholars supplying their own needs or those of their friends, may occasionally have occurred.[95] No dated Latin astrolabes have survived from the 12th or 13th centuries, but a few undated examples have been ascribed

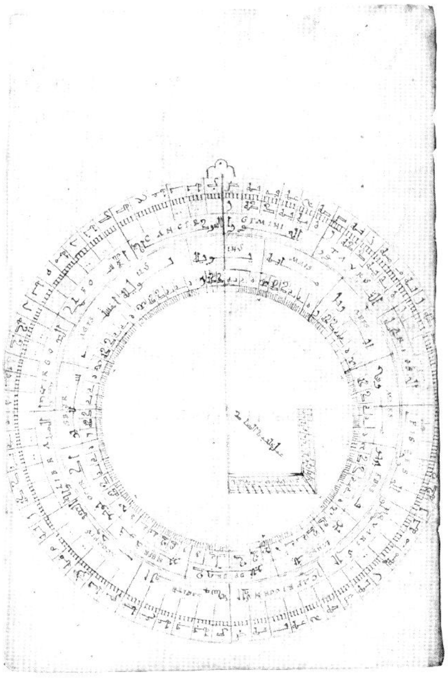

figure 23. Back of an Hispano-Moorish astrolabe, 11th century A.D. Paris, Bibliothèque Nationale Ms Lat. 7412 fol. 23v. (Photograph: Bibliothèque Nationale)

90. Kaye (I), 32 & pl. ix; Gunther (V), i, 214. Some additional details are given by Sharma.
91. An even wider significance is given to the astrolabe by Pedersen (II), 68, who suggests that not only did it force Western scholars to acquire the mathematical knowledge necessary to understand it, but that 'it brought Medieval astronomy out of its impasse, providing a new link between observation and theory as well as the new awareness of the fact that mathematics is a valuable key to nature.' Clear evidence of the use of astrolabes for making observations is rare, but see the comments by Levi ben Gerson (1288–1344) translated in Goldstein.
92. For examples, see the early 4th/10th-century astrolabe traditionally associated (alas without evidence) with Gerbert, now in the Museo di Storia della Scienza, Florence, Inv. 1113, IC/CCA no. 101, to which a zodiacal calendar and a shadow square have been added. See Gunther (V), i, 230–1; Bonelli, 163; Garcia, 133–52; Maddison & Turner, 41; and an astrolabe by Ibrāhīm b. Sa'īd of Toledo A.H. 460 (A.D. 1068), Mayer (I), 51; Gunther (V), i, 253–6; Maddison & Turner, 47, now in the Museum of the History of Science, Oxford (IC/CCA no. 118). An unsigned example is shown in Gunther (V), i, 244 (IC/CCA no. 110).
93. Bibl. Nat., Paris, Ms Lat 7412 ff 19v–23v. Reproduced in Poulle (VIII), pl. I–IV.
94. Copernic, 71–2; Estreicher, 19–21.
95. As an example of scholarly craftsmanship we may cite Gerbert making a celestial globe for his correspondent Remi of Trier. Lattin (III), no. 156. The 'Carolingian' astrolabe from the Ripoll-Barcelona region mentioned above should also be recalled in this context.

to this period.[96] Their number is, however, very small and it is clear that the astrolabe was not yet widely known in Europe outside communities of scholars. The contrast in this respect between Christian Europe and Islam is marked, but it may be explained by the different role played by the astrolabe in each society. In Islam the astrolabe was a long-established instrument used regularly in the determination of the hours of prayer and for astrological prognostication. In Christian Europe, by contrast, the astrolabe had no such social purpose. It was still a relatively new instrument, the use of which was restricted to those few scholars and their pupils who concerned themselves with geometry, astronomy, and astrology. Whereas in Islam the *muwaqqit* of every local mosque might use an astrolabe, in Christian Europe the number of men who even knew what such an instrument did was severely limited. This situation, after the cultural and educational developments of the 12th and 13th centuries, began to change.

The standardization of descriptions of the construction and use of the astrolabe in the 13th century has been discussed above. Following the standardization, we can see the beginnings of dissemination. Already, at the beginning of the 12th century, Héloise and Abelard had named their son 'Astrolabe,' perhaps reflecting a diffusion of the instrument in the university at Paris. From the end of the century, however, as Poulle has pointed out,[97] the astrolabe appears with increasing frequency in cathedral sculpture and miniature paintings, symbolizing astronomy or an astronomer. Such depictions are rarely accurate. They show, however, that the astrolabe was no longer entirely unknown, and knowledge of it spread through two channels – the universities and the princely courts.

From the 13th century onward, the astrolabe became increasingly studied in the universities, where a formalization of astronomical teaching was taking place. Beginning at Paris in the middle of the century, a corpus of texts for teaching astronomy was established centering upon three works by Sacrobosco, the *Algorismus*, the *Computus*, and *de Sphaera*. To these a treatise on the astrolabe was often added,[98] and the instrument itself was placed in university libraries where it could be consulted together with the books. Indeed, it could even be borrowed, as the loans register of the Sorbonne library shows.[99] It is, therefore, perhaps not too extravagant to sup-

96. IC/CCA nos. 546, 457, 547, 290, 198, 428, 2041.
97. Poulle (I), 89. For reproductions of two sculptures from the cathedrals of Sens and Laon, see García, 34, although as Poulle (I), 89, n. 3, notes, the descriptions have been interchanged.
98. Beaujouan; Pedersen (II). 74 ff. An idea of the penetration of astrolabe treatises into the curriculum is given by the considerable number of such works surviving among the 185 astronomical manuscripts in the Jagellonian Library, Cracow, which are known to have been used for teaching the astronomy course at Cracow in the 15th century. For a detailed analysis, see Rosińska.
99. Copernic, 72; Vieillard, *passim*. Only the register from 1403–1530 survives, an earlier volume having been destroyed. The practice, however, does not seem to have been novel. Astrolabes were included among the library equipment given by Charles V to his foundation, the Collège de Maître Gervais at Paris. '. . . et leur donna belle librairie bien garnie des livres, speres, astrolabes, saphées et autres instrument servons a ladicte science [astrology].' Wickersheimer, 4. (. . . and [he] gave them a fine library well-stocked with books, spheres, astrolabes, sapheas [?Azarchelis], and other instruments useful for the said study [astrology]).

figure 24. Goat with an astrolabe, which is suspended from a fixed point; it is not supported by the goat. From the position of his trotters, he is perhaps adjusting a rule which has not been drawn in. He is not using the astrolabe for observation, but is perhaps using the unequal-hour scale. This is drawn incorrectly, only four hour lines being shown on each side, and is upside-down in the lower quadrant. Detail from a marginal illustration to an early-14th-century north French copy of the *Almagest.* London, British Library Ms Burney, 275 fol. 390v. (Photograph: British Library)

pose that there can have been few students of astronomy and mathematics in the universities of the later Middle Ages who did not learn something of the astrolabe and its use. The treatises used were those standardized in the mid-12th century; they, like the single-latitude astrolabe itself, changed hardly at all during the next five centuries.[100] What did change, however, was the number of universities in Europe, the number of men attending them, and in consequence the number of men who knew something of the astrolabe. It was often only a superficial knowledge, for those who acquired a university education in the later Middle Ages did so less for love of learning than to make a reputation and so win preferment in the church, in the royal administration, or in both. Even so, this development entailed a further degree of popularization in royal and princely courts, which were the obvious centres of promotion. Educated men flocked to them, and, inevitably, their presence affected the life of the courts. Patronage of learning and culture became in

100. For editions of treatises on the astrolabe written before 1500, see Poulle (VIII), 870–1.

the later Middle Ages as much a mark of position and authority as was the ability to find clients and protect their interests. Occasionally patronage was allied to a genuine intellectual curiosity and at the courts of such men as the Emperor Frederick II (1194–1250), Alfonso X of Castile (1226–84), or Charles V of France (1337–80) learning was encouraged and the astrolabe was certainly esteemed. As early as 1142–46 Adelard of Bath dedicated his treatise on the astrolabe to Henry Plantagenet (the future Henry II of England).[101] Two other characteristics of the period also had consequences for the astrolabe. One was the great growth of writing in the vernacular, the other was the permeation of political and social life by astrology.

Dependence on astrology in later medieval Europe was sometimes abject, and it was far more widespread than it had been in the 11th and 12th centuries.[102] The astrologer (with his instruments) became a familiar figure at court, called upon for advice in illness,[103] for explanation of misfortune, for prediction of the weather, or for advice on the most propitious time to travel, wage battle, marry, or purchase a house. Examples of court astrologers and their activities are many. Michael Scot in Sicily, with Frederick II;[104] Thomas de Pisan at Paris; with Charles V, for whom he conjured the English out of France;[105] Symon de Phares at the court of Charles VIII;[106] and the Italian, William Parron at the court of Henry VII of England.[107] We are not here concerned with the activities of these men and the hundreds of their fellows,[108] but that they brought the astrolabe into cognizance and into relatively common use. In his *Liber Particularis*,[109] Michael Scot listed the astrolabe among the necessary instruments. Charles V of France owned many astrolabes, a fact not unconnected with his heavy dependence on astrology.[110] The astrological interests of Arnaud of Brussels (*fl.* 1464–77), a Flemish humanist scribe and printer who worked in Naples, led him to collect or copy several astrolabe treatises. Among those he copied was

101. Haskins 28–9. See also Cochrane.
102. The career of Thomas de Pisan as sketched by Solente, i, i–xv, strikingly illustrates astrology's fundamental importance.
103. Medical astrology was of the greatest importance – so much so that the Paris medical school was known as the faculty of medicine and astrology, Lewis (II), 24. For general introductions to the subject, see Cardoner, Cosman, Thorndike (I), Wedel, and White (although he seems to exaggerate the role of medical astrologers in the development of instrument-making).
104. Thorndike (III), *passim*.
105. Lewis (II), 24–5, who gives a general sketch of the importance of astrology in France.
106. Symon's work *Recueil des plus célèbres Astrologues* (ed. E.Wickersheimer, Paris, 1929) is replete with examples of astrologers attached to great men and with specimens of their advice and successes.
107. Armstrong. See Germana Ernst's descriptions of two of his manuscripts in Zambelli, *fl.* 371.
108. For general and detailed discussion, see Thorndike (I), esp. iv. For attacks upon astrology by the orthodox in France, see Coopland.
109. Haskins 291, n. 118. According to Scot, knowledge of the astrolabe was won by master Gislebertus (= Gerbert) from demons whom he had conjured and forced to explain it to him. What they told him he wrote down and the knowledge thus became available to many. Thorndike (III), 94. Elsewhere in the *Liber Particularis* he says that the astrolabe is sometimes used for conjuring evil spirits, *Ibid*, 117.
110. Charles was, according to Christine de Pisan, '. . .des haultes choses de philosophie comme d'astrologie, tres expert et sage en ycelle (. . .of the high things of philosophy like astrology, very skilful and learned therein), and governed his life thereby. Solente. ii. 16–17. *Cf.* the citation from Christine's *Livre du corps de policie*, and other references provided by Solente. For Charles' astrolabes, see below.

figure 25. Astrolabe signed 'Blakene me fecit anno dōi 1342.' London, British Museum, Department of Medieval and Later Antiquities, Register no. 53 11-41 (Ward, 113). (Photograph: British Museum)

that of Raymond of Marseilles, which is unknown in any other copy.[111]

The teaching of the astrolabe in the universities and the use of astrology in European society entailed two new developments for the astrolabe in the West. One was a demand for vernacular treatises. This development may be seen as resulting partly from the general growth of vernacular literature during this period, but it also reflects an increase of interest in the astrolabe itself. The second was that a new demand was felt for the instruments themselves, a demand sufficient to sustain a few commercial workshops. These, albeit scattered throughout Europe and somewhat sporadic in their production, produced sufficient instruments for a few examples to have survived into the 20th century.

The first dated astrolabe surviving from Western Europe is an unsigned instrument of 1326, now in the British Museum.[112] The earliest known instrument that is both signed and dated is an English astrolabe inscribed 'Blakene me fecit anno dōi 1342'(fig. 25).[113] Al-

111. Poulle (VII), 872–3; Poulle (IV), 61–72. If the audience for Chaucer's poetry was expected to notice and appreciate the astrological inner structure recently elucidated in North (III), then the fact is a further testimony to the importance of astrology among the educated classes.
112. IC/CCA no. 91. Gunther (V), 465–7; Ward 112 & pl. LI.
113. IC/CCA no. 292. Gunther (V), 468–9; Morley (I), 45. This astrolabe is also in the British Museum. Ward 113 & pl. LII.

though only three other dated examples are known from the 14th century, and none from the first half of the 15th, several astrolabes have been ascribed to this early period, three to the 13th century, twenty-one to the 14th, and eighteen to the first half of the 15th.[114] In themselves these figures mean nothing, but compared with each other they do suggest that there was a considerable increase in the number of astrolabes made in Christian Europe. In addition, there were many Islamic instruments in circulation, and other instruments have, doubtlessly, been lost or melted down. In 1380, twelve astrolabes, including one made of gold and two of silver, were listed among other instruments in the inventory of the possessions of Charles V of France.[115] None of them is now known to survive. That Charles V owned so many astrolabes was partly, but not entirely, ostentation. In 1362, he commissioned Pelerin de Prusse to write a treatise on the instrument in French. This, the first French vernacular treatise, was perhaps not widely known,[116] but its successor by Jean Fusoris in *c.* 1407 was more widely copied and was partly, but not faithfully, translated into Italian.[117] A further French work on the astrolabe was written by Jehan de Bregny *c.* 1435 for René of Anjou and Sicily.[118]

Preceding these among vernacular treatises, however, was Geoffrey Chaucer's English version (with some additions) of Pseudo-Māshā'allāh (1391–2), which was evidently popular – some twenty-four manuscripts of it survive[119] – and remains the best account of the subject in English, despite the fact that it is unfinished.[120] These manuscripts are the more interesting for preserving several of the illustrations to which Chaucer made repeated reference in his text. These illustrations are of some importance, for the design of the *rete* of the astrolabe there shown (fig. 26a) and the scales on the back of the instrument (fig. 26c) resemble closely a number of surviving astrolabes which may thus be given an approximate date. To conclude that these surviving astrolabes (of 'Y-type Gothic' form,

114. IC/CCA no. 44. In these totals Hispano-Moorish instruments have been ignored.

115. Labarte, nos. 1990, 2072, 2216, 2270, 2427, 2714, 2817, 3119, 3121.

116. Poulle (V), 7. Charles V's own copy is now at St. John's College, Oxford, Ms. 182, fols. 111–9. The volume contains other astrological treatises written by Pelerin de Prusse for Charles. For a detailed analysis of the volume, see Delisle, i, 266–9; *Fastes*, n. 289; and for the miniature of Charles which it contains, Sherman, 22 & 74. For the little that is known about Pelerin or Pelegrin de Prusse, Delachenal, i, i, n. 1; ii, 279, 295, & 368.

117. Ms Vatican lat 1732. Poulle (V), 12–3.

118. According to Symon de Phares, Jehan gave many examples and marked stars which were not in the treatise of Fusoris. Wickersheimer, 255.

119. To the 22 manuscripts listed and analyzed by Skeat may be added that edited by Pintelon from Mss 4862–4869 of the Bibliothèque Royale, Brussels, and Aberdeen University Library Ms 123 fols. 10^{v} *et seq*, briefly described by North (III), 432, n. 1.

120. For discussion of the date of the treatise, see North (III), 452–3 and for the contents of the missing parts *Ibid*, 436–7. The classic edition is that of Skeat, reprinted without apparatus in Gunther (III). The most recent and convenient text is that of Robinson. For other editions, see Hammond and Pintelon. Chaucer's source in Messahalla was first pointed out by the jurist and Orientalist John Selden (1584–1654) in his preface to his notes to Michael Drayton's chorographic poem *Poly*-olbion (1613). '. . . by his treatise of the Astrolabe (which, I dare sweare, was chiefly learned out of Me*ssahalah*) it is plain he was much acquainted with the Mathematiques, and amongst their Authors had it.' (1622 edit., A3v.) A Maghribi astrolabe which had belonged to Selden passed with his books to the Bodleian Library in 1659. It is now in the Museum of the History of Science, Oxford (Gunther (I), 194). For definitions and descriptive astronomy, Chaucer also used Sacrobosco.

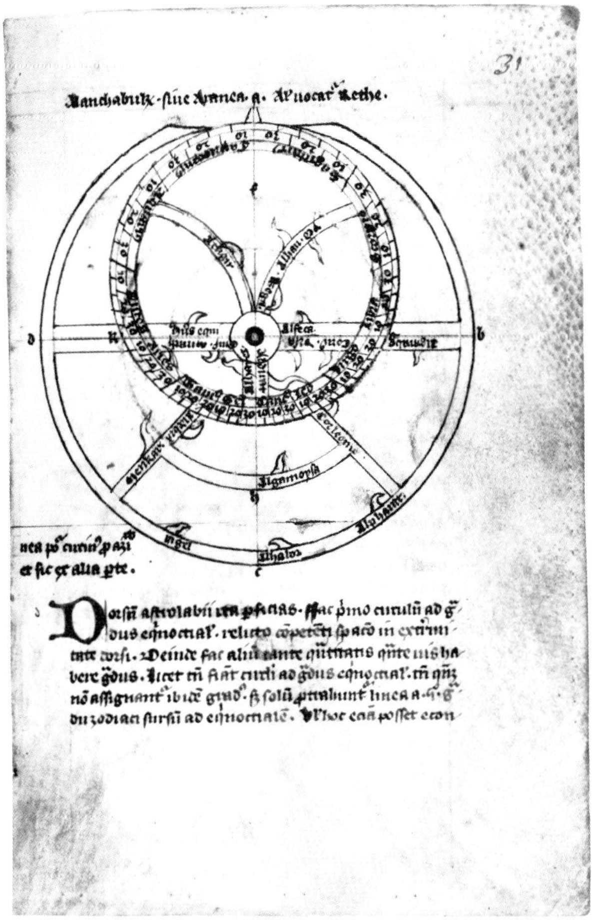

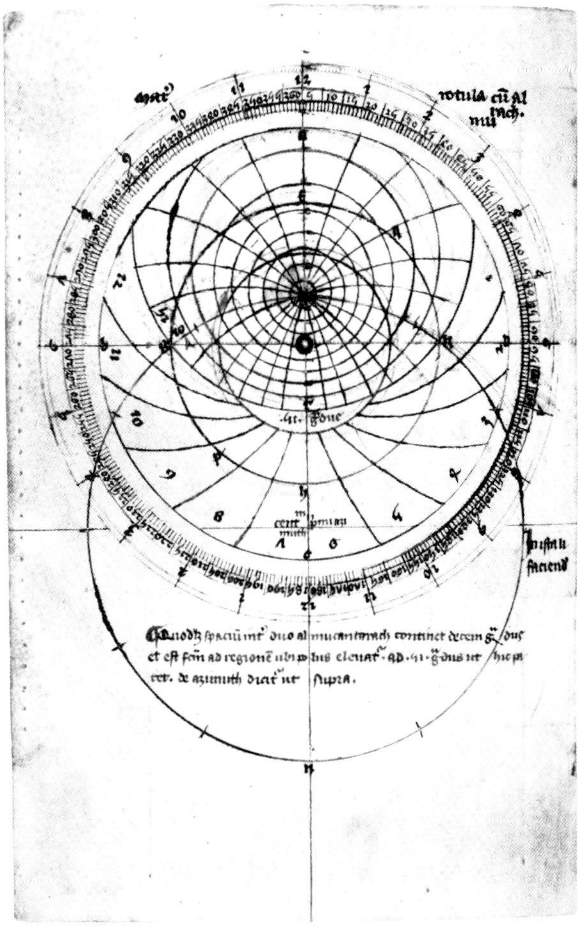

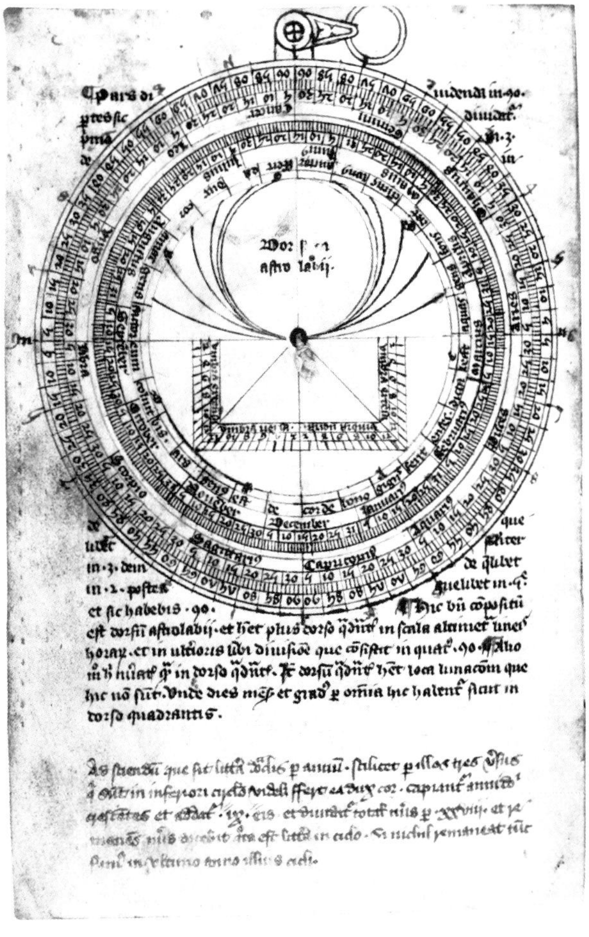

figure 26. *Rete* (26a, upper left), plate (26b, upper right), and back (26c, left) of an astrolabe. Illustrations to an anonymous treatise on the astrolabe, *incipit* 'Astrolabium ita construitur. Recipe tabulas de auricalco ad hoc aptas.' Drawn in an English ms *c.* 1400. Oxford, Bodleian Library Ms Bodl. 68 fol. 31r (26a), 30v (26b), 33v (26c). (Photograph: Bodleian Library)

figure 27a. Moorish Gothic *Rete*.

figure 27b. Trefoil/Quatrefoil *Rete*.

figure 27. *Rete* designs of medieval astrolabes.

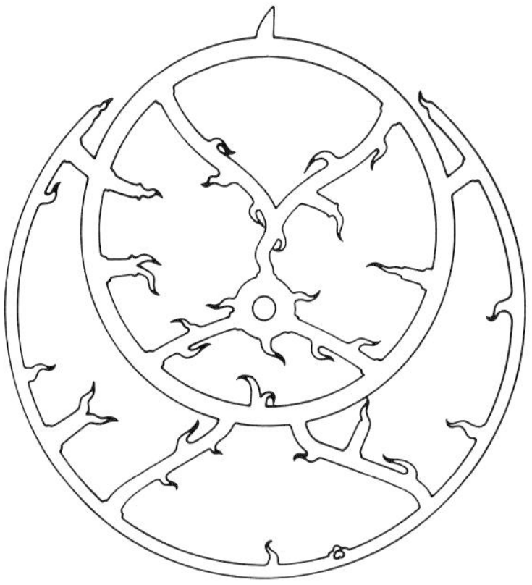

figure 27c. Y-type *Rete*.

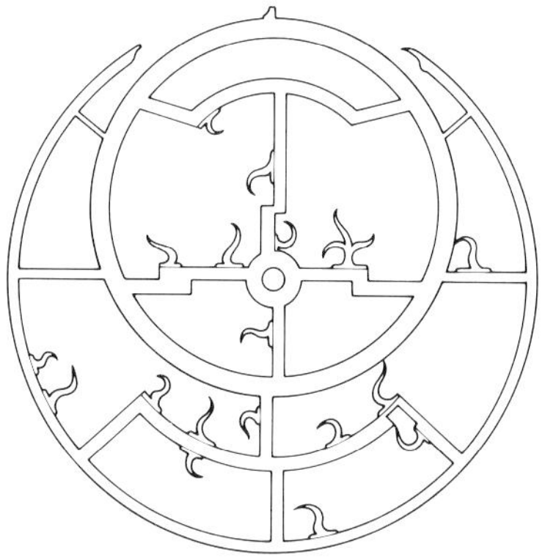

figure 27d. Late Gothic *Rete*.

fig. 27c) are therefore of English manufacture would be to press the evidence too far.

From among the small number of surviving medieval Western astrolabes, four basic types have been discerned and given chronological arrangement (fig. 27).[121] There is, however, a considerable overlap between them, both in style and in dating. This may result, as Michel suggested,[122] from the itinerant and international nature of astrolabe production. With few masters available, those who wished to learn the craft of making an astrolabe were forced to seek out, wherever they might be, such men as could teach them. Conversely, the makers of astrolabes themselves tended to migrate to those places that offered plentiful raw materials, promise of good markets, or a rich patron. In consequence, the astrolabes of Christian Europe show less marked geographical or regional characteristics than those of Islam. The stylistic similarities which appear are caused by the succession of master and pupil who, being itinerant, worked in the same style in widely different parts of Europe. But

121. By Price (III), 247–8.
122. Michel (III), 149.

this is hypothetical. What is definite is that by the 15th century the astrolabe was known and made throughout Europe from the Atlantic seaboards of England and Spain to the Central European plains of Bohemia and Poland. Theory, however, was considerably in advance of practice, and some of the astrolabes produced were poorly executed.[123] That many of their owners noticed may be doubted.

If in the 15th century there was no change in the theory and little improvement in the construction of astrolabes, there was, however, development in the manner of their production. If the surviving Gothic astrolabes with Y-type *retes*, similar to that of the astrolabe illustrated in Chaucer manuscripts suggest a common source, we know nothing about the workshop that produced them and, therefore, nothing of its size and organization. From approximately the same period, however, there is a group of fifteen or more similar astrolabes, described by Price as 'Late Gothic,' ascribed to the workshop of Jean Fusoris.[124]

Born at Giraumont in the Ardennes *c.* 1365, the son of a pewterer, Fusoris studied arts and medicine at Paris, attaining both the master's and doctor's degrees. He was thus an educated man, not purely an artisan. To seek to win a reputation, and so the possibility of a successful career in the church or royal bureaucracy, by way of mathematical, philosophical or medical studies, was not unusual. Obvious examples of philosopher-churchmen are Nicholas Oresme and Nicholas of Cues.[125] The invention and making of instruments was somewhat more unusual, but likewise not without precedent. For example, Richard of Wallingford made instruments.[126] What set Fusoris apart from his fellows was that besides inventing new devices and writing treatises about them, he set up a commercial workshop of instruments. Until the time he went to England, attempting to obtain payment from a recalcitrant client, and – as a result of the renewal of warfare between France and England – was arrested as a spy on his return to France, he was very successful. Even after his sentence to perpetual residence at Mezières-sur-Meuze, he continued to receive important commissions.

At once a scholar and a craftsman, Fusoris, in the normal academic manner, received pupils such as Henri Arnault of Zwolle and Jean de Berle, and had at least one assistant. Astrolabes were probably the major part of his production of astronomical instruments and clocks, although their production ought always be thought of in this wider context. Characterized by clear engraving, a convenient

123. From an analysis of the existing treatises and of 22 surviving astrolabes from before 1500. Poulle (II), 125, concluded: 'Perfect in theory, the making of an astronomically exact astrolabe by artisan's methods ran into some important technical difficulties. There were so many opportunities for errors and for approximations, intentional or not, that the best astrolabist could make a very lovely instrument, but unusable for astronomy.'
124. Poulle (V), 20–1. It should be noted that not all the astrolabes described as late Gothic can be ascribed to Fusoris' workshop.
125. *D.S.B.*, x, 223–9. For Oresme, see Coopland, 2, who suggests that Oresme received the bishopric of Lisieux 'perhaps as a reward for his translations of Aristotle's works into French. . .'
126. North (IV), *passim*.

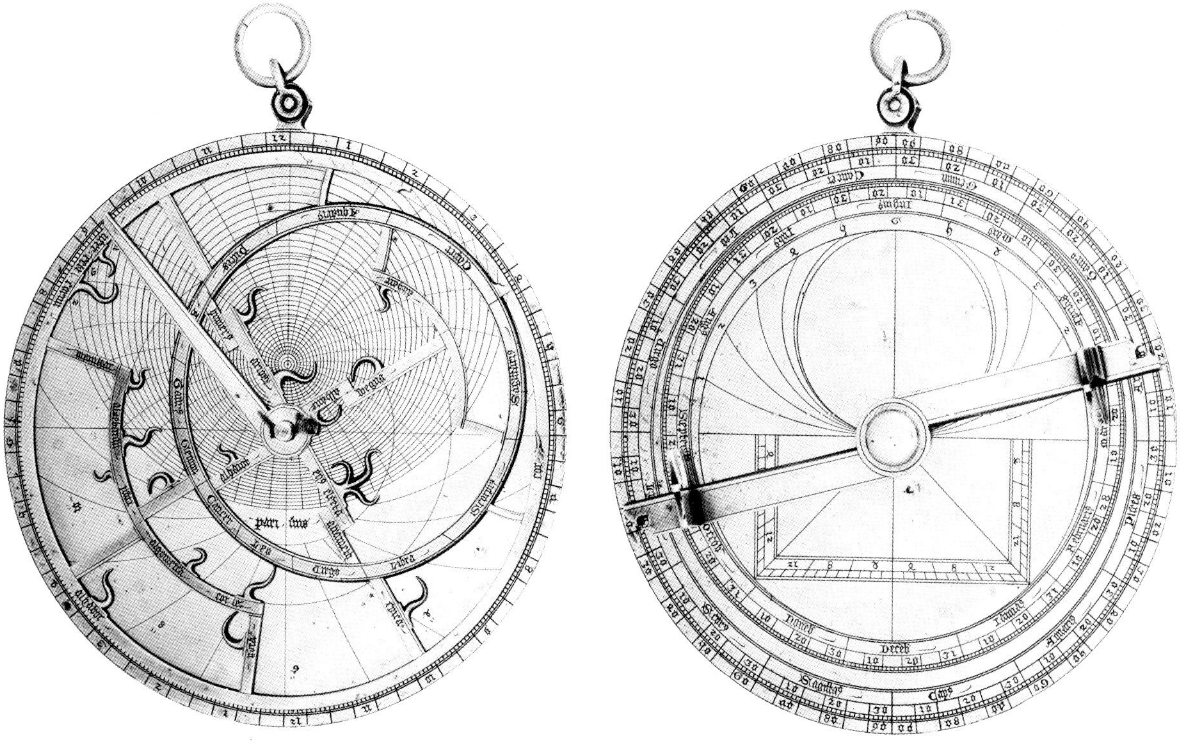

figure 28. A late-Gothic astrolabe which may be ascribed to the workshop of Jean Fusoris. Front (above) and back (right). Private Collection.

size, a simple yet elegant *rete* and great clarity, Fusoris' astrolabes (fig. 28) presumably exhibited the innovations which, in his treatise on the subject, he claimed to have made. Fusoris' claimed innovations were a modification to the sight vanes of the alidade, the addition of an *ostensor* (rule) to the face of the astrolabe, and the addition of an equal hour scale to the limb. In addition to this, his description of the horizon plate is the first such description to be found in Western treatises.[127]

Fusoris, part medieval clerk, part entrepreneur, part designer, and part craftsman, is the forerunner of a series of scholar-craftsmen-merchants whose work, both theoretical and commercial, would transform the making of scientific instruments during the Renaissance, from the unavoidable but occasional occupation of a scholar interested in astronomy who needed some instruments, to a broadly based commercial activity meeting demands from widely differing sections of society. At first, like that other new intellectual 'trade' of the 15th century – printing[128] – trade in astrolabes stayed close to the universities, where its basic market was to be found, as the example of Fusoris shows. Soon, however, thanks perhaps to the influence of astrology, a taste for astronomy and for complicated but beautiful astronomical/astrological devices and machines per-

figure 29. (opposite top) An astrolabe from a 14th-century copy of the treatise on the astrolabe of Pseudo-Māshā'allāh. London, British Library Egerton Ms 844. Fol. 64r: *rete* (left). Fol. 59r: back (right). (Photograph: British Library)

figure 30. (opposite bottom) An astrolabe from the astrolabe tract included in a computus manuscript, ?14th century. The similarity of the representation with that shown in figs. 29a and 29b, will be noted. London, British Library Harleian Ms 3647. Fol. 68v: *rete* (left). Fol. 64r: back (right). (Photograph: British Library)

127. For all which concerns Fusoris, see Poulle (V), *passim*.

128. And indeed like the writing shops which preceded it. Febvre & Martin 22 ff, 255–7.

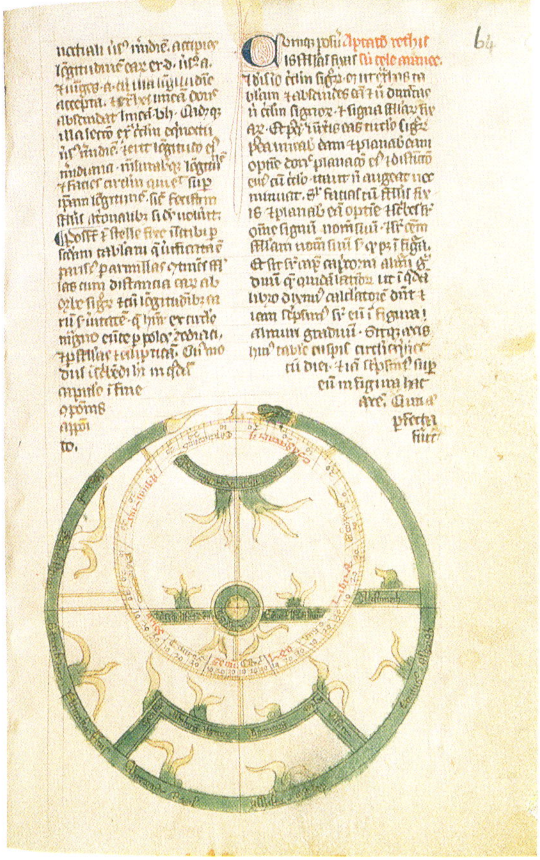

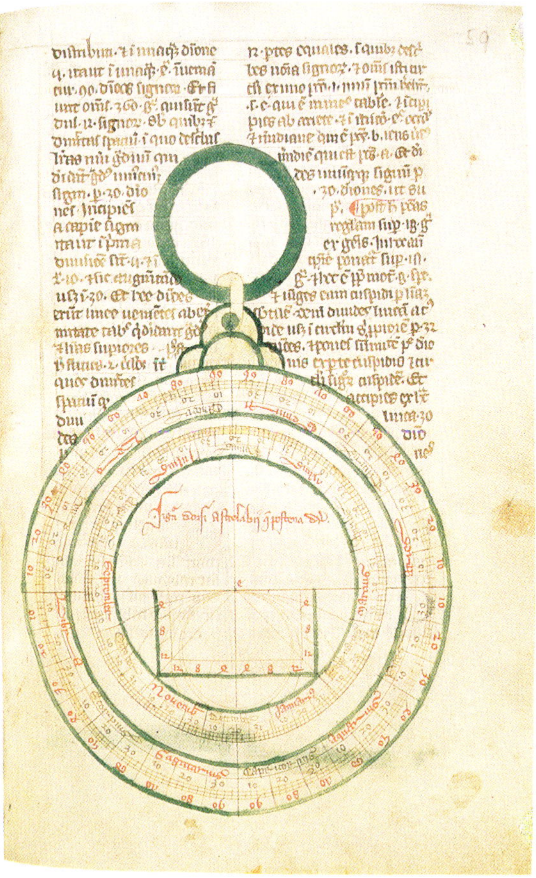

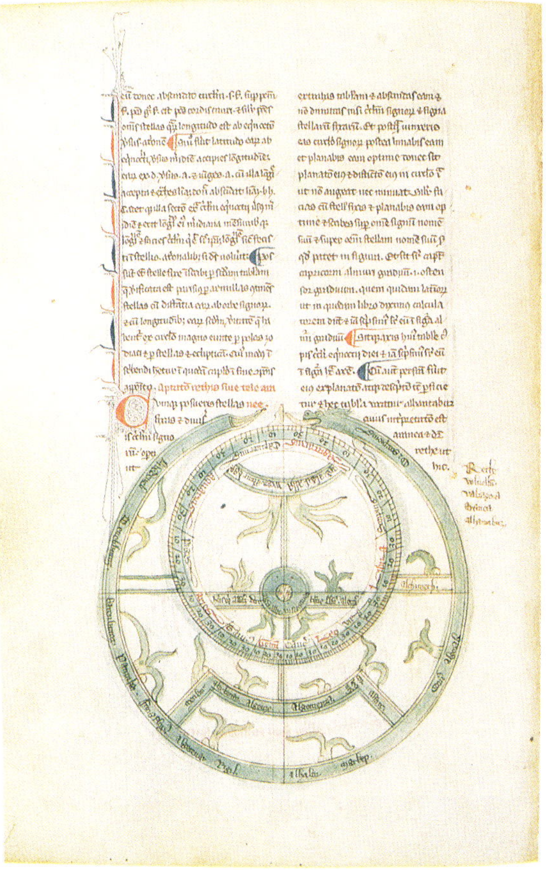

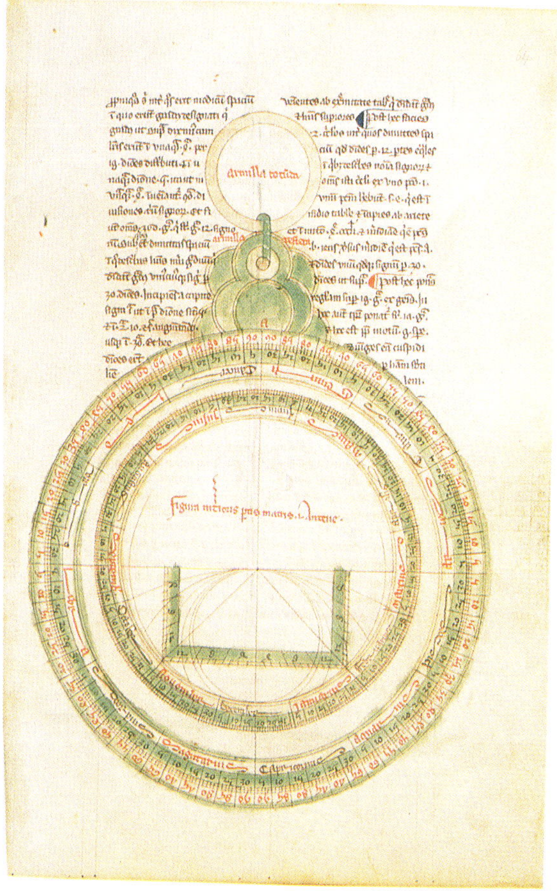

meated the culture of the wealthy, the noble, and the leisured classes. The process was actively encouraged by the solicitations of entrepreneurs like Fusoris, ever seeking munificent patrons to whom they could dedicate the products of their ingenuity. Instrument-making thus became caught up in the complex world of patronage, prestige, and courtly pleasure, but this meant that it could extend its commercial base. Other developments also played their part. In the maritime nations of Europe, the rapid development of oceanic, astronomical navigation from the mid-15th century onwards created a demand for new instruments in quantity. With the geometrization of warfare, cartography, and land surveying, further markets would be created during the 16th century.

The astrolabe lay at the heart of this development, for not only was it a key instrument of the teaching of geometry and astronomy, and possibly a useful device for time-finding and altitude measurement, but numerous other instruments could be derived from it. The late 15th and 16th centuries, therefore, saw a popularization in which the instrument reached the apogee of its career in Europe. The pace of production intensified. More astrolabes and more books about them were produced than ever before, use and comprehension of the instrument spread more widely through society, and the endeavours of gilders and engravers gave to the instruments an artistic quality commensurate with their mathematical sophistication. Symbol of astronomy and tool of astrologers, for the 16th century the astrolabe was truly a 'mathematical jewel.'[129]

If from the time of Fusoris onward, the anonymity which had largely covered the world of instrument makers and craftsmen in the Middle Ages began to lift, development at the time was still slow. Although we need not think that Fusoris' workshop was unique, it seems clear that the survival of workshops was still a matter of great uncertainty, and no strong tradition of instrument-making established itself in France or Burgundy as a result of Fusoris' activities. Henri Arnault of Zwolle, who had been a pupil of Fusoris, did, it is true, continue the practice of mathematics and instrument-making and became court astrologer and physician, first to the Dukes of Burgundy, for whom he made at least two sundials, carried out land surveys, and designed astronomical clocks, and later to Charles VI.[130] Nothing, however, is known of Fusoris' other apprentices – such as Jean de Berle – and even the work of Arnault seems to have left little trace at Dijon, where he passed much of his time, or elsewhere in the Burgundian dominions.

Further south, at Carpentras, stimulated by the patronage of René of Anjou, the Zeelander Guillaume Gilliszoon of Wissekerke

129. So John Blagrave (1585) entitled his treatise on a form of universal astrolabe, '. . . a Mathematical Iewel, of no small vertue and efficacie, to furnish the willing wits of this our age . . .' (fol 2r). The phrase, however, is considerably older.

130. Poulle (VI), 27; Zinner (II), 235–6; Omont, 127–8. For Arnault's work in other areas of technology such as musical instruments, gem-polishing machines and siege engines, see White, 310; LeCerf & Labande, xi–xiv. Richard.

produced a number of instruments which included clocks, portable sundials, and two 'astrolabes' which showed the movements of the moon and the planets.[131] These are listed in René's accounts between 1476 and 1480. Later he devised an equatorium, which has survived,[132] and wrote an astronomical treatise first printed in 1494. The treatise received several impressions.[133] But the evidence for instrument-making in France, Burgundy, and indeed most of Western Europe during the 15th century suggests that it was carried out on a very small scale, and was heavily dependent upon the patronage of court and university. Only in Central Europe, especially in the region of Nuremberg and Augsburg, was production more advanced and established on a permanent commercial basis.

On 4 July 1471, the scholar and mathematician Regiomontanus (1436–76) wrote in a letter that he had now settled in Nuremberg '. . . not only on account of the availability of instruments, particularly the astronomical instruments on which the entire science of the heavens is based, but also on account of the very great ease of all sorts of communication with learned men living everywhere since this place is regarded as the centre of Europe because of the journies of the merchants.'[134] Clearly Nuremberg was already a centre for the manufacture of instruments. In 1441, Nicholas of Cusa brought an astrolabe, a torquetum, and a large celestial globe there,[135] and the influence of the city was beginning to spread into central Europe. The town, as Regiomontanus indicated, possessed exceptional advantages of communication while also being conveniently close to rich sources of copper and other minerals. A centre of printing from its early days, it contained craftsmen skilled in metal-working, the cutting of dies and punches, and engraving on wood and (later) metal. At the same time, although there was no university at Nuremberg, it possessed excellent schools, and its long tradition of wide-ranging, large-scale commerce meant that not only was there a high level of everyday commercial mathematics but that there was also a receptivity of mind which could grasp and quickly exploit new ideas and possibilities.[136] The arrival of Regiomontanus in Nuremberg reinforced the mathematical and astronomical studies which had been established in the city during the second quarter of the 15th century. These were largely inherited from the lively mathematical tradition of Vienna, where Regiomontanus had studied with Georg Peurbach (1423–61). Peurbach, himself a

131. Poulle (III), 225.
132. For equatoria, instruments which facilitated calculations concerned with the position of the planets, see Poulle (XII). Gilliszoon's device is now in the Bibl. Nat., Paris, Ms Latin 7276B.
133. Poulle (III), 226–7, and references there given.
134. Cited from Rosen in *D.S.B.* xi, 351. 'Nuperrime Norimbergam mihi delegi, domum perpetuam, tum propter commoditatem instrumentorum maxime astronomicorum quibus tota sederalis innititur disciplina, tum propter universalem conversationem facilius habendam cum studiosis viris hic citam degentibus. quod locus ille perinde centrum Europae, propter excursum mercatorum habeatur.'
135. Durand, 88–9 who claims that Cusa's *torquetum* was probably constructed at Nuremberg *c.* 1434.
136. For Nuremberg, see Steinberg, 58–60, 137–8; Gallois, 5–7, 71–2; Lunardi, 1–38; Durand, 87–90, who stresses the dependence of the city on outsiders for scientific activity.

graduate of the university there, had devoted some attention to mathematical instruments. Among his writings is a treatise on the astrolabe.[137] An astrolabe dated 1457 carrying scales specifically mentioned in Peurbach's treatise is preserved in the Germanisches National Museum, Nuremberg.[138] To conclude that it was therefore made by Peurbach, however, would be premature.

Regiomontanus lived in Nuremberg for only four years; his influence on the city, therefore, should not be exaggerated. It was rather the combination of his advanced mathematical and astronomical studies with the availability of skilled craftsmen, astute merchants, and the excellent communications of Nuremberg with Italy, Spain, and Portugal,[139] Central Europe, and the Low Countries which made Nuremberg, as well as Augsburg, pre-eminent in Europe for the manufacture of astrolabes and scientific instruments. Nuremberg remained prominent even after its supremacy began to be challenged in the later 16th century. The surviving astrolabes which may be ascribed to Nuremberg *ateliers* about 1500 are generally smaller and a little more decorative than those made earlier by Fusoris. The layout of their *retes* looks back to the trefoil and quatrefoil Gothic forms of the 13th and 14th centuries. Peurbach's treatise and the astrolabe made after it give us some idea of the prevailing style, and it is interesting to note how closely this astrolabe resembles that made at Vienna in 1486 by Hans Dorn for Martin Bylica. The similarity is not astonishing, for Dorn had been a pupil of both Peurbach and Regiomontanus.[140] An astrolabe dated 1462,[141] and usually ascribed to Regiomontanus, shares the prevailing style (if with some calligraphic peculiarities), but is closer in the design of its *rete* to the later work of Georg Hartmann.

Georg Hartmann (1489–1564)[142] was born at Eggolsheim. He studied mathematics and theology at Cologne and, during a visit to Italy in 1518, became friendly with Copernicus' brother Andreas. While in Italy he began to design sundials and to study the phenomenon of magnetic dip. Moving to Nuremberg, where he was vicar of St. Sebald's from 1518–44, Hartmann established a commercial manufactory of instruments which was evidently of some size, for over one-hundred instruments signed by him have survived (for one, see catalogue no. 17). Although it is certain that some of these instruments are 19th-century fakes, this still represents a considerable output. Hartmann's astrolabes became widely distributed throughout Europe. One, for example, dated 1527, was owned by Archbishop Laud, who in 1634 gave it to St. John's College,

137. Ms Vin 4782 fols. 225^{a}–270^{b}; Ms Vin 5176 fols. 156^{a}–162^{b} *Canones astrolabii*; Ms Vin 5176 fols. 43^{b}–47^{a} Compositio *quadrantis astrolabii*. For a brief account of his life, see C. Doris Hellman & Noel M. Swerdlow in *D.S.B.*, xv, 473–9; Zinner (II), 463–5.

138. The astrolabe is described and its face illustrated in Urschlechter 71, no. 15. The back is illustrated, also with a description in Zinner (II), 465 & pl. 50.

139. For trade relations between Germany and Portugal, see Bensaude, 22.

140. Zinner (II), 292.

141. National Maritime Museum, Greenwich. No. A56/58–20, IC/CCA no. 640. See Price (IV).

142. For his life, see Lucille B. Ritvo in D.S.B., vi, 143–5; Zinner (II) 357–61.

Oxford.[143] Hartmann evidently felt that he was, to some degree, working in the tradition of Regiomontanus. In 1526 he stimulated Johann Schöner to publish Regiomontanus' *Problemata XXIX Saphea* and it seems probable that he consciously copied his *rete* design from that of Regiomontanus. Although a learned man who published several works in astronomy, Hartmann was also a skilful merchant. He produced fine instruments for noble patrons such as Duke Albrecht of Prussia and Duke Ottheinrich, who in 1544 ordered two ivory dials, a brass astrolabe, and a brass armillary sphere.[144] Hartmann's instruments had sufficient prestige for one to be depicted in a painting by Lucas de Heere (1534–84) (fig. 31). At the same time he also developed inexpensive, yet potentially profitable, instruments for everyday use. Hartmann was the pioneer of the cheap astrolabe, being perhaps the first to print astrolabe parts on paper which could be cut out and pasted onto pieces of wood to make a serviceable and inexpensive instrument. At least five such instruments by Hartmann have survived,[145] with dates ranging from 1531 to 1540, but there surely were many more. Indeed, since

figure 31. *The Liberal Arts in Time of War* by Lucas de Heere (1534–1584). (Photograph: Galleria Sabauda, Turin)

143. Gunther (V), ii, 213–15.
144. For Regiomontanus' *Saphea*, see Zinner (I), no. 1573, 183. For orders from patrons, see Voigt, 277–96.
145. Lewis Evans Collection, Museum of the History of Science, Oxford; British Museum, London (2); Kestner Museum, Hamburg; Mensing Collection, Gunther (V), ii, 438–40; IC/CCA nos. 263–7.

several other paper instruments are found throughout the 16th and 17th centuries, we may speculate that it was in this form that the astrolabe was most widely circulated.[146]

In marked contrast was the astrolabe-clock, which, far from being intended for popular consumption, was a prize which could be acquired only by the richest of patrons. Use of the astrolabe as the dial of a clock was an old and continuous tradition in the making of public clocks which may eventually find its source in the anaphoric water-clock of Vitruvius. During the 16th century, however, astrolabe dials were incorporated in sumptuous spring-driven striking-clocks which might also incorporate automata or other astronomical indications. Clocks of this type were difficult to make – they were often prescribed as masterpieces by the south German clock-making guilds – and very expensive. They were not widely diffused, and their place in the history of the astrolabe is a relatively minor one.[147]

By the time of Hartmann's death, the astrolabe was more readily available in Europe than it had ever been before. For the first time, perhaps, it reached the same level of familiarity and widespread use in European society as it had for so long enjoyed in Islam. If the immediate reason for this popularity was the ease with which information about the astrolabe could be obtained from printed books,[148] the basis of its appeal and demand, as in Islam, lay in astrology. Occult study and prediction were fundamental parts of the cosmology and belief of Renaissance Europe. Astrology entwined itself into every aspect of life, influencing action and belief to a quite remarkable degree. The astrolabe was presented as the indispensable tool of astrology, aiding the calculation of horoscopes, finding the positions of sun and stars, and even, used in conjunction with an ephemerides, the positions of the planets. In 1545 this function of the astrolabe was clearly described by Dominicque Jacquinot in his *L'Usage de l'astrolabe,*

> 'Which instrument . . . I will dare say and affirm is the most perfect, reliable and necessary [one] there is among the all other astronomical instruments, as much for the observation of celestial movements and to know the course of the stars as for many other strange things belonging to the divine science of Astrology.[149]

Two years earlier, in 1543, Vincentius Campanatius had aided Luca Gaurico to determine the propitious moment for laying the cornerstone of a building near the Church of St. Peter's, Rome, by inspect-

146. Paper astrolabes are known to have been produced by such makers (among others) as Johan Krabbe, Ioan Blaeu, Phillippe Danfrie, Henry Sutton, John Prujean, Nicholas Bion. The method was also used for a wide range of other instruments. For astrolabes, see Gunther (V), ii, 220. For paper instruments in general and the impact of printing on science and instrument-making, see Poulle (XII), 73 ff. and index references under 'Papier,' and Turner (II).

147. For a detailed discussion of Renaissance astrolabic clocks, see the forthcoming volume of this catalogue on Renaissance clocks by J.H. Leopold. For a useful selection of examples, see Maurice & Mayr, nos. 12–46.

148. See below,

149. 'Lequel instrument . . . i'oze bien dire & asseurez estre le plus parfaict, certain & necessaire, qui soyt entre tous autres organes astronomicques, tant pout l'observation des mouvements celestes, & cognoistre le cours des astres, que plusieurs autres singularitez appartenantes à ceste Science divine d'Astrologie.'

ing the sky with an astrolabe.[150] He thus supplies us with a striking and concrete example of astrological use, even if we may suspect he employed his astrolabe as much for show as for its practical advantages. If this were not enough to ensure the popularity of the instrument, it was, in addition, always useful for altitude measuring and as a pedagogic device. Astrolabes, therefore, were widely used. Demand for them in 16th-century Europe was greater than ever before, but thanks to the existence of commercial workshops making a wide range of instruments, this demand could easily be met.

The supremacy of the Nuremberg-Augsburg region of southern Germany in the manufacture of astrolabes and other instruments was a function of that region's general pre-eminence in metal-working, in turn a reflection of the rich natural deposits of copper and other minerals available in the Austrian Tyrol. During the late 15th century the mines of Hungary at Neusohl and elsewhere – and so commerce in copper and its alloys – had increasingly been controlled by the association of Augsburg merchants led by the Fugger banking house. This control, thanks to the widespread commercial interests of the Fuggers, also extended to the Vienna copper market and spread throughout Europe, giving the Fuggers an effective monopoly of the supply of brass and copper.[151] At Augsburg and Nuremberg and the Central European Habsburg dominions centering on Vienna and Prague, brass and copper could be readily obtained, as at Antwerp, where the Fuggers maintained an entrepôt. Elsewhere, however, supply was more difficult and this may partly explain why so little trace is found of any extensive instrument-making trade outside these two areas during the early 16th century.[152]

Throughout the first half of the 16th century, the south German instrument-making *ateliers* were virtually without rival for the supply of astrolabes and other instruments, except for one or two makers such as members of the della Volpaia family in Italy, and perhaps a few craftsmen in France. From about 1530 onward, however, a new centre began to develop in the Low Countries, in the Antwerp-Louvain region, significantly close to the source of copper and brass supplies. Like the enterprises of Fusoris and Georg Hartmann, the leading workshop in this development was the inspiration of a learned man. Gemma Frisius (1508–55)[153] (fig. 34), like his predecessors, had received a university training. Unlike them he remained attached to the university as a professor of medicine at Louvain. Possibly it was by his studies of astrology in relation to medi-

150. For the occult and astrology in Renaissance thought, see Evans, and for England, Thomas, ch. 10–12. *Cf.* John Aubrey's remark *c.* 1680 that every schoolboy should have a paper astrolabe 'to teach him to erect a Scheme presently' (i.e. to draw up a horoscope). Bodleian Library Aubrey ms 10 f 109, see Turner (I), 65. For an astrological astrolabe perhaps made for Queen Elizabeth, see Gabb. Astrological scales were not infrequently added to his astrolabes by Habermel. For Campanatius, see Thorndike (I). v. 259, and for the detection of thieves with an astrolabe, *Ibid*, 384.
151. Klarwill (I), xxiii; Hay. 371–2; For the Fuggers and England, Hamilton (II), 3–4; for a general survey with numerous references, Kellenbenz.
152. Michel (IV), *passim*.
153. For the life of Gemma, see van Ortroy, and more briefly, de Smet (I).

Qui varium cœli morem penitusq;
remoti
Mortales docuit ætheris
ire vias:
Quiq; nouos Medicis meditando
repperit vsus,
Hes Gemæ effigies, hic fuit
oris honos.
EFFIGIES GEMMÆ FRISII OBIIT ANNO 1555 Ætatis suę 47

cine that Gemma was drawn more generally into the study of astronomy and mathematics.[154] Whatever the case, he quickly began to exploit an aptitude for the design of astronomical and mathematical instruments, first by commissioning the engraver Gaspar van der Heyden (Gaspar a Myrica)[155] to construct the terrestrial and celestial globes that he designed, then by establishing a full-scale workshop producing instruments on a commercial basis. That Gemma ever constructed an instrument himself seems unlikely, but he was extremely fortunate in the workmen he employed and perhaps trained. In 1536, Gemma's second terrestrial globe was made jointly by Gaspar van der Heyden and Gerard Mercator,[156] who was later to become famous throughout Europe for his accurate and beautifully engraved maps. Mercator seems to have made a wide range of instruments during his years at Louvain, and never entirely abandoned their construction.[157] In 1552, however, he left Louvain for Duisburg, when his place as engraver/maker at Louvain was taken by a nephew of Gemma, Gualterus (Walter) Arsenius.[158]

Under the direction of Arsenius, the manufactory inspired by Gemma grew and prospered throughout the second half of the 16th century despite Gemma's early death in 1555. Of its prestige and leading position in northwest Europe there was and is no question. Orders flowed into the workshop from Spain, France, England, and elsewhere. Arsenius, like his uncle, was a skilful tradesman and established relations with merchants who could relieve him from the burden of direct marketing. Instruments from the workshop of Gemma had first been distributed by the printer Bartholemew Graevius,[159] but other agents were used by Arsenius, such as the printer and publisher Christopher Plantin at Antwerp.[160] Arsenius' instruments were of a very high standard, greatly esteemed and much sought after.[161] That they made so strong an appeal is perhaps a reflection of the fact that unlike most contemporary German instruments, which had their numerals and inscriptions punched in somewhat squat Roman characters, the products of the Louvain *atelier* were engraved throughout in a fine, elegant, Italian italic or humanistic script developed by Mercator (figs. 32 & 33). From 1535 onwards, he adopted this script as the best suited, in clarity, compactness, and decorative possibilities, for the lettering of Latin.[162]

Although we know extremely little about the size and structure of

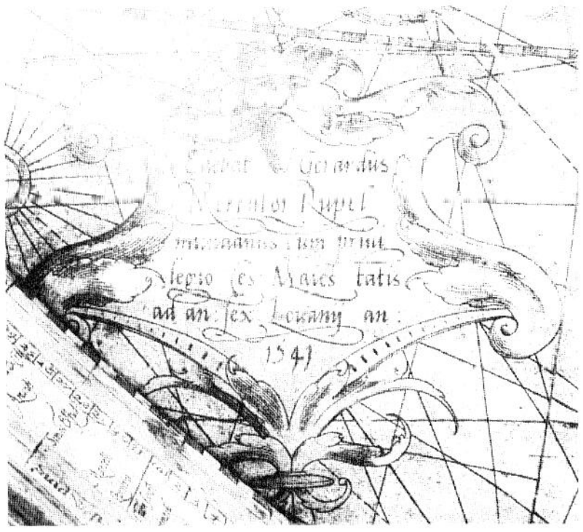

figure 32. Mercator's signature on his terrestrial globe of 1541. Greenwich, National Maritime Museum G. 96. (Photograph: National Maritime Museum)

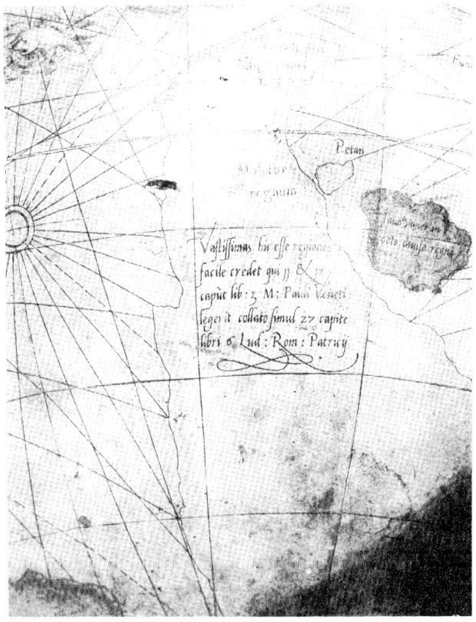

figure 33. Lettering by Mercator on his celestial globe of 1551. Greenwich, National Maritime Museum G. 97. (Photograph: National Maritime Museum)

figure 34 (opposite). Portrait of Gemma Frisius engraved by Jan van Stalburch in 1557, two years after Gemma's death. Oxford, Ashmolean Museum. (Photograph: Ashmolean Museum)

154. Van Ortroy, 23. Gemma's son, Cornelius Gemma, considered his father's most important work to be a 2-volume medical treatise, the first volume of which listed his patients, their illnesses, and their cures, and the second volume the positions of the stars and the daily changes of the wind. *Ibid.*, 55.
155. For whom see de Smet (II).
156. de Smet (III), 34. For Mercator's life, see Osley.
157. '. . . from time to time was fashioning and constructing scientific instruments (i.e. spheres and astrolabes), astronomer's rings, and similar apparatus in bronze.' Walter Ghim's *Life of Mercator* in Osley, 185. Likewise in 1553, Pierre Beausard remarked of Mercator that he could think of no one who excelled him in Europe for the making of instruments. Michel (IV), 16.
158. See cat. no. 18.
159. Osley, 91.
160. Clair 198–9; van Ortoy, 102–4.
161. Van Ortroy, 101.
162. Osley, 50–1.

the workshops of Gemma Frisius, Mercator, and Arsenius, it is clear from the number of surviving instruments signed by Arsenius that their output, at least in the later period, must have been considerable. Some forty astrolabes signed by Arsenius are known;[163] they raised the art of astrolabe-making in Europe to new technical and aesthetic heights. Gemma and Arsenius made several innovations in the astrolabe: Mercator's legible and decorative italic script; the introduction of a pleasing balanced *rete* of a characteristic design with up to fifty or sixty stars; the insertion of a small compass in the suspension mount to allow horizontal angles to be measured;[164] and the inclusion, either on the back or on one of the plates, of Gemma's 'universal' projection,[165] and of other scales such as the *quadratum nauticum* for use in navigation.[166] All these innovations distinguished the astrolabes of the Arsenius workshop and the adoption of some or all of them in the work of later makers illustrates the influence that they had. That Ferdinand Arsenius[167] should have produced pieces almost identical with those of Gualterus is perhaps not astonishing, nor is it astonishing that other Flemish makers such as Coignet, Piquet, Lambert Damery, or Adrian Zeelst, who may have been apprentices or journeymen in the workshop, employed these characteristic designs. What is more striking is that through the Protestant expatriate engraver Thomas Gemini (*c.* 1524–70),[168] the Louvain tradition was to influence the development of instrument-making in England.

Gemini, who may have been banished from his native region as a heretic in 1533, had emigrated to England by 1540. Here he worked as a copper-plate engraver and instrument-maker. Certainly he is the first instrument-maker of whom anything is known in England, and it was probably he who re-introduced knowledge of astrolabe-making, passing it on to his workman (and, perhaps, apprentice) Humphrey Cole (*c.* 1520–91), 'a englishman born in ye north,' as he declared in one of his signatures.[169] Of Cole's successors we know little, but his contemporary, the engraver and instrument-maker Augustin Ryther (*fl.* 1576–95), also used a Mercatorian italic script and was familiar with Dutch cartography.[170] Possibly, Ryther had

163. IC/CCA no. 63. About 30 of them are listed by Zinner (II), 237–9. In both cases I have counted the instruments signed Gualterus Arsenius and instruments signed Regnerus Arsenius as the work of the same man. For discussion of this problem, see below,
164. This was not a totally new idea, for Hans Dorn had similarly incorporated a compass in the astrolabe he completed for Martin Bylica in 1486. For Gemma's intentions, see Pogo.
165. See cat. no. 18.
166. See cat. no. 16.
167. Van Ortroy, 364–5.
168. For Gemini, who was born Thomas Lambert or Lambrichts at Lixhe near Markt-Weset in the bishopric of Liège, see Gunther (VI); Michel (V); Taylor (II), 165–6.
169. On a map of Palestine engraved in 1572 for the printer Richard Jugge. The contention in Michel (V) that this emphatic phrase is to be interpreted as meaning that Cole was an Englishman by adoption and not by birth, need hardly to be taken seriously. For a general account of Cole's work, see Gunther (II).
170. In 1549, 20 shillings was paid for an astrolabe by All Soul's College, Oxford, but it is not known to whom, Rogers, iii, 573. On Ryther, see Taylor (II), 179; Hind, i, 138–49; Brown (II), 24, 58–9; for his cartographical work Tyacke & Huddy, 35–7, 41, who comment on the influence of Flemish cartographical techniques on the Saxton world map of 1583 which Ryther is presumed to have engraved.

worked with Gemini or Cole; possibly, he had himself been in the Low Countries. In any case, Ryther's surviving work may be placed in the Louvain tradition which thus—through his apprentice Charles Whitwell, in turn the master of the leading English instrument-maker of the early 17th century, Elias Allen—influenced the whole development of English instrument-making.

The influence of the Louvain school of astrolabe making was not confined to the northwest of Europe. Via the Frisian Hugo Helt (*c.* 1525–94/5) and the Spaniard Juan de Rojas[171] (*fl.* 1544–50) it spread to the Iberian peninsula, whither several Arsenius instruments were also despatched by Arias Montanus during the years he spent in Flanders (1568–72) supervising the production of Plantin's polyglot bible.[172] A more indirect, but equally striking example of influence is provided by Erasmus Habermel's use of a *rete* design which closely resembles that of Arsenius and of scales such as the *quadratum nauticum* which had originated with Gemma Frisius. That a leading maker in central Europe, whom one might expect to have followed in the earlier Nuremberg styles, was, instead, engraving astrolabes in the fashion of Louvain is a striking testimony to the prestige of Arsenius' instruments and also to the failure of the south German workshops to maintain their prestige in the face of the competition from Louvain. There is no evidence, however, to suggest that Habermel ever had any direct contact with Arsenius, nor that he ever visited Louvain. His calligraphy, moreover, although italic, is unlike the Mercatorian italic used by Arsenius and his followers.[173]

Ten astrolabes signed by Erasmus Habermel have survived into the 20th century.[174] For no other astrolabe-maker after Arsenius do we have so much material. From the later decades of the 16th century, however, we have the names of far more makers of whose work one to five examples have survived. This is one of the most striking contrasts between the 16th and 17th centuries. Whereas in the earlier century the bulk of the known astrolabes were produced by the two large workshops of Hartmann and Arsenius, with new but smaller workshops becoming established in the closing decades of the 16th century in England (Gemini and Cole), Italy (Egnatio Danti and the della Volpaia family),[175] and France (Phillippe Danfrie),[176] in the 17th century many more makers each produced a small number of instruments. This fragmentation of astrolabe production seems to have occurred from about the 1580's onwards, and may in part represent a geographically more widespread demand for instruments, as well as an increase in the number of competent workers. Thus alongside the Arsenius workshop we find such

171. See below, for Helt & Rojas.
172. Clair, ch. 4; van Ortroy, 101. For a brief account of Arias Montanus, see Bell.
173. Osley, 99, n. 4. For Habermel in general, see Eckhardt.
174. IC/CCA no. 65.
175. For Danti, see Gunther (V), ii, 332–3; for della Volpaia, Maccagni.
176. Gunther (V), ii, 358–64.

figure 35. Portrait of Johann Stöffler. From J. J. Boissard, *Bibliotheca, sive Thesaurus... Illustrium Eruditione & doctrina Virorum... Effigies & Vitae...*, pt. II, Frankfurt, 1630, 143. Engraved by Theodore de Bry.

lesser makers as Coignet and Lambert Damery,[177] and in Germany good makers such as Johann Krabbe at Minden and Caspar Vopel at Cologne. Nonetheless, from the 17th century approximately only one-third as many astrolabes have survived as from the 16th. In part, this difference may be explained by the continued use of second-hand astrolabes. Robert Gordon of Straloch, for example, in the mid-17th century was using an astrolabe made three centuries earlier by Jean Fusoris.[178] An even more striking example is provided by the astrolabe signed by Muḥammad b. Fattūḥ al-Khamā'irī at Seville in A.H. 619 (A.D. 1223/4) which had its *rete* and one plate replaced in the workshops of Arsenius in *c.* 1560 (catalogue no. 2). Perhaps a second reason is to be found in the widespread circulation of astrolabes printed on paper and pasted onto wood or cardboard. At least eight makers (Hartmann, Danti, Danfrie, Krabbe, Blaeu, Sutton, Prujean and Bion) are known to have produced such instruments, and John Aubrey recommended their educational use in *c.* 1683.[179] Being easily perishable, few of them have survived. Even with these qualifications made (and both apply, to some extent, to both centuries), the difference in the number of astrolabes surviving probably reflects a gradual decline in overall demand for the instrument during the 17th century. Before considering this phenomenon, however, we may first glance at the history of the printed literature of the astrolabe.[180]

The astrolabe, as already noted, had been much studied and much explained in the Middle Ages. It would be astonishing, therefore, if works upon it had not quickly appeared in print. By 1500, at least nine treatises had been printed, and it is not unlikely that there were more. The treatises printed before 1500 are listed in Table I. What is perhaps astonishing is that no version of Pseudo-Māshā'allāh was available until 1510, when it was issued by J. Grüninger.[181] Two years later, John of Seville's medieval Latin version was included in a reissue of Gregor Reisch's *Margarita Philosophica*.[182]

In 1512/3 one of the most popular and beautiful of all books on the astrolabe appeared in Germany – Johann Stöffler's *Elucidatio fabricae ususque astrolabii* (figs. 35 & 36). The influence of this work, which was itself based on Pseudo-Māshā'allāh, may be measured by the fact that 'Stöffler's astrolabe' became a synonym for the 'particular' or 'fixed latitude' astrolabe when it was necessary to distinguish this from the universal forms. Thomas Blundeville, for example, remarked that 'of Astrolabes I have never seen but three sorts, First, that of *Stofflerus*, which for these hundred years past or there-

177. For whom, see Michel (I) & (III), 165, 166.
178. Now in the Royal Scottish Museum, Edinburgh. See Cumming; Hutchieson.
179. Turner (I), 65.
180. This account is largely based on the bibliographies given in Houzeau & Lancaster; Klebs; Poulle (I), 870–1; Gunther (V).
181. Only the copy in the Crawford Collection at Edinburgh seems to be known. See Grassi, 38.
182. A highly popular scientific encyclopedia which was first printed in 1503 and not, as often stated, in 1496 (Stilwell, 284). It was several times reprinted, e.g. in 1508, 1512, 1515, 1517, 1535, and 1583.

figure 36. Astrolabe. From Peter Jordan, *Coelestium Rerum Disciplinae, atque totius Sphaericae peritissimi, Ioannis Stoflerini… Astrolabiorum compositionem seu fabricam…*, Mainz, March, 1535. *Rete* (upper left), *mater* with plate (above), and back (left). These pages, which are not called for in the collation, were intended to be removed from the work and mounted on wood or pasteboard to make a usable astrolabe. Oxford, Museum of the History of Science. (Photograph: Museum of the History of Science)

Table 1. The Earliest Printed Treatises of the Astrolabe[1]

Year	Author	Short title and incipit
8 July 1475	Andalo di Negro[2] (Pietro Buono Avogaro, ed.)	*Opus praeclarissimi astrolabii...* a) 'Si astrolabium facere volueris primo et ante omnia fac tabulam' b) 'Nomina instrumentorus astrolabii sunt hec; primum est annulus'
c. 1477	[Robert of Chester][3] (Ulysses Lanciarinus, ed.) published with:	*de astrolabio canones* 'Cum (or quia) plurimi ob nimiam quandoque accurationem...'
c. 1477	Prosdocimo de Beldomandi	*Compositio astrolabii et operatio secundem novam et veterem compositionem* 'Quamvis de astrolabii compositione tam modernorum quam veterum...
24 December 1484	Henry Bate	*Magistralis compositio astrolabii* 'Universorum entium radix et origo Deus qui...'
?[*c.* 1490]	[Robert of Chester][4]	*Astrolabii quo primi mobilis motus deprehendentur Canones* 'Sphera solida que et astrolabium sphericum appelatur...'
1492/93	Bonet de Lates	*Anuli per eum compositi super astrologicam utilitates*[5]
[1497–19]	[Robert of Chester][6]	*Astrolabii... canones* above, no. 5.
30 September 1498	Anon[7]	*...Nicephori Astrolabi Expositio...* 'Si astrolabi peritiam tenere volueris ipsam hoc modo...'
30 September 1498	Nicephor Gregoras[8]	*de Structura Astrolabi...* 'Exterius conceptaculum quod tympana comprehendit...'
No date[9]	Faber Barduvicensis	*Compositio Astrolabii* 'Astrolabium ut Abraham judeus inquit omnium mathematicalium instrumentorum...'

Notes to Table 1

1. Despite its title, the *Astrolabium Planum*... of Johan Engel [Angelus], is a purely astrological work, and has here been omitted.
2. Reprinted with introduction by G. Bertolotto, 'Il Trattato sull'Astrolabio... riprodotto dall' edizione Ferrarese del 1475', *Atti della Societa Ligure di Storia Patri*, XXV, 1892, 49–142.
3. Both parts of the work were published by Lanciarinus as being by Robertus Anglicus, to whom they are still often ascribed. For the ascription to Robert of Chester, see Thorndike (II), 467, n.l.; and for Prosdocimo di Beldomondi, Poulle (VIII), 871, n. The text breaks off in mid-sentence on f20v.
4. No copy of this issue has been located. For the ascription to Robert of Chester, see n.5.
5. The work describes a small astrolabe mounted in a finger ring (fig. 43). It was dedicated to

Imprint & References	Contents
Ferrara, Picard: BMC vi, 608; Hain, 967; Klebs, 63:1; Stilwell, 808	a) Construction b) Uses
Perugia: BMC, vi, 879; Coppinger, 5 34; Klebs, 850:1; Stilwell, 884	Uses
ff. 27r-42r	Composition
Venice, E. Ratdolt: BMC, v, 291; Hain, 21; Klebs 4:1; Stilwell, 823	Uses (astrology)
Augsburg: Houzeau & Lancaster 3253	Description Uses Mensuration
Rome: Houzeau & Lancaster 2304; Gingerich, 16	
Venice, Paganinus de Paganinis: BMC, v, 458; Klebs 119:1	Description Uses Mensuration
Venice, Simon Bevilaque: BMC, v, 523; Klebs, 1012:1	Construction Uses
Venice, Simon Bevilaque: BMC, v, 523; Klebs, 1012:1	Construction
Klebs, 386:1	Construction

Pope Alexand VI.
6. The text of this work, which is anonymous, is identical from f 2r, 1, 18 to f 15r, 1. 1, with that of Robert of Chester's treatise (2, above). This point was noted by an anonymous annotator of BMC, v., 458.
7. This work, which is generally attributed to Nicephoros Gregoras, is not part of his treatise. It is perhaps a translation from an as yet unidentified original. See Segonds (II), 83–84, 111. Since it often occurs in manuscripts following the treatise of Philopon, it acquired the title *Explicatio altera*.
8. A *scholium* of Mararios Hieromonachos is interpolated by Valla into this translation. Segonds, (II), 111.
9. Klebs, 135, conjectures 1499.

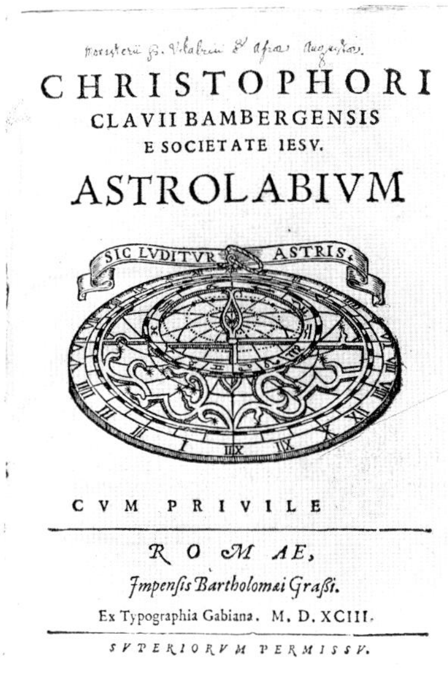
CHRISTOPHORI
CLAVII BAMBERGENSIS
E SOCIETATE IESV.
ASTROLABIVM
SIC LVDITVR ASTRIS
CVM PRIVILE
ROMAE,
Impensis Bartholomæi Grassi.
Ex Typographia Gabiana. M. D. XCIII.
SVPERIORVM PERMISSV.

figure 37. Title-page, Christopher Clavius, *Astrolabium*, Rome, 1593.

The conclusions of the Astro labie.

figure 38. Title-page, Chaucer's treatise on the astrolabe, from the 2nd collected edition of his works, 1532.

abouts hathe beene had in most price and estimation, as an instrument containing all the uses, or at least the most part of all other Mathemeticall Instruments. . . .'[183] Stöffler's work indeed had an enormous success, passing through at least sixteen editions before 1620, as well as two editions of a French translation[184] and a German edition. Stöffler's book was published by his friend Jacob Köbel (1460/5–1533). Köbel himself, a few years later, wrote a treatise on the astrolabe[185] which was also popular, passing through some twelve editions, and which was often issued together with that of Stöffler upon which it was based. A third widely used work was *de Usu Astrolabii Compendium*. . . , by Johan Martin Poblacion, issued by the Paris press of Estienne in 1518. Before 1550, when it was translated into Spanish and published at Valladolid, this received some twelve editions. The edition of 1546 included the tracts of Pseudo-Proclus and the *Explicatio altera*, incorrectly ascribed to Nicephoros Gregoras (Table I, no. 9). No other work was as popular as these three. But there was no shortage of books on the astrolabe, nor of accounts of it in general mathematical works. Of these, the most important was Gregor Reisch's *Margarita Philosophica*(1503 and nine further editions in the 16th century) already mentioned. The most comprehensive treatment of the subject, by Christopher Clavius[186] (fig. 37), had only one edition.

Literature in the vernacular was also not lacking. Although the best treatise on the subject in English, that by Geoffrey Chaucer, was not issued as a separate work until 1870,[187] it was, nonetheless, available in the second collected edition of Chaucer's works issued in 1532 by William Thynne and Thomas Godby (fig. 38) and remained available in this form throughout the 16th and 17th centuries. Of other printed works on the single-latitude astrolabe in English, however, there were none before Robert Tanner's translation of Köbel's treatise as *The Traveller's Joy and Felicitie or a Mirror for Mathematics*. . . , 1587, and, indeed, of the other five English books on the astrolabe issued during these centuries, only Thomas Oliver's *A New Handling of the Planisphere*. . . , (1601) concerned itself with the single-latitude astrolabe.

The small number of books in English contrasts markedly with the position in Germany, where Johann Copp's *Wie man diss hoch-*

183. Blundeville, 280. *Cf.* Valerius Regnartius, *Astrolabiorum seu Utriusque planispherii universalis et particularis Usus*, Rome, 1610, 31. 'Astrolabii Ptolemai quod stoflerini vocant' (Ptolemy's astrolabe, which they call Stöffler's), and throughout the book. Nearly another century later the phrase was still in use. John Aubrey, for example, used it in his educational treatise completed *c.* 1683/4 (Turner (I), 65 & 74). Pepys also used the phrase in his 'Fragments of English Shipwrightry' (Pepys Library Ms 2820, cited in Gunther (V), 518).

184. The basis of the French edition, Sig A3v, was an old manuscript translation of the second part (Uses) which was acquired by G. de Bordes and prompted him to translate the first part. The second part was revised and expanded at his request by his friend, Jean Pierre de Mesmes.

185. *Astrolabii declaratio ejusdemque usus mire jucundus*. . . , Nuremberg, 1517. For Köbel's life, see Kurt Vogel in *D.S.B.*, vii, 418–20.

186. *Astrolabium*, Rome, 1593. The verbose paper by Mascart gives a hostile account of the work and its author.

187. By Brae. It was followed 2 years later by Skeat's remarkable edition. Both editors acknowledge the merits of the edition prepared by Walter Stephens in 1553–5, which unfortunately was never published.

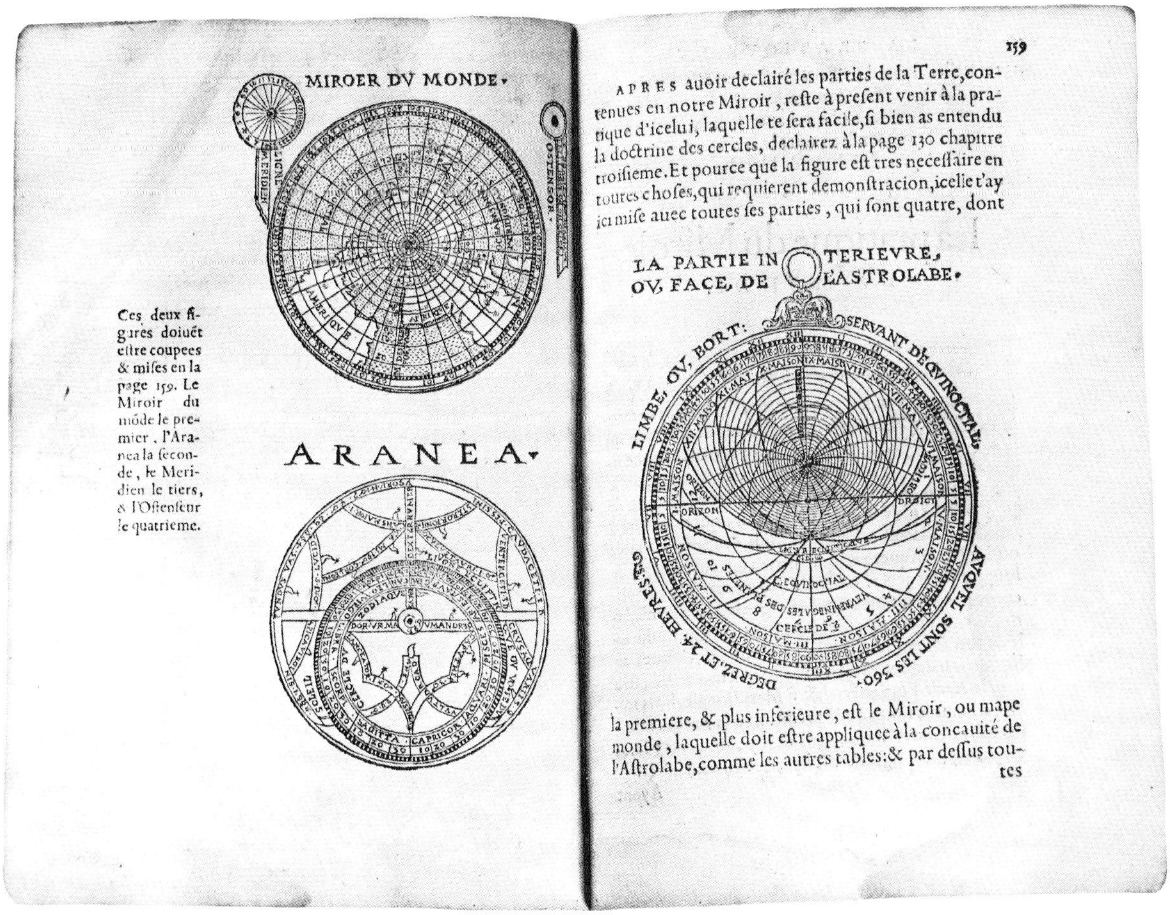

figure 39. Geographical plate and *rete* for an astrolabe (left) intended to be cut out and fixed to the *mater* (right). From Jacques Focard, *Paraphrase de l'astrolabe* (rev. by J. Bassentin), Lyons, 1555, fol. 158v–59r.

berumpt astronomischer und geometrischer kunst Instrument Astrolabium brauchen soll appeared simultaneously at Bamberg and Nuremberg in 1525 and passed through four more editions by 1600. In three of them the work was augmented by Zacharias Bornmann. Andreas Schöner included a succinct account of the astrolabe in a work on dialling in 1562[188] and also gave instructions for the construction of a rectangular or column form of astrolabe.[189] In the 17th century, three further German treatises—by Johann Krabbe (1608, 1625, 1656), by Adam Olearius (1632), and by Franciscus Ritter (1599, 1613, 1640, 1650)—were published. In Italy, the works of Egnatio Danti (1562)[190] and Giovani Paolo Gallucci (1597)[191] provided excellent treatments, as in France a little earlier did the treatises of Dominicque Jacquinot (1545)[192] and Jacques Focard (1546)[193] (figs. 6, 7, 8, & 39). In Spain, the translation of Poblacion in 1550 remained somewhat isolated, and works on the astrolabe in Dutch were late to appear. Adrian Metius' *Astrolabium; Hoc est astrolabii utriusque accurata Descriptio* (1626) was translated in the following

188. *Gnomonice Andreae Schonerii Noribergensis. Hoc est de Descriptionibus Horologiorum*. . ., Nuremberg, 1562, fol. xciiii, ff.
189. *Ibid*, ciiii^{r}–cviii^{r} 'Astrolabii columnaris proiectio.' It should, however, be noted that this seems to have nothing in common with the linear astrolabe. See below.
190. *Trattato dell Uso et della Fabrica del'Astrolabio*. . . Florence, 1562.
191. *Della Fabrica et use di diversi Stromenti di Astronomia et Cosmografia*. . ., Venice, 1597. This was basically a translation of Pseudo-Māshā'allāh's treatise.
192. *l'Usage de l'astrolabe avec un traicte de la Sphere*, Paris, 1545.
193. *Paraphrase de l'astrolabe*. . ., Lyons, 1546.

year, and in 1628 it was joined by Philip Lansberg's *Astrolabium datis. . .Verclaringhe de platte sphaere. . .anders Astrolabium genaemt. . . .* There were three further editions (in 1635, 1655, and 1680), and two additional Dutch books on the astrolabe were issued before the end of the 17th century.

Just as the production of astrolabes began to slacken markedly in the middle and later 17th century, so too did that of books about them, less than a dozen appearing after 1650. In 1702, Nicholas Bion published an excellent short account of astrolabes, probably to accompany the wood-and-paper instruments which he issued at about this time,[194] and in 1712, F. Bottens published *Magnum et universale astrolabium notam faciens*. This, however, was the last work on the subject until, toward the end of the century, an interest in the astrolabe as an antiquity stimulated new publications.[195] The wood-and-paper astrolabes produced by Bion were also, it would seem, the last astrolabes to be produced as part of the normal stock-in-trade of a professional instrument-maker. With the exception of Islam, these astrolabes mark the end of a long tradition, the end of the period of about 1500 years during which it can be claimed that the astrolabe had a fundamental scientific and social function.

What were the causes of this demise? Why, after being established for so long should the astrolabe have disappeared from use in the short space of about three or four decades?

Answers to this question can, in the nature of things, only be impressionistic. They are to be sought partly in the nature of the instrument itself, partly in the broader development of astronomy, of mathematics, and of society itself. In the first place, the astrolabe in its origin and use was the product of particular cultures and societies. These societies were the Hellenistic Roman world and its successor – Islam. Their cultures rested on a geometrically enunciated, geocentric cosmology that the astrolabe partially represented. From the moment when this cosmology, with its concomitant symbolism and astrological arguments, was destroyed in the West at the hands of Copernicus, Galileo, Kepler, and Newton, the astrolabe became increasingly irrelevant.[196] Much of the appeal of the astrolabe during the Middle Ages and early Renaissance lay in the fact that it was an economical device. One object combined the functions of observing instrument, celestial computer, and model of the heavens. Because it did so, it had obvious attractions for the rela-

194. One example of Bion's pasteboard astrolabes is now in the National Maritime Museum, Greenwich, A24/36-6c. It is illustrated in Hamilton (I), no. 7, being then in the Ganze collection. See also Proctor (I)

195. Nonetheless, it is interesting to find a copy of Bion's *l'Usage des astrolabes* being supplied in 1788 to Cassini for use in the Paris Observatory at a cost of 2 livres (Observatoire de Paris Ms D. 538: I am indebted to Mlle J. Alexandre for this reference). The interest of the book for Cassini was, we may suppose, more antiquarian than practical.

196. It is striking, for example, to see how in his *Lexicum Technicum* (first edit. 1704; fifth edit. 1736), John Harris ignored the astronomical astrolabe as an instrument entirely, describing only the mariner's astrolabe. For him it existed only as an example of projection. 'There are also some Projections of the Sphere which are called by this Name, as that of *Gemma Frisius* and *Stoffler*.' Of only these 2 projections does he give a brief description.

tively poor, underdeveloped societies of Europe. From the moment, however, that a commercial manufacturing trade in instruments developed and demand began to be made for precision of measurement and ease of use, the astrolabe began to disappear. The multiplicity of purposes to which it could be put meant that its precision was low. The complexity of its theory meant that it was often inaccurately made. If an astrolabe in the late 16th or 17th centuries was really to be of use, it needed to be large and heavy. As an instrument for time-finding, therefore, it became inconvenient and was passed over in favour of the many forms of small pocket dial which were then available. As a device for travellers it was equally inconvenient, and for sea travel – exposed to high winds – quite impossible. To use the astrolabe correctly required long training and much practice. It was, moreover, difficult to make, and its manufacture required considerable time. Therefore it was expensive.

If these general considerations meant that the astrolabe was becoming less and less useful, the demand for greater precision, combined with the desire of commercial instrument-makers to demonstrate their skills by the invention of specific instruments for specific tasks – which also gave them more objects to sell – meant that its very universality told against the astrolabe in the course of the 17th century. As an observing instrument, an astrolabe small enough to carry conveniently could not compete with an instrument such as a Holland Circle, graphometer, or a quadrant in accuracy or convenience, while the introduction of telescopic sights after *c.* 1667/68 made it almost entirely redundant. Displaced as a time-finder by portable sundials and watches, as a star map and model of the heavens by printed celestial maps, by armillary spheres and planetaria, it was in its didactic and astrological aspects that the astrolabe survived longest. A diminution in the influence of astrology, however, occurred in the early 18th century. Didactic models that could be used to teach the wonders of the Newtonian world system were required, rather than the simple geocentric universe shown by the astrolabe which, in so far as it was needed, could just as easily be demonstrated on a globe. Obsolete as a practical and scientific tool, outmoded as a cosmological model, and deprived of a social function with the decline of astrology, the astrolabe quietly disappeared. Only when the attention of Orientalists was drawn to the relics of early Islam and a new interest developed in the West in the history and antiquities of science, did the astrolabe once more come to have a place in European culture.[197]

BONETI DE
LATIS HEBRÆI ME-
DICI PROVENZALIS ANNVLI
Aſtronomici vtilitatũ liber ad Alexandrum ſextum Pontificem maximum.

Idetur mihi compoſitio iſtius Annuli multò facilior Aſtrolabij cõpoſitione, & quadrati Iſraelis, & aliorũ inſtrumẽtorũ aſtrologiæ, & minus tædioſa, maximè quia eſt ornamentũ nobile, ac ſemper viſibile in manibus, ad ſciendũ neceſſaria, & eſt inſtrumentũ nouum, ideo adaperire debeo ſuos vſus & vtilitates, quia plures ſunt. Imprimis autem deo creatori cœli gratias agimus.

DE

figure 40. Astrolabe mounted on a finger-ring. From Bonetus de Latis, *Annuli Astronomici...*, in J. Dryander, *Annulorum trium diversi generis instrumentorum astronomicorum componendi ratio atque usus*, 1537.

197. For an interesting account of making a modern astrolabe with 18th-century techniques, see Honig.

Astrolabe Catalogue
Planispheric Astrolabes

1. Eastern (? Persian) Islamic Astrolabe

?late 4th/10th century or early 5th/11th century
Brass, of high copper content
Diameter 3½ in. (89 mm)
Signed: [198]
(Made by al-Muḥsin b. Muḥammad the physician)

Inventory 507

The *rete* carries eighteen named stars, has dagger-shaped pointers and a straight east-west bar; the zodiacal signs are named on the ecliptic circle (*al-ḥūt* for Pisces). There is one *mudīr.* The *mater* is composed of a back sheet rivetted to the limb, which carries a 360° scale divided and numbered in groups of 5°. The *kursī* is plain and lobed, with two holes, and carries a shackle and ring. The interior of the *mater* is blank, as is the reverse of the instrument, except for a degree scale in the upper-left quadrant and the maker's name on the *kursī*. There is one plate, with the almucantars drawn for every six degrees for latitudes 31° and ?34°, pierced with a small hole that fits over a pin in the *mater* to retain the plate in place. The plate is also engraved on each side below the horizon with unequal-hour lines. The alidade, which is unusable and probably a replacement, is straight, with two pin-hole sights. The instrument is engraved throughout in a plain *kufic* script.

Signed by an otherwise unrecorded maker, this astrolabe should be compared with the earliest surviving astrolabes from the 3rd/9th and 4th/10th centuries.[199] With these instruments it shares a family likeness, being particularly close in style to three of them. These are an astrolabe by Khafīf (late 3rd/9th century), now in the Museum of the History of Science, Oxford[200] an astrolabe by Basṭūlus dated A.H. 315 (A.D. 927/8), now in a private collection;[201] and an astrolabe by Aḥmad b. Khalaf *c.* A.D. 950, now in the Bibliothèque Nationale, Paris.[202] In all four instruments the ring and shackle are virtually identical and there are only minor differences in the form of the *kursī*, that of al-Muḥsin's astrolabe being very slightly more curved and decorative in form than the other three, and distinctly less angular than that of Khafīf. The single *mudīr* on the *rete* of Khafīf's astrolabe is in the same position as it is on al-Muḥsin's while the *rete* design and form of the star pointers are closely similar for all four and, indeed, do not differ greatly from the non-Persian astrolabes in this group. The astrolabe of al-Muḥsīn, however, has one more star on the *rete* than the other three. Each of the astrolabes is signed on the back of the *kursī* in an unadorned *kufic* script,

198. Pointing added.
199. For these, see above, n. 28.
200. Maddison (I), no. 155 & pl. XXIIIa.
201. Brieux (I), 6–9; Maddison & Brieux, *passim*.
202. Gunther, (V), i, no. 99, 230.

figure 41. Catalogue no. 1: Astrolabe by al-Muḥsin b. Muḥammad. Front.

figure 42. Catalogue no. 1. Back.

and the back of each instrument is very plain. The instrument by al-Muḥsin is marked only with one scale of degrees in the upper-left quadrant. Aḥmad b. Khalaf's has scales of degrees in both upper quadrants as originally had the instrument by Khafīf. The astrolabe by Basṭūlus is fully engraved with degree scales and has in addition a scale of co-tangents in the lower-right quadrant.

While the strong similarities between al-Muḥsin's astrolabe and the surviving 4th/10th-century instruments suggest that it also should be ascribed to the 4th/10th century, it does differ from most of the instruments in the group. First, it is appreciably smaller and secondly, it uses the term *al-ḥūt* (a direct translation from the Greek) for the zodiacal sign *Pisces* instead of the older Arabic term *as-*

Samaka,[203] which is used on the astrolabes of Khafīf, Basṭūlus, and Aḥmad b. Khalaf. *Al-ḥūt* came into general use after the appearance of *az-Zīj aṣ-Ṣābī* (The Sabian Astronomical Tables) of al-Battānī (A.D. 858–929), a Ṣābian from Ḥarrān. That there was very much time lag in the adoption of the new term seems unlikely given the fact that both terms are used in aṣ-Ṣūfī's book on the constellations, *Kitāb ṣuwar al-kawākib ath-thābita* (A.H. 355 [A.D. 965/6]), and *al-ḥut* is used on the surviving astrolabes of al-Khujandī (A.H. 374 [A.D. 984/5])[204] and of Aḥmad and Muḥammad b. Ibrāhīm (A.H. 374 or 394 [A.D. 984/5 or 1003/4]).[205] A date in the late 4th/10th century A.D. for this instrument is, therefore, not improbable, although it could have been made in the 5th/11th century or later, although in what would then have been an antiquated style.

Provenance: Collection of K.F.S.D. Gilkes, Sussex, acquired from an old excavation site near Rayy.

Sotheby & Co. Ltd., 19 December 1966, lot 71.

IC/CCA nos.: 3522, 3527, 3904, 3919.[206] Brieux-Maddison, MHSN MHMD TBIB 1.

figure 43. Catalogue no. 1. Plate for latitudes 31° (top) and ?34° (bottom).

203. Kunitzsch (I), 21–3, 68, 104.
204. Brieux (I), 10–3.
205. Gunther (V), i, 114–5.
206. The 4 separate IC/CCA nos. for this one instrument arise partly from the fact that the signature has been incorrectly read in the past (as al-Mushi b. Muḥammad aṭ-Ṭushi) and partly from another entry (as al-Mushi b. Muhammad aṭ-Ṭabib) having been duplicated.

2. Andalusian Astrolabe with Flemish Rete and Plate

A.H. 619 (A.D. 1222/3) and 3rd quarter of 16th century
Brass
Diameter 7¾ in. (199 mm)
Signed:
(In the name of God. Made by
Mu[h]ammad b. Fattūḥ al-Khamā'irī
in the city of Seville in the year
619 of the Hijra)

Inventory 3407

The *rete*, which is gilt, is a replacement made in the 16th century in the Low Countries. Although it is not signed, the elegant tracery and fine lettering are characteristic of the workshop of Gualterus Arsenius at Louvain *c.* 1560–70.[207] There are 43 named stars with their magnitudes and astrological information, and three unnamed star pointers.[208] The limb and backplate were separately cast and rivetted together to form the *mater.* The *kursī* is low and triangular, with decorative piercing along the edge. The reverse side carries the maker's inscription with date. The limb is engraved with a scale of degrees to which western 'Hindu-Arabic' numerals have been added at 10° intervals, as have some, but not all, of the equal-hour numerals, at 15° intervals. The *mater* is inscribed with a circular table giving the triplicities for day and night and the terms and faces of the planets according both to Egyptian tradition and to that of Ptolemy. On the back, in the circumference of the upper two quadrants, is a degree scale, and in the lower two quadrants, scales of co-tangents. Within these is a zodiacal calendar (0° Aries = 13 ¾ March) and three concentric circles which form a table for finding the day of the week on which the Christian year begins.[209] Within these are a simple form of *dastūr* diagram and a shadow square. There is no alidade. The present counter-changed rule is a later replacement. There are seven plates, six of them (a-f) original, engraved:

a. for latitudes 31°30′ and 37°30′
b. for latitudes 34°30′ and 40°30′
c. for latitudes 21° and 25°30′
d. for latitudes 33° and 39°
e. for latitudes 30° and 36°
f. for latitude 45° and with a tablet of ecliptic co-ordinates
g. for latitude 51° with lines for the 12 astrological houses; on the other side a tablet of horizons, shadow square, and equal-hour and unequal-hour diagram.

207. For whom, see cat. no. 18.
208. The single name of Ursa major, however, accounts for 4 stars.
209. Such tables long continued to be placed on Maghribi astrolabes. *Cf.* the example on an astrolabe by Muḥammad b. Aḥmad al-Baṭṭūṭī A.H. 1136 (A.D. 1723/24) in *Instruments Scientifiques*, no. 173, 110–1.

figure 44. Catalogue no. 2: Astrolabe
by Muḥammad b. Fattūḥ. Front.

figure 45. Catalogue no. 2. Back.

Lines for the Muslim prayer times and for dawn and dusk are drawn on both sides of plates a-f. The seventh plate (g) was made in the Arsenius workshop and, like the *rete*, is gilt.
Brieux & Maddison MHMD FTUH 7.

Muḥammad b. Fattūḥ al-Khamā'irī is one of the last known astrolabists of Islamic Spain. His work was of high quality and thirteen other pieces (11 astrolabes and 2 *saphea*) have survived, with dates ranging from A.H. 604 (A.D. 1207/8) to A.H. 634 or 638 (A.D. 1236/7 or 1240/1). He was thus working a few years before the fall of Seville, capital of the Almohad Andalusian kingdom, to Ferdinand III of Aragon after a siege of seventeen months.

This remarkable instrument, in which the work of two of the finest makers of their respective eras is combined, provides striking evidence for the continued use of Islamic instruments in the West, even in quite late periods. Although the zodiacal calendar would have been useless because of the precession of the equinoxes, the rest of the astrolabe (*rete* excepted) could perfectly well have been used, even by someone who did not understand Arabic. A late-16th-century owner, therefore, realizing this, modernized the instrument by ordering a new *rete* and a plate appropriate to his latitude from the leading instrument-maker of his time. Whether he did this because the instrument had already lost these parts, or as up-to-date replacements of those that were still there, cannot now be determined. In either case, he acquired an elegant, usable instrument at a lower cost than he might otherwise have expected.

figure 46. Catalogue no. 2. *Mater.*

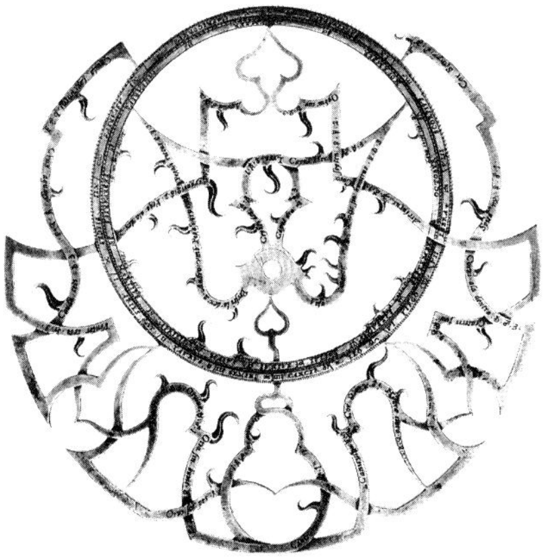

figure 47. Catalogue no. 2. *Rete.*

figure 48. Catalogue no. 2. Plates.

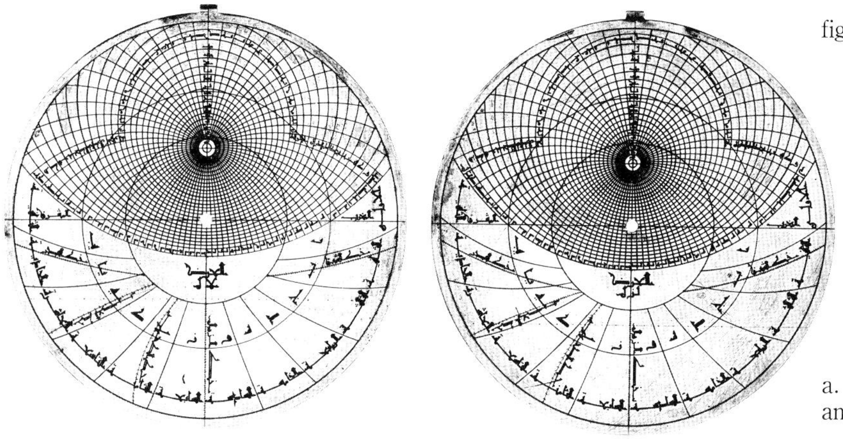

a. For latitudes 31°30′ (left) and 37°30′ (right).

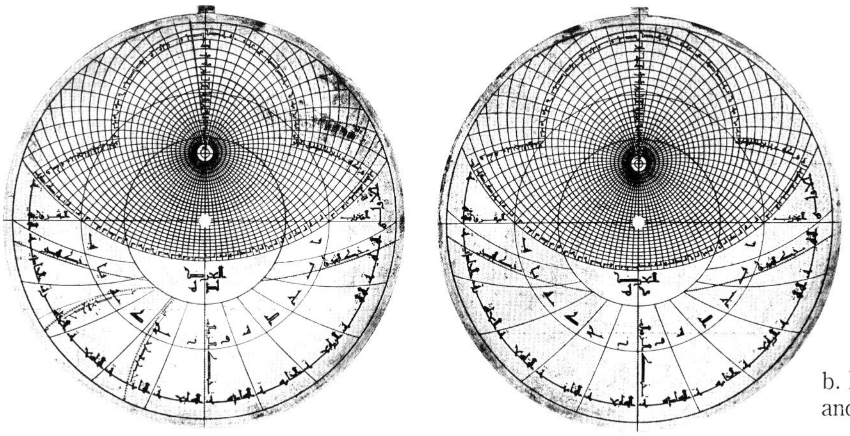

b. For latitudes 34°30′ (left) and 40°30′ (right).

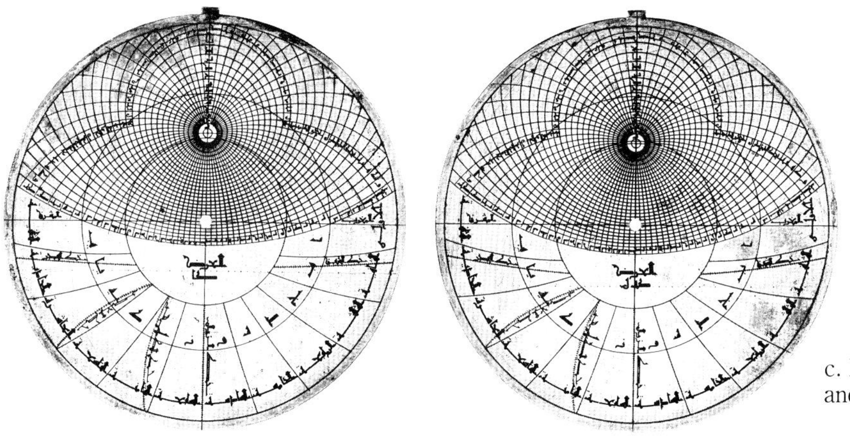

c. For latitudes 21° (left) and 25°30′ (right).

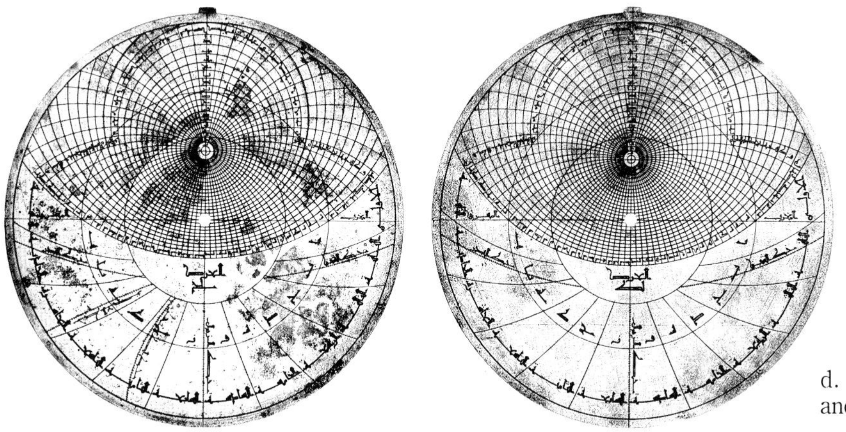

d. For latitudes 33° (left) and 39° (right).

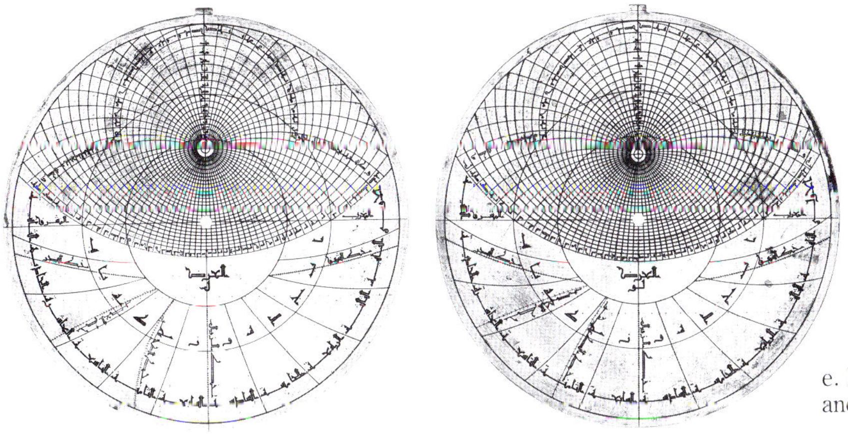

e. For latitudes 30° (left)
and 36° (right).

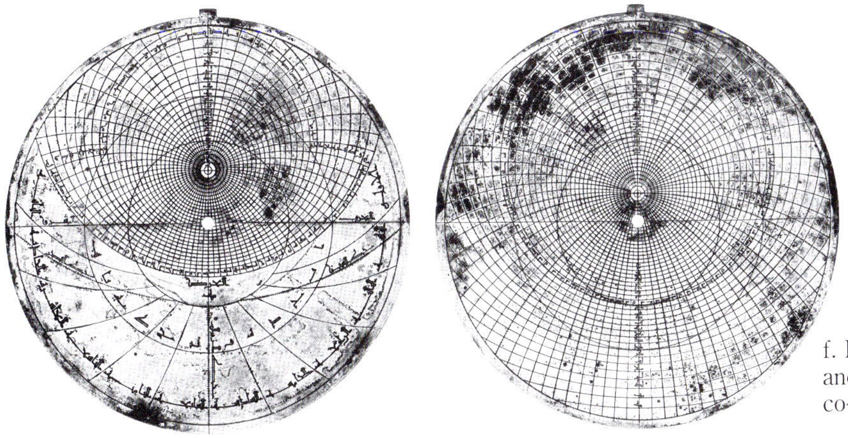

f. For latitudes 45° (left)
and with a tablet of ecliptic
co-ordinates (right).

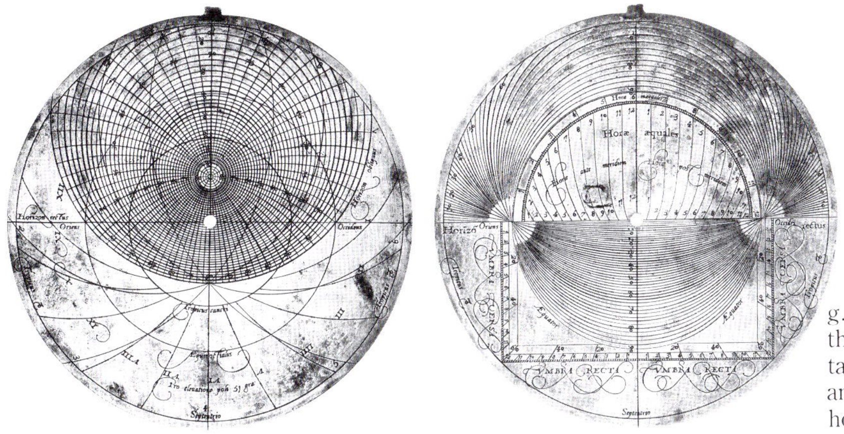

g. For latitude 51° and
the 12 astrological houses, (left) and
tablet of horizons, shadow square,
and equal-hour and unequal-
hour diagram (right).

3. Maghribi Astrolabe

?mid 8th/14th century–9th/15th century
Brass
Diameter 4⅜ in. (111 mm)
Not signed
Inventory 1433

The *rete* carries twenty-six named stars, which originally were probably inset with silver knobs. There are narrow, slightly curving, star pointers arising from circular bases, and curved tracery in the *rete* with a Moorish trefoil arch beneath the arc of the equator. The east-west bar is counter-changed. The ecliptic circle has named signs, and there are three *mudīrs*. The limb and backplate were separately cast and rivetted together to form the *mater*. The *kursī* is low and triangular, decorated with an ogee (a moulding convex above and concave below), on each side and at the top, with a semicircular moulding between them. There is a shackle, a suspension ring, and an *'ilāqa* of red cord. The limb is graduated with a 360° scale reading to 2° in groups of five. The *mater* is plain but inscribed with three concentric circles evidently intended as the tropic of Cancer, the Equator, and tropic of Capricorn for a latitude plate that was not completed. On the back, in the circumference of the upper two quadrants, is a degree scale (0°–90°–0°). The inner rings provide a zodiacal calendar (0° Aries = 13 March). There is a double shadow square divided by twelve. The alidade is straight and has a replacement horse and pin. There are two plates, with the almucantars drawn for every 6°:

a. for latitudes 37° and 31°30′ (Meknès)
b. for latitudes 37°30′ and 21°40′

Plate a) is engraved by a different hand from plate b) and the rest of the astrolabe, and is probably a later replacement. The almucantars are numbered, as are the azimuths and the unequal-hour lines drawn beneath the horizon. In addition, each plate is engraved with a dotted arc running horizontally across the instrument as far as the tropic of Cancer. This is the crepuscular line marking the period of twilight and thus the hours of morning and evening prayer. There are also two dotted vertical arcs among the unequal-hour lines on the east side of the plate. These are the *khaṭṭ aẓ-ẓuhr* (shortly after midday), signifying the time of the midday prayer, and the *khaṭṭ al-'aṣr* (afternoon), which indicated the time of the afternoon prayer. The lower part of the meridian line is also dotted (*khaṭṭ az-zwāl*) to signify the moment of true noon.[210] The instrument is engraved throughout in Maghribi *kufic* script.

210. Morley (I), ii. According to Sédillot, 170, n. 6, the *khaṭṭ aẓ-ẓawāl* marks the moment of true noon and the *khaṭṭ az-zuhr* also marks the hottest period of the day. A method of constructing the prayer lines with a description of their use is given by al-Bīrūnī in his *Ifrād al-maqāl fī amr aẓ-Ẓilāl*, see Kennedy (II), i, 233–5, ii, 145.

figure 49. Catalogue no. 3: Maghribi astrolabe. Front.

figure 50. Catalogue no. 3. Back.

This is a well-made astrolabe which closely resembles an instrument found in Egypt by Jean-Joseph Marcel (1776–1854) during the Napoleonic occupation[211] and taken by him to France where it was already lost by 1840.[212] A detailed description of it was, however, given by Sédillot.[213] In design, the two instruments are virtually identical, marking the same twenty-six stars on the two *retes* which have identical designs. The layout of scales on the back of the instrument and on the plates is similar and the *kufic* script has marked resemblances. Despite a number of minor differences (the Marcel astrolabe evidently had a shackle and a swivel-ring suspension, a single shadow square, and more plates), the two instruments clearly derive from the same workshop.

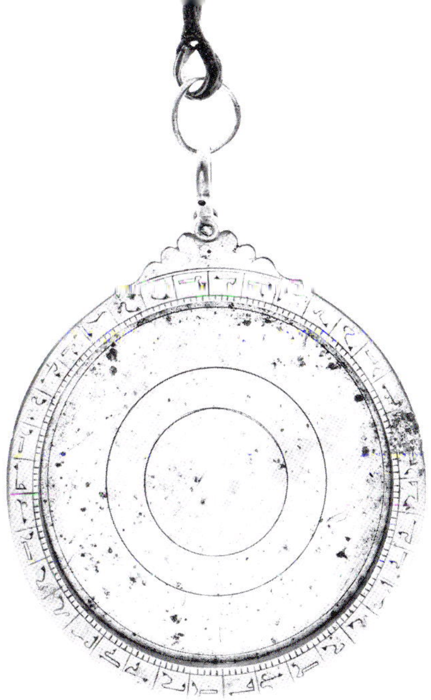

figure 52. Catalogue no. 3. *Mater*.

figure 51. Catalogue no. 3. Plates.

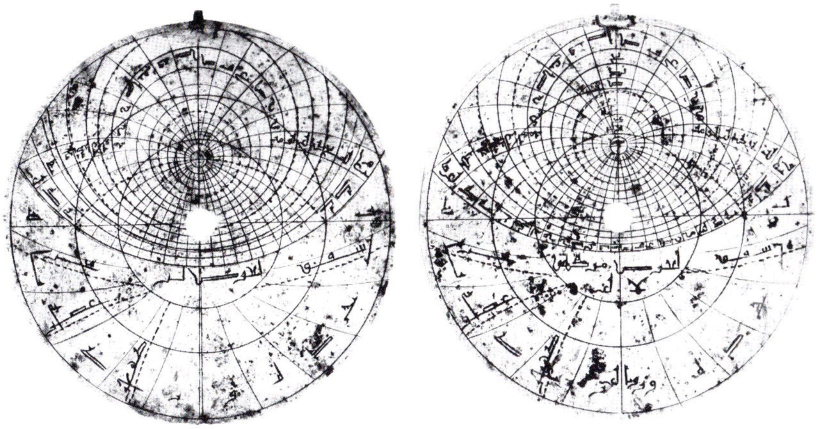

a. For latitudes 37° (left) and 31°30′ (right).

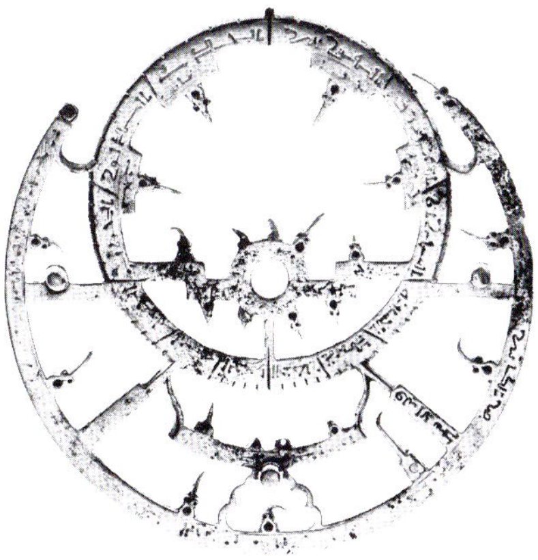

figure 53. Catalogue no. 3. *Rete*.

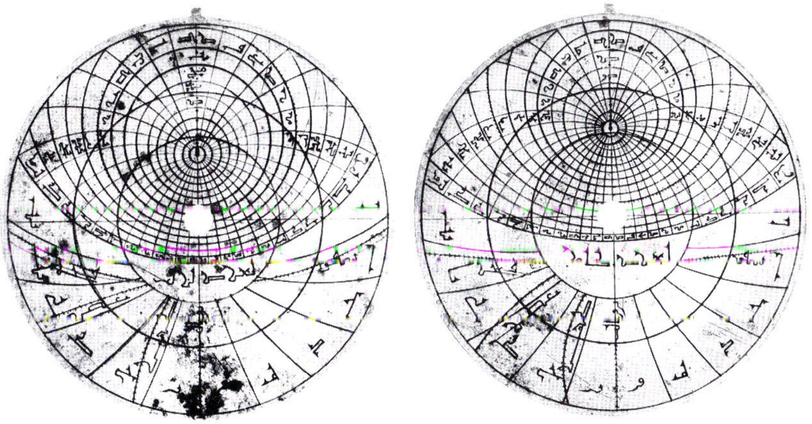

b. For latitudes 37°30′ (left) and 21°40′ (right).

211. See Jomard, ii, pl. HH. Marcel intended to give a detailed description of it in the final volume, but this was never achieved. See Sédillot, 166, n. 2.
212. Sédillot, 166, n. 2.
213. *Ibid*, 166–72. There is a shorter account in Gunther (V), i, 282–3.

4. Indo-Persian Astrolabe with Northern and Southern Projections

11th/17th century
Brass
Diameter 5 in. (127 mm)
Not signed, but from the school of Lahore
Inventory 308

There are two *retes*. The *rete* projected from the south celestial pole (fig. 54) provides the map of the northern hemisphere and has thirty-six named stars; that projected from the north celestial pole (fig. 55), and so showing the southern hemisphere, has twenty-seven. The stars are indicated by leaf-shaped pointers rising from flowing tracery. The ecliptic circle on each rete is divided in groups of 6°, but on that for the southern hemisphere the zodiacal divisions are reversed and a declination scale, such as normally would be engraved on the alidade (e.g. in catalogue no. 6), is engraved on a meridian line within the ecliptic circle. The *rete* also has an east-west bar with a single counter-change, whereas that for the northern hemisphere has three counter-changes. Since the *retes* are placed one on each side of the instrument, there is no *mater*, but the *retes* and plates are held in the limb by four rotatable lugs set on each side of the instrument. The limb is engraved on each side with a degree scale in four quadrants, reading to 1° in groups of three. The *kursī* is high and pierced, with decorative outline engraving, and is cast in one piece with the limb. The shackle suspension piece carries an *ʿilāqa* of chain which may not be original. There are seven plates, with the almucantars drawn for every 2°:

a. for latitudes 19° (north) and 22° (south)
b. for latitudes 32° (north) and 40° (south)
c. for latitudes 27° (north) and 36° (south)
d. for latitudes 11° and 29° (north)
e. for latitudes: 30° (south) and 0°; and for latitudes: 34° and 38° (south)[214]
f. with a tablet of horizons and a tablet of ecliptic co-ordinates
g. which carries the scales normally found on the back and *mater* of a conventional astrolabe

This backplate (g) is engraved on one side with a sine graph; a single shadow square divided by twelves; the arcs of the signs of the zodiac; and sigmoid graphs of the meridian altitudes of the sun for latitudes 19°, 27°, 32° (see Glossary);[215] astrological tables for the mansions of the moon, the signs of the zodiac, and the limits of the

214. 'Double plates with co-ordinates for two sets of latitudes engraved on the same side are not uncommon among Indian [*sic*] instruments.' Gunther (V), i, 183. This section largely follows Kaye (I).
215. See fig. 193. 'The graphs take a characteristic sigmoid form and are diagnostic of Indian [*sic*] astrolabes.' Gunther (V), i, 184.

figure 54. Catalogue no. 4:
Indo-Persian astrolabe.
Northern hemisphere projection.

figure 55. Catalogue no. 4. Southern hemisphere projection.

planets. On the other side, it has a circular gazetteer of the longitudes and latitudes of seventy-seven places in the northern hemisphere,[216] arranged according to the seven *climata* derived from ancient Greek geography, which divided the earth into seven latitudinal regions counted from the equator towards the north pole.[217] The starting place could, however, vary. In addition, all the latitude plates are marked with the length of the longest day in hours and minutes, and also in *ghaṭi*.[218] The astrolabe is held together in the usual manner with a pin and horse and is furnished with a rule that can be used on either side.[219] Since the lugs holding the plates and *rete* in place would impede the use of a conventional alidade, there are two sighting pinnules set on each *rete*, one at each end of the east-west bar.

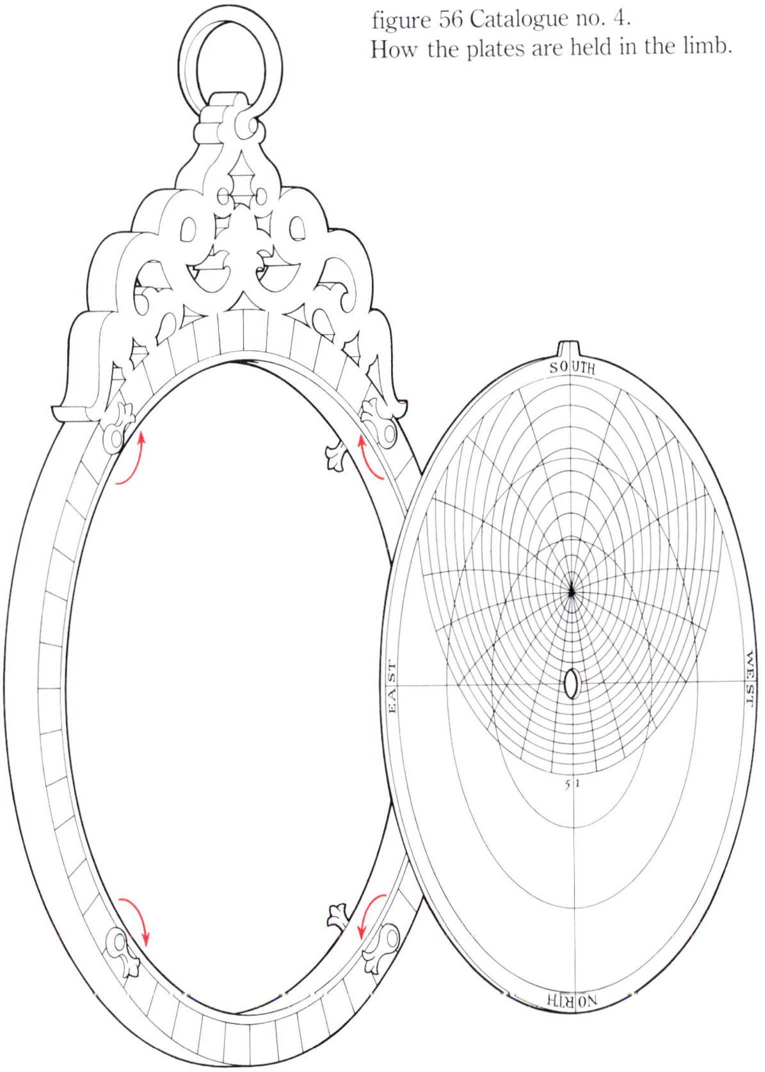

figure 56 Catalogue no. 4.
How the plates are held in the limb.

figure 57. Catalogue no. 4. Limb and *kursī*.

216. For a complete listing, see Brieux (III). I am greatly indebted to M. Brieux for generously placing at my disposal his entire dossier on this instrument.
217. Gunther notes the existence of a table of climates on the Jaipur A astrolabe (Kaye (I), 24) but knew of no other example. Gunther (V), i, 185 & 205.
218. Indian unit of time measurement in which 60 *ghaṭi* = 24 hours.
219. *Cf.* the similar sighting arrangement employed in the astrolabe with geared calendar movement by Muḥammad b. Abī Bakr ar-Rāshidī al-Ibari al-Iṣfahānī, A.H. 618 (A.D. 1221/2), where an alidade could similarly not be used. Gunther (V), i, pl. XXV.

Although not signed, this instrument may be attributed to the workshops of Lahore on general stylistic grounds (the pierced *kursī*, the numerous leaf-shaped star pointers, the abundance of scales all being characteristic of the Lahore makers). Although it has a family resemblance with the astrolabes of Muḥammad Muqīm b. Īsà and Jamāl ad-Dīn b. Muḥammad, it may be more closely associated with five other instruments from the same school. The first of these is an astrolabe now in the Museum of the History of Science, Oxford,[220] which contains a plate engraved in the normal fashion, but with the addition of a zodiac circle, and with the stars marked on the *rete*. These are indicated by a dot in a circle together with their names. This astrolabe, like the double astrolabe, has a pierced *kursī* decorated with outline engraving. The second instrument is another double astrolabe signed by Ḍiyā' ad-Dīn Muḥammad b. Qā'im Muḥammad (*fl. c.* 1646–80/1) and dated A.H. 1085 (A.D. 1674/5) (fig. 58). It is composed of a single plate engraved on both sides as a latitude plate but like the plate described in the first astrolabe, also with a zodiac circle (northern on one side, southern on the other), and with each named star marked by a dot in a circle (fig. 58). On one side, the instrument has a *rete* (reduced simply to the zodiac circle, a meridian, and a cross bar; it has no stars) and a rule; on the other side, an alidade. Unlike the other two instruments, however, the *kursī* is relatively low and solid.[221] The third instrument[222] is also a double astrolabe with north and south zodiac circles engraved on each side of a single plate. The stars are marked in the same way, as dots in circles with their names. All this is superimposed on a latitude plate, but in this case on one side a double plate is engraved and on the other a quadruple (i.e. with co-ordinates for four latitudes) plate. The *kursī* of this instrument is more elaborately and delicately pierced than the others, being indeed very similar in style to the *kursī* of several of Muḥammad b. 'Īsà (*fl. c.* 1622–38) and Muḥammad Muqīm b. 'Īsà's (*fl. c.* A.D. 1609–60) astrolabes with which it also shares a similar central motif.[223] This motif should be compared with that on the Rockford double astrolabe (figs. 54 & 55). It should also be noted that heavier *kursī* similar to this one are also found on astrolabes by Muḥammad Muqīm. The fourth astrolabe is an unsigned instrument (fig. 59)[224] in which the limb is so placed that it overhangs on each side, thus creating recesses on both faces of the instrument. There are two *retes*. One of them is of the conventional form for the northern hemisphere. The other has no stars marked and is similar in form to the so-called 'ship' *rete*.[225] The

220. See Gunther (V), 1, 218–20.
221. This instrument (IC/CCA no. 2703; Brieux & Maddison MHMD QAIM MHMD 36) is now in the Jaipur Observatory.
222. Victoria & Albert Museum, London, 809–1889.
223. See, for example, those illustrated in Gunther (V), i, 193, fig. 94 & 199, fig. 99.
224. Formerly in the Mensing Collection, now in the Adler Planetarium, Chicago, M 39.
225. *Cf.* Gunther (V), no. 90.

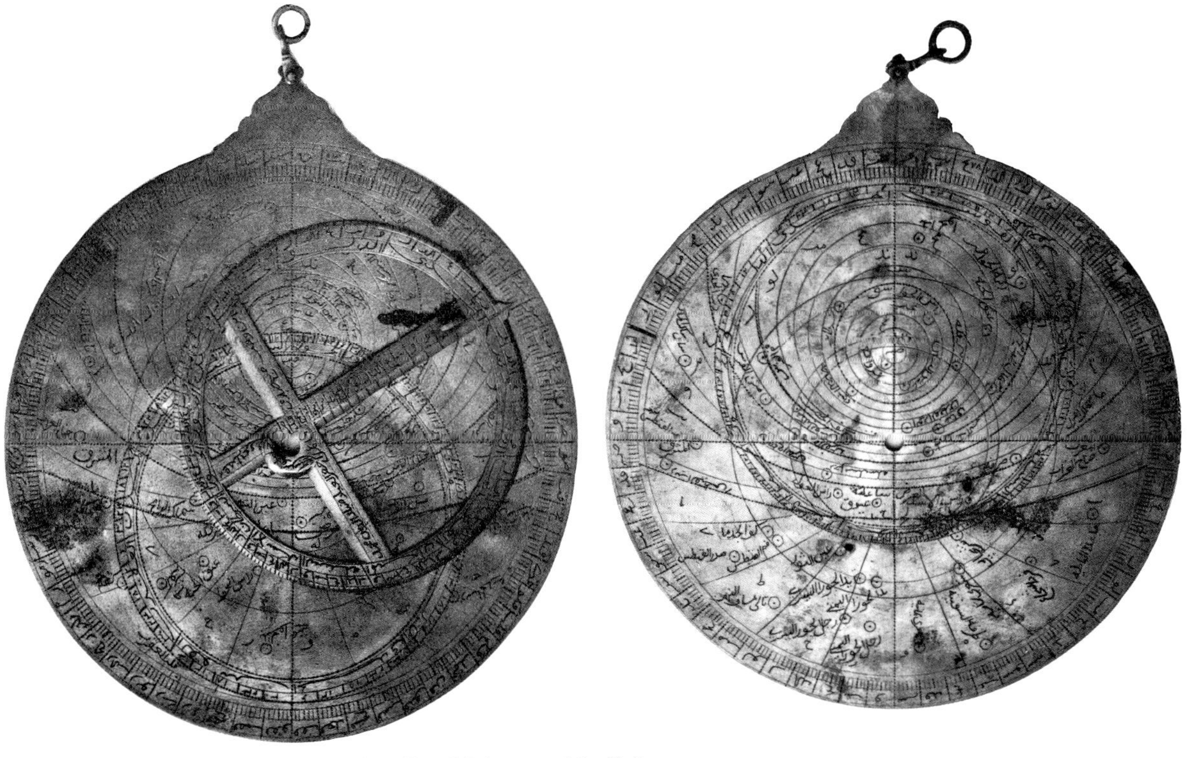

figure 58. Double astrolabe by Ḍiyā' ad-Dīn Muḥammad b. Qā'im Muḥammad (left). Reverse (right). Jaipur, The Observatory. (Photograph: Alain Brieux, Paris)

figure 59. Double astrolabe, not signed (left). Reverse (right). Chicago, Adler Planetarium, M. 39.

figure 60. Catalogue no. 4. *Retes*. Northern hemisphere (left) and southern hemisphere (right).

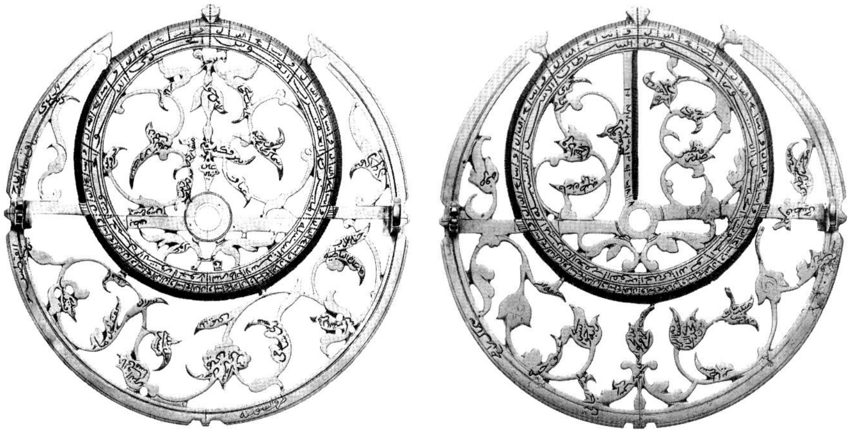

figure 61. Catalogue no. 4. Plates.

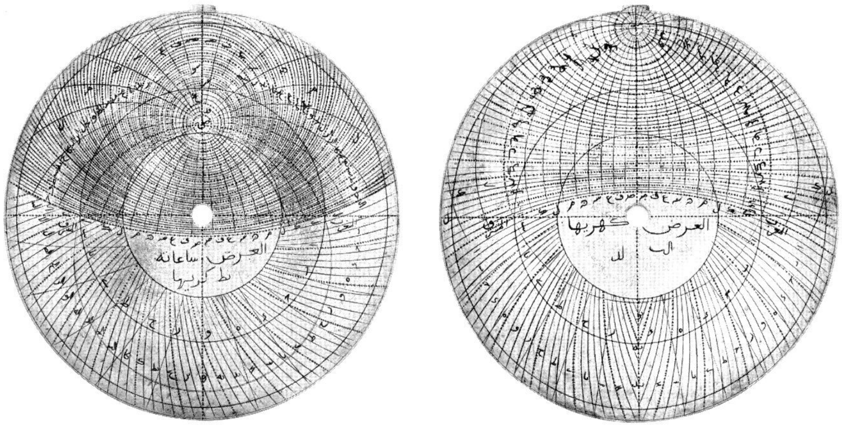

a. For latitudes 19° north (left) and 22° south (right).

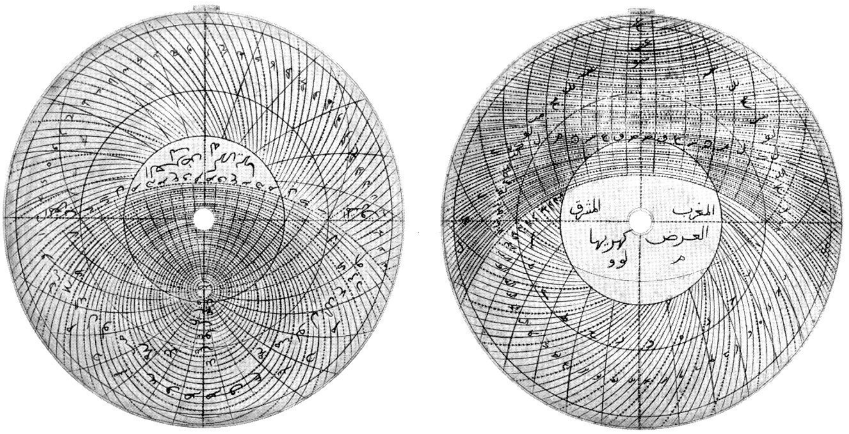

b. For latitudes 32° north (left) and 40° south (right).

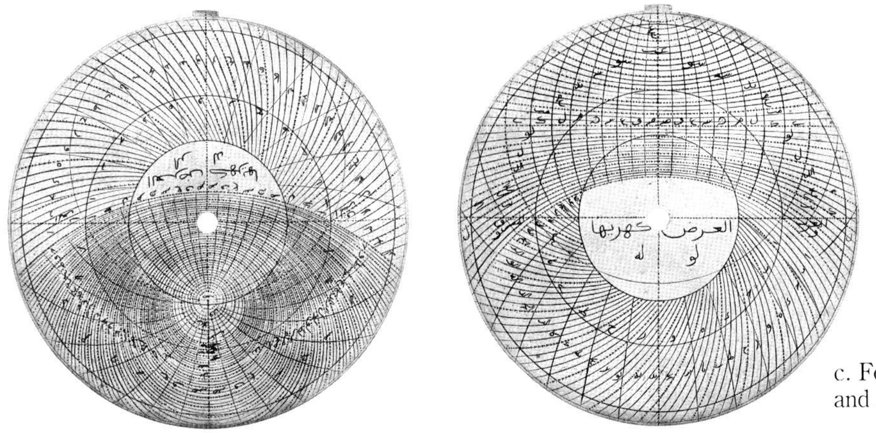

c. For latitudes 27° north (left) and 36° south (right).

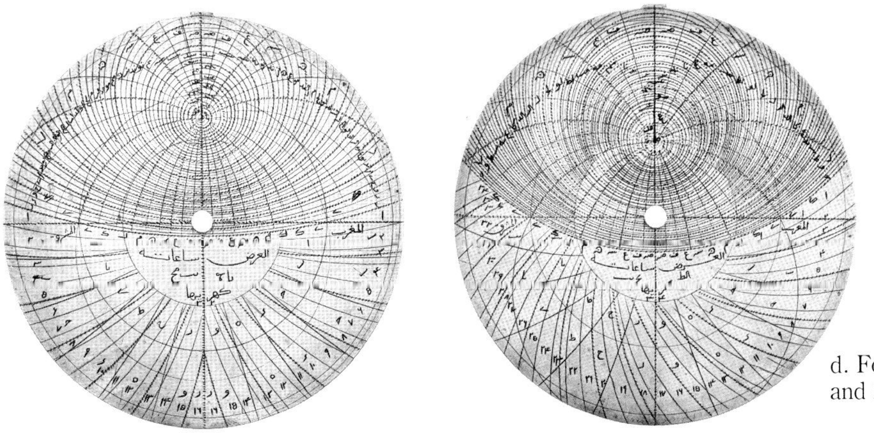

d. For latitudes 11° north (left) and 29° north (right).

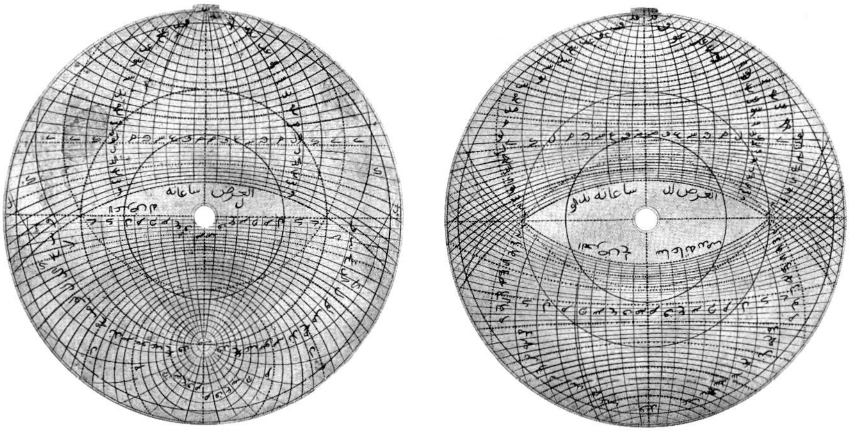

e. For latitudes 30° south and 0° (left) and 34° and 38° south (right).

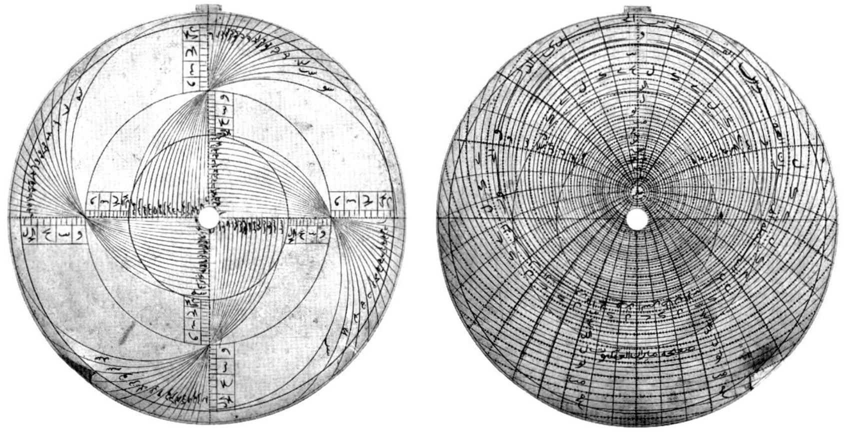

f. With tablet of horizons (left) and tablet of ecliptic co-ordinates (right).

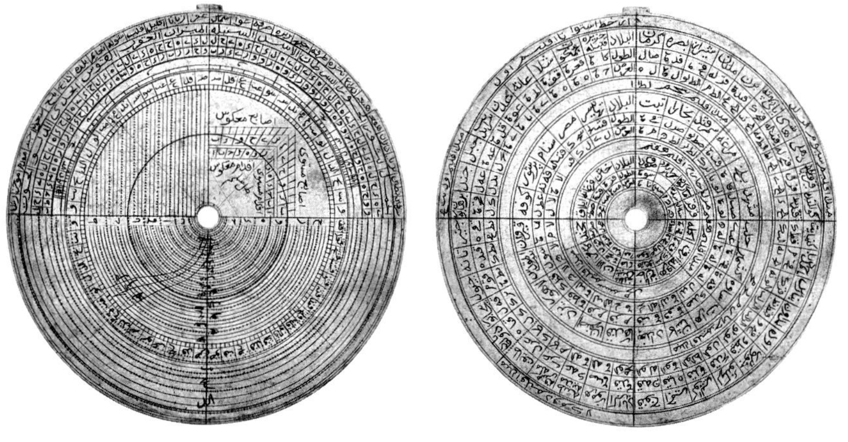

g. Scales normally found on the back and *mater* of a conventional astrolabe.

instrument is thus double in having two *retes*, but does not use them to show the two different hemispheres. The *kursī* is high and pierced. The plates are silvered and both the layout of the *rete*, and the lettering show distinct similarities with those of the north/south astrolabe. The fifth instrument,[226] also unsigned, is an otherwise conventional astrolabe for which two *retes* – both northern – have

226. IC/CCA no. 2569 now in the Smithsonian Institution, National Museum of American History. Gibbs & Saliba, 165–9.

been provided, although one does not extend beyond the ecliptic.

The multiplication of complexity and a delight in the unusual, seem to be typical of the astrolabe-makers of Lahore. In an otherwise conventional astrolabe dated A.H. 1044 (A.D. 1634/5), Muḥammad b. ʿĪsà included two double plates, a plate for latitude 0°, and a composite plate with a miniature projection for latitude 42° at the centre surrounded by a diagram named on the instrument as 'A table of the projection of rays' (al-Bīrūnī's 'equalization of houses').[227] It is, therefore, not astonishing that astrolabes with fanciful *retes* and astrolabes doubly projected from the north and south pole were produced there. Although the exact sources of information available in Lahore are not known, most of the complexities introduced by Lahore makers into their instruments are mentioned in the early treatises of aṣ-Ṣūfī[228] and al-Bīrūnī,[229] the latter's section on unusual astrolabes being related to the slightly earlier and more extensive work of as-Sijzī.[230] There is a later summary by the 13th-century astronomer Abū l-Ḥasan al-Marrākushī.[231] Many of the ingenious *retes* there shown are partially drawn for a northern astrolabe, partially for a southern.[232] Possibly it was by such early texts and these, or others based on them, that the Lahore makers were stimulated to make their complicated astrolabes. There was, after all, very little else in the astrolabe which after the 13th century offered much scope for originality.[233]
IC/CCA no. 3917.

227. Now in the Lewis Evans Collection, Museum of the History of Science, Oxford. Gunther (V), i, 191–7.
228. *Kitāb al-ʿamal bi-l-asṭurlāb* (Book of the Making of the Astrolabe), a work which ran to 378 chapters.
229. *Kitāb fī istīʿāb al wujūh al-mumkina fī ṣanʿat āl-asṭurlāb* (Book of Detailed Treatment of All Possible Ways of Making an Astrolabe).
230. *c.*A.H. 334–*c.* 411 (A.D. *c* 945–*c.* 1020). For his life and bibliography, see *D.S.B.*, xii, 431–2. Sezgin, 224–6. King (III), 117.
231. *Jāmiʿ al-mabādiʾ wa-l-ghāyāt* (The Whole of the Beginnings and the Ends), Sédillot, 181–4.
232. Michel (III), 69.
233. It is interesting to note that a few decades earlier in Western Europe the Belgian Jesuit Odon van Maelcote (1572–1614) had devised a combined northern and southern instrument with a single *rete* of which an example was made by Lambert Damery, who described it in his work *Astrolabium Aequinoctiale Odonis Malcotij*. . ., Brussels, 1607. Michel (III), 71–2, 174, 176; van Ortroy, 81, n. 2. A further description of the instrument was given by Regnart, in 1610. Further editions followed in 1613 (Rome) and 1652 (Breslau). An earlier Western design for a double astrolabe, however, is included in a manuscript written by Henri Arnault, a pupil of Jean Fusoris, with whom the design should perhaps be associated (Bibl. Nat. Ms. Lat 7295 f 19v: Poulle (I), 94–5; Poulle (V), 25–26).

5. Indo-Persian Astrolabe *Mater*

A.H. 1071 (A.D. 1660–61)
Brass
Diameter 4½ in. (114 mm)
Signed in the shadow square:
(Work of the least of the servants Ḍiyā ad-dīn Muḥammad b. Qā'im Muḥammad b. Mullā 'Īsā b. Shaikh Ilāh-dād, the imperial astrolabist of Lahore in the year 1071 of the Hijra)

Inventory 3172

The mater is cast in one piece with the limb and *kursī*, which is engraved with foliate decoration. Engraved on the limb is a scale of degrees in four quadrants (90°–0°, four times) and in the centre of the *mater* is a gazetteer of the latitudes, longitudes, and *inḥirāf* (direction of Mecca) for 54 places. On the back are scales of degrees in the upper quadrants (0°–90°–0°) surrounding (left) a sexagesimal sine graph and (right) the arcs of the signs of the zodiac crossed by two sigmoid graphs of meridian altitudes of the sun. In the lower quadrants is a double shadow square surrounded by a scale for correlating the signs of the zodiac with the 28 mansions of the moon, and co-tangent scales used in conjunction with the shadow square.

Published in Wynter & Turner, 19.

Provenance: Sotheby & Co., London, 26 February 1968, lot 12.

IC/CCA no. 2607. Brieux & Maddison MHMD QAIM MHMD 24.

Ḍiyā' ad-dīn Muḥammad b. Qā'im (fl. 1646–80/1) was a member of the well-known family of astrolabists established at Lahore.[234] At least 39 instruments (24 astrolabes and 15 globes) by him have survived,[235] two of which were made for the Mughal Emperor 'Alamgīr I, surnamed Abū-l-Ja'far Mūḥyi ad-Dīn Muḥammad Aurangzīb (A.H. 1028–1118 [A.D. 1618/9–1706/7]).

234. See p. 26.
235. They are listed in Brieux & Maddison under the entry MHMD QAIM MHMD.

figure 62. Catalogue no. 5: Astrolabe by Muḥammad b. Qā'im Muḥammad. Front.

figure 63. Catalogue no. 5. Back.

6. Persian Astrolabe

Late 11th/17th century
Brass, with silver mudirs *and inlaid turquoise*
Diameter $1^7/_8$ in. (48 mm)
Signed:
(from the poor [in god] man Muḥammad Mahdī)

Inventory 1849

The *rete* is symmetrically designed and marks twenty named stars indicated by leaf-shaped pointers springing from a very restricted amount of symmetrical tracery. The ecliptic is divided to 6° with the signs of the zodiac named and with each division marked by a turquoise (nine missing). The east-west bar is straight and there is one *mudīr*. The *mater* is cast in one piece with the high, four-lobed *kursī* which bears an inscription on each side and carries a shackle, suspension ring, and *'ilāqa*. The inscription reads:

> The one who has strength, beauty, power, perfection, authority with splendour.

Inscribed around the edge of the instrument (fig. 64) is the famous 'throne' verse from the Qur'ān (II: 256–7):

> God
> there is no god but He, the
> Living, the Everlasting
> Slumber seizes Him not, neither sleep;
> to Him belongs
> all that is in the heavens and the earth.
> Who is there that shall intercede with Him
> save by His leave?
> He knows what lies before them
> and what is after them,
> and they comprehend not anything of His Knowledge
> save such as He wills.
> His Throne comprises the heavens and earth;
> the preserving of them oppresses Him not;
> He is the All-high, the All-glorious.[236]

figure 64. Catalogue no. 6. Inscription on edge.

236. Translation from Arberry, 37.

figure 65. Catalogue no. 6: Astrolabe by Muḥammad Mahdī.

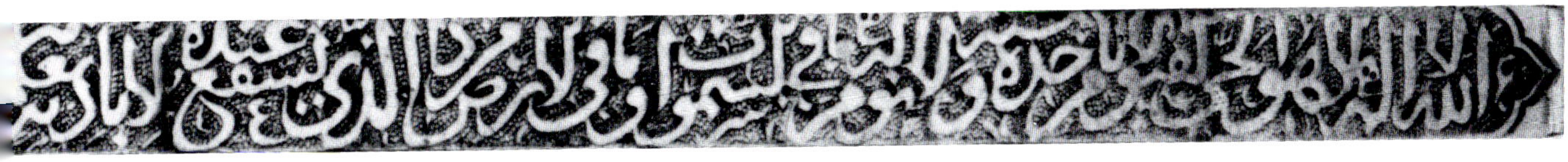

figure 67. Catalogue no. 6. Back.

figure 68. Catalogue no. 6. *Mater* (left) and back without alidade (right).

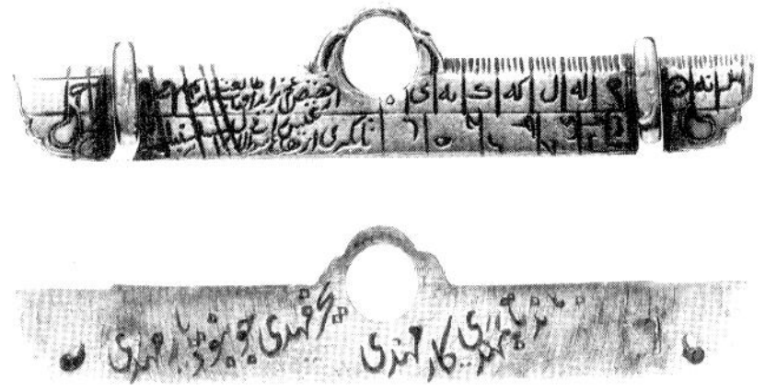

figure 66. Catalogue no. 6. Alidade. Front (top) and back (bottom).

The limb is engraved with a degree scale, each fifth degree indicated by a dot on the division graduation: The edge of the back of the instrument is marked in the upper two quadrants with a degree scale (0°–90°–0°) surrounding, in the upper-left quadrant, ninety horizontal lines drawn from each degree with dotted radials used in conjunction with the sexagesimal division on the edge of the alidade as a sine graph and, in the upper-right quadrant, a graph of the arcs of the signs of the zodiac together with the arcs of the solar altitudes in the azimuths of the *qibla* for Baghdad, Isfahan, and Tus. In the lower two quadrants is a double shadow square of seven and twelve divisions, within which is a table of triplicities. Beneath the shadow square, in a cartouche, is the maker's signature and surrounding this is a series of astrological tables. The limb is marked with scales of tangents and co-tangents. At the base of the astrolabe in a small cartouche is a quotation from the *Gulistān* of Sā'dī, which may be translated as 'The object of this engraving is that it should outlast us.' On the *umm* a list of thirty-one places is engraved with their latitude, longitude, and *inḥirāf*, while at the base there is a small projecting lug that fits into a slot in the plates to prevent them from turning with the *rete*. There are five plates (a-e), with the almucantars drawn for every 6°:

a. for latitudes 32° and 26°
b. for latitudes 36° and 38°
c. for latitudes 34° and 22°
d. for latitudes 39° and 45°
e. for latitude 41° and with a tablet of horizons marked with the length of the longest day

Plates a) and e) have background decoration in the latitude panel

figure 69. Catalogue no. 6. *Rete.*
Front (left) and back (right).

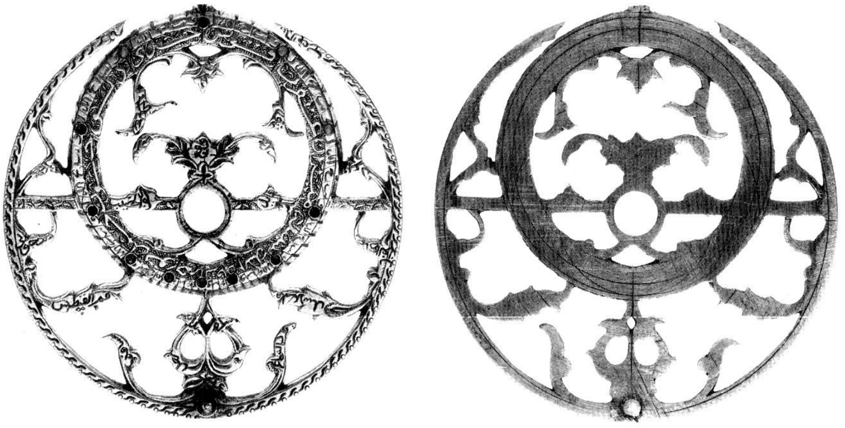

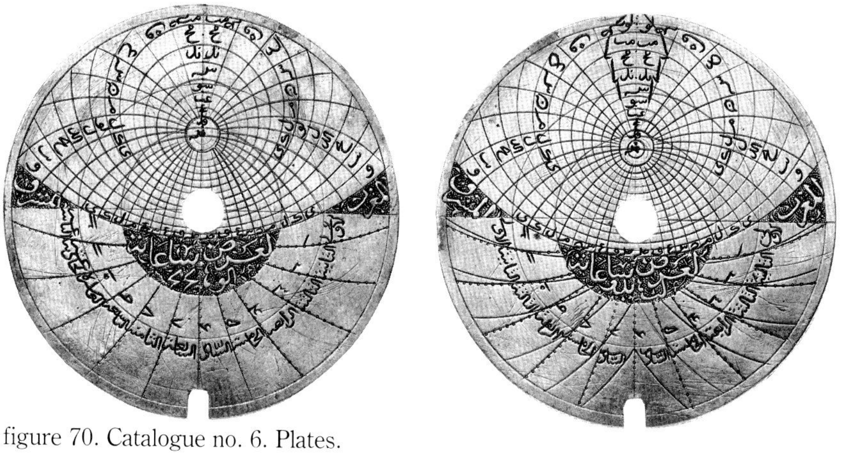

a. For latitudes 32° (left) and 26° (right).

figure 70. Catalogue no. 6. Plates.

and where the east and west points of the horizon are named. On plate a) the circles for the equator and the tropic of Capricorn, usually drawn on the plates, have been marked but not fully engraved. In addition to the sexagesimal scale, the straight alidade is engraved with an unequal-hour scale and with declination scales. The horse, pin, and washer are original.
Brieux & Maddison MHMD MHDI YZDI 21.

This is a remarkably small astrolabe that displays well the virtuosity of its maker, one of the best reputed, prolific, and prestigious astrolabists of Safavid Persia. Muḥammad Mahdī al-Khādim al-Yazdī, the son of Muḥammad Amīn[237] is known by twenty-three astrolabes that have survived, besides others decorated by him for ʻAbd al-A'imma, Muḥammad Khalīl, and Muḥammad Muqīm al-Yazdī. Few of his astrolabes are dated but he was working in A.H. 1070 (A.D. 1659/60) when he made an astrolabe for Muḥammad Bāqir Iṣfahānī,[238] and another for Ṣafī Qulī Beg.[239]

237. Possibly the maker encountered by Sir John Chardin at Isfahān. See above, p. 24; Mayer (I), 70–1 & 63.
238. *Ibid*, 70. Brieux & Maddison.
239. Now in the National Maritime Museum, Greenwich (A64/69–6). See Maddison & Turner, no. 56. Brieux & Maddison.

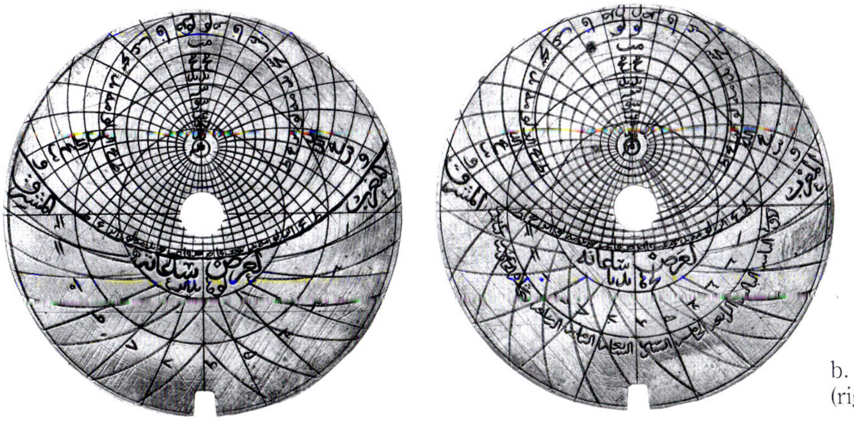

b. For latitudes 36° (left) and 38° (right).

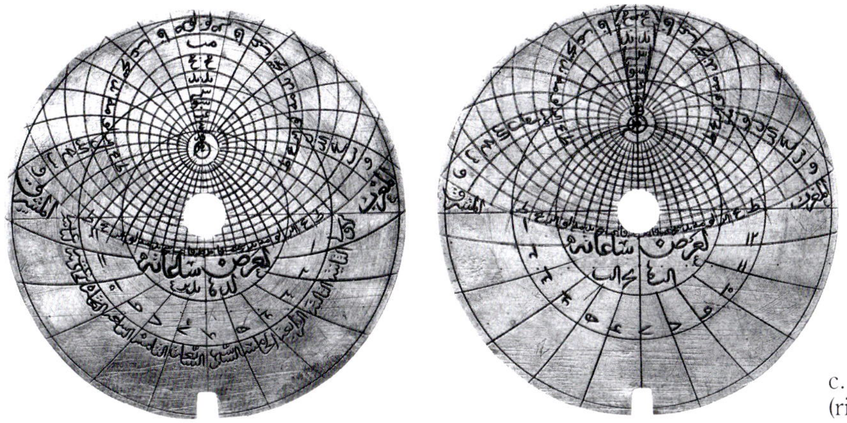

c. For latitudes 34° (left) and 22° (right).

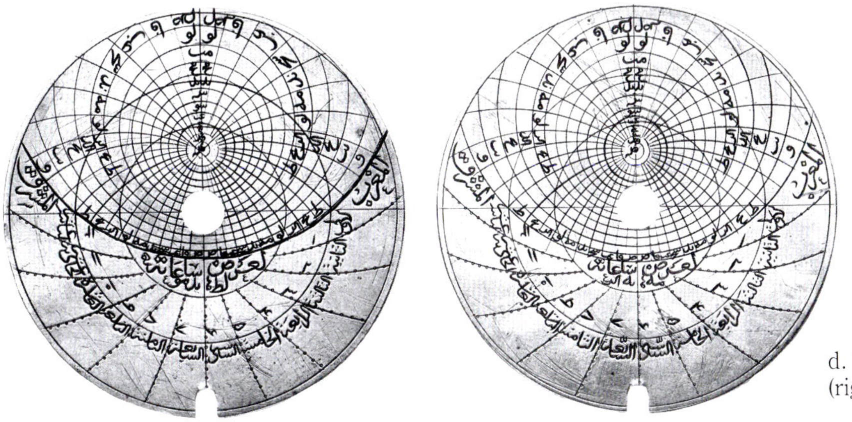

d. For latitudes 39° (left) and 45° (right).

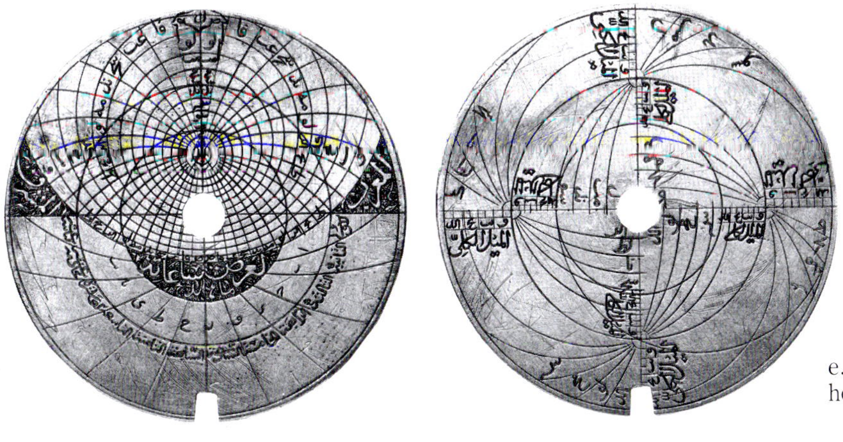

e. For latitude 41° (left) and tablet of horizons (right).

7. Persian Astrolabe

A.H. 1194 (A.D. 1780)
Brass with inlaid turquoise
Diameter 3 ⅝ in. (93 mm)
Signed:
(Its designer and its maker is Muḥsin ibn Muḥammad 'Alī ash-Sharīf al-Kirmānī – 'may their sins be pardoned' – in the year 1194 of the Hijra *and 2091 of Alexander*[240] *and 1149 of Yazdigird*[241] *and 702 of the Jalālī[era])*[242]

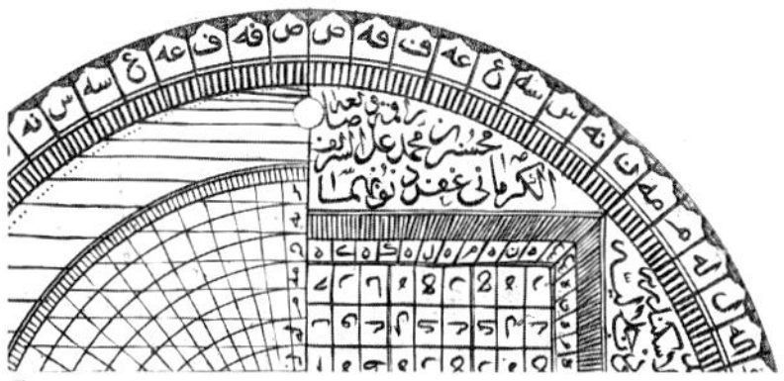

Inventory 1436

figure 71. Catalogue no. 7. *Mater.*

The *rete* marks the positions of twenty-seven named stars represented by star pointers in the form of leaves or birds. It is symmetrically patterned with flowing tracery and a straight east-west bar. The ecliptic has named signs, divided to 6°. Neither the equator nor the tropic of Cancer are shown. The high, scrolled, open-work *kursī* is decoratively engraved and cast in one piece with the *mater*. There is an engraved pierced shackle, a suspension ring, and *'ilāqa* plaited from blue, red, green, and yellow cord. Engraved on the limb is a degree scale (360° reading to 1° in groups of five), each numeral set against an arched background. Engraved on the *mater* is a gazetteer of thirty-eight towns with their latitude, longitude, and *inḥirāf*. On the back in the upper-left quadrant is a scale of degrees (0°–90°) surrounding a graph of horizontal lines drawn from every fifth degree of the scale on the limb. Overlaying this is a sexagesimal sine graph with radial lines. In the upper-right quadrant a scale of degrees (90°–0°) surrounds a single shadow square of twelve divisions each divided to five. On the limb the lower quadrants carry a scale of cotangents used for a direct reading of the length of a shadow of a vertical gnomon in an horizontal plane,[243] and within are a series of astrological tables.[244] The alidade is short and of straight-bar form. The *faras* is in the form of a horse. There are four plates (with the almucantars drawn for 6°) each separated by a later pink paper buffer. The plates are:

a. for latitudes 34° and 36°
b. for latitudes 38° and 40°
c. for latitudes 32° and 30°
d. for latitude 22° and with a tablet of horizons

Each, except for the tablet of horizons, is also engraved below the horizon with an unequal-hour scale, azimuth lines, and (in dotted lines) an equal-hour scale reading 1–12 from the west side of the instrument.

240. More properly known as the era of Seleucides, this era is counted from the year 311 B.C. and therefore probably relates to the death of Alexander IV (Aegus). Taqizadeh 124–30.
241. This era counts from the accession of Yezdegred III to the Persian throne 16 June A.D. 632. Nicolas, 20. Taqizadeh, 117–8.
242. The reformed Persian calendar introduced during the reign of the Seljūk Malik Shāh I at the instigation of his vizier, the great statesman Niẓām al-Mulk, and calculated with the advice of several astronomers among whom was the poet and mathematician 'Umar Khayyām. The alāli era commenced A.H. 471 (15 March A.D. 1079). For details, see Taqizadeh, 108–17.
243. Michel (III), 40. *Cf.* the description of this scale and its construction by al-Bīrūnī in Kennedy, i, 111–15; ii, 54–5.
244. For details, see Michel (III), 42; Morley (I), 30.

figure 72. Catalogue no. 7: Astrolabe by Muḥammad Muḥsin b. Muḥammad. Front.

figure 73. Catalogue no. 7. Back.

figure 74. Catalogue no. 7. Back of alidade, pin, and horse.

Provenance: Bahari Collection.
Alain Brieux, Paris.
IC/CCA no. 3521. Brieux-Maddison, MHMD MHSN AL 1.

Nothing is known of Muḥammad Muḥsin except that in the year after he made this instrument, he signed an astrolabe, which belonged to Luṭf'alī b. 'Abd an-Nabī, as its engraver.[245] This latter instrument was in the Chadenat Collection and was exhibited in 1936 in the Musée des Arts Décoratifs.[246]

245. Mayer (I), 74.
246. *Exp. des Inst.*, 35, no. 263.

figure 75. Catalogue no. 7. Plates.

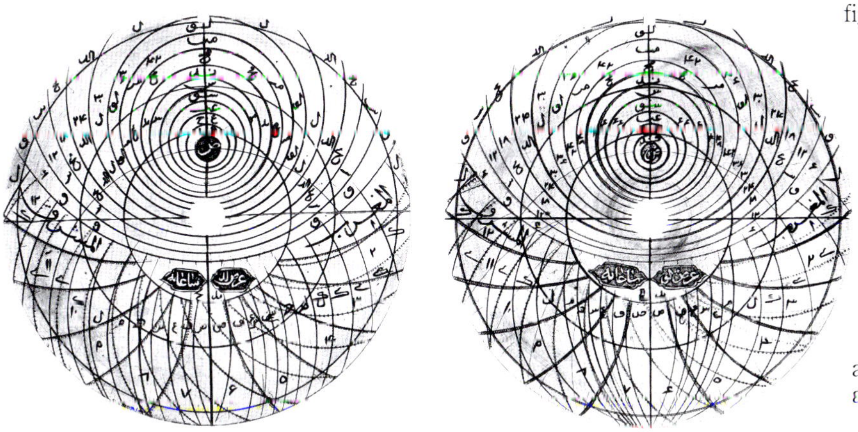

a. For latitudes 34° (left)
and 36° (right).

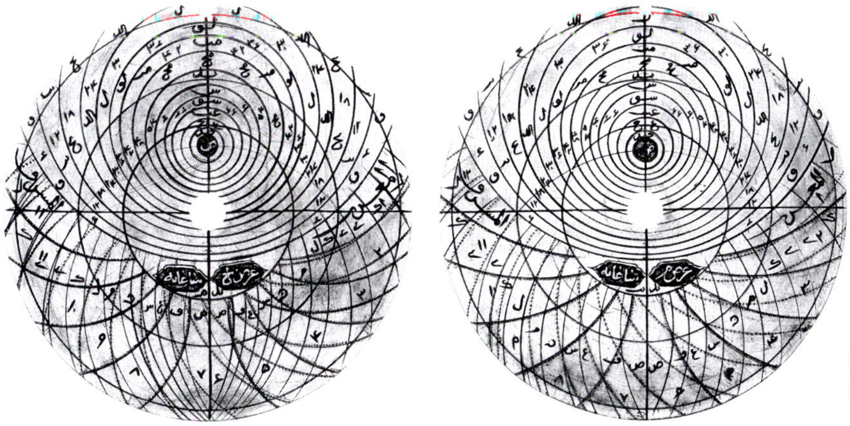

b. For latitudes 38° (left)
and 40° (right).

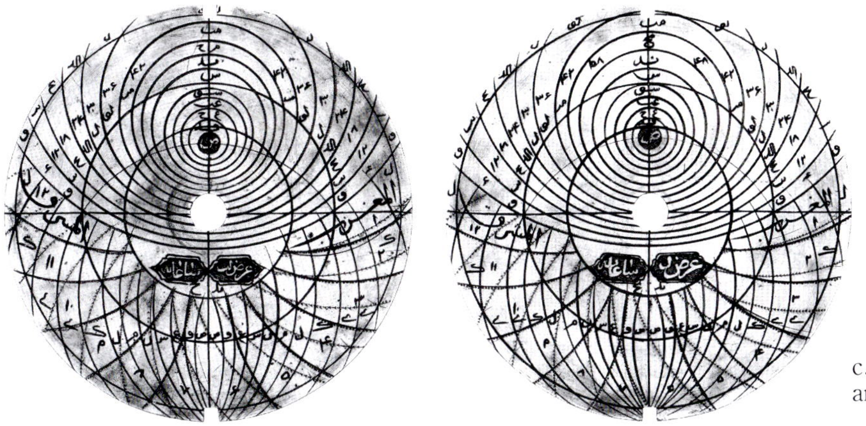

c. For latitudes 32° (left)
and 30° (right).

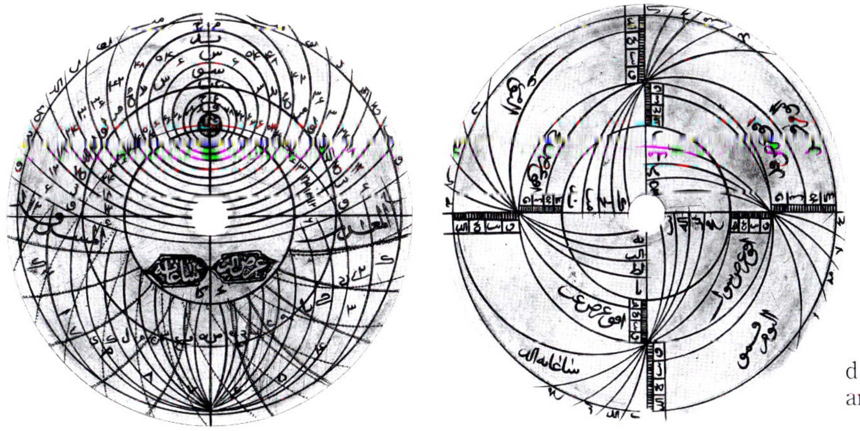

d. For latitude 22° (left)
and tablet of horizons (right).

8. Persian Astrolabe

[A.H.] 12[00] (A.D. 1785/86)
Brass, diameter 3½ in. (89 mm)
Signed, on the back in a cartouche in the shadow square:
(Made by Ḥājjī 'Alī)

and in a second cartouche at the base of the instrument:
(work three 12[00])

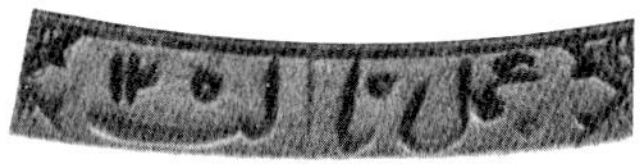

Inventory 1437

figure 76. Catalogue no. 8. *Mater.*

The *rete* marks the positions of twenty-nine named stars indicated by squat leaf-shaped pointers springing from flowing tracery. The ecliptic circle is divided to 6°, with signs of the zodiac named, and contains a counter-changed east-west bar which does not extend beyond the ecliptic circle. There is one *mudīr.* The limb is cast in one piece with the *mater* and *kursī* and is divided with a scale of 360°. The *kursī* is high and lobed, with a symmetrical leaf pattern, and carries a shackle with suspension ring. The *umm* is engraved with the latitude, longitude (measured from the meridian of the Fortunate Islands), and *inḥirāf* of thirty-four places[247] and has a projecting lug at the bottom which fits into a slot in the plates to prevent them from turning. On the back, in the upper-left quadrant, is a degree scale (0–90°) reading from the east point, surrounding a sexagesimal sine and cosine graph.[248] In the upper-right quadrant is a degree scale (0–90°) reading from the west point, surrounding the arcs of the signs of the zodiac. The arcs crossing them from left to right are the arcs of the azimuth of the *qibla* for Shiraz, Baghdad, Isfahan, and Tus.Crossing both these scales from beside the centre of the instrument toward the *kursī* are a series of arcs of the meridian altitude of the sun for latitudes 28°, 30°, 32°, 34°, 36°, 38°, and 40°. In the lower quadrants are, on the limb, scales of tangents and co-tangents surrounding a series of astrological tables[249] which surround a double shadow square with seven and twelve divisions. Within this is a table of the triplicities.[250] There are four plates with the almucantars drawn every 6°. The plates are:

a. for latitudes 30° and 36°
b. for latitudes 37° and 38°
c. for latitudes 32° and 34°
d. for latitude 29° and with a tablet of horizons

each of which is also marked with the length of the longest day. Each plate is engraved beneath the horizon with lines for the equal and unequal hour, both of which are counted from the western (right) horizon. The straight bar alidade is engraved with declination scales.

Provenance: H. Swainson Cowper, who purchased the instrument

247. For a table of these, see Cowper, 472.
248. Gunther (V), i, 185. For a detailed description, see Hartner (I), 2546 and (II), 303.
249. Described by Michel (III), 42; Morley (I), 30.
250. For a detailed description, see Morley (I), 19; Hartner (I), 2546 and (II), 303.

figure 77. Catalogue no. 8: Astrolabe by Hājjī ʿAlī. Front.

figure 78. Catalogue no. 8. Back.

in Baghdad in the early 1890's. He was told that the instrument had come from Kerbala.[251]

Devlay Collection, Paris.

Alain Brieux Collection, Paris.

IC/CCA no. 1163. Brieux-Maddison ALI.

A late representative of the Persian astrolabe-making tradition, Ḥājjī 'Alī fully maintained the standards of his predecessors. Other astrolabes signed by him range in date from A.H. 1203 (A.D. 1788/9) to A.H. 1210 (A.D. 1795/6).[252] Unique among astrolabists, he numbered some of his astrolabes. Astrolabe no. 13 dated A.H. 1208 (A.D. 1793/4) is now in the National Maritime Museum, Greenwich;[253] astrolabe no. 15 (not dated) was formerly in the S.V. Hoffman Collection.[254] Astrolabe no. 20 (not dated) was formerly in the Linton Collection.[255] The highest number known is 24.[256] His instruments resemble each other closely in style, in decoration, and in the scales on them.

251. Cowper, 75.
252. Brieux & Maddison, s.v. ''Alī.'
253. Mayer (I), 39, no. III.
254. Gunther (V), i, 155–6; Christie, 31 no. 102.
255. *Instruments Scientifiques*, 120, no. 178.
256. Brieux & Maddison, who list 13 astrolabes by 'Alī.

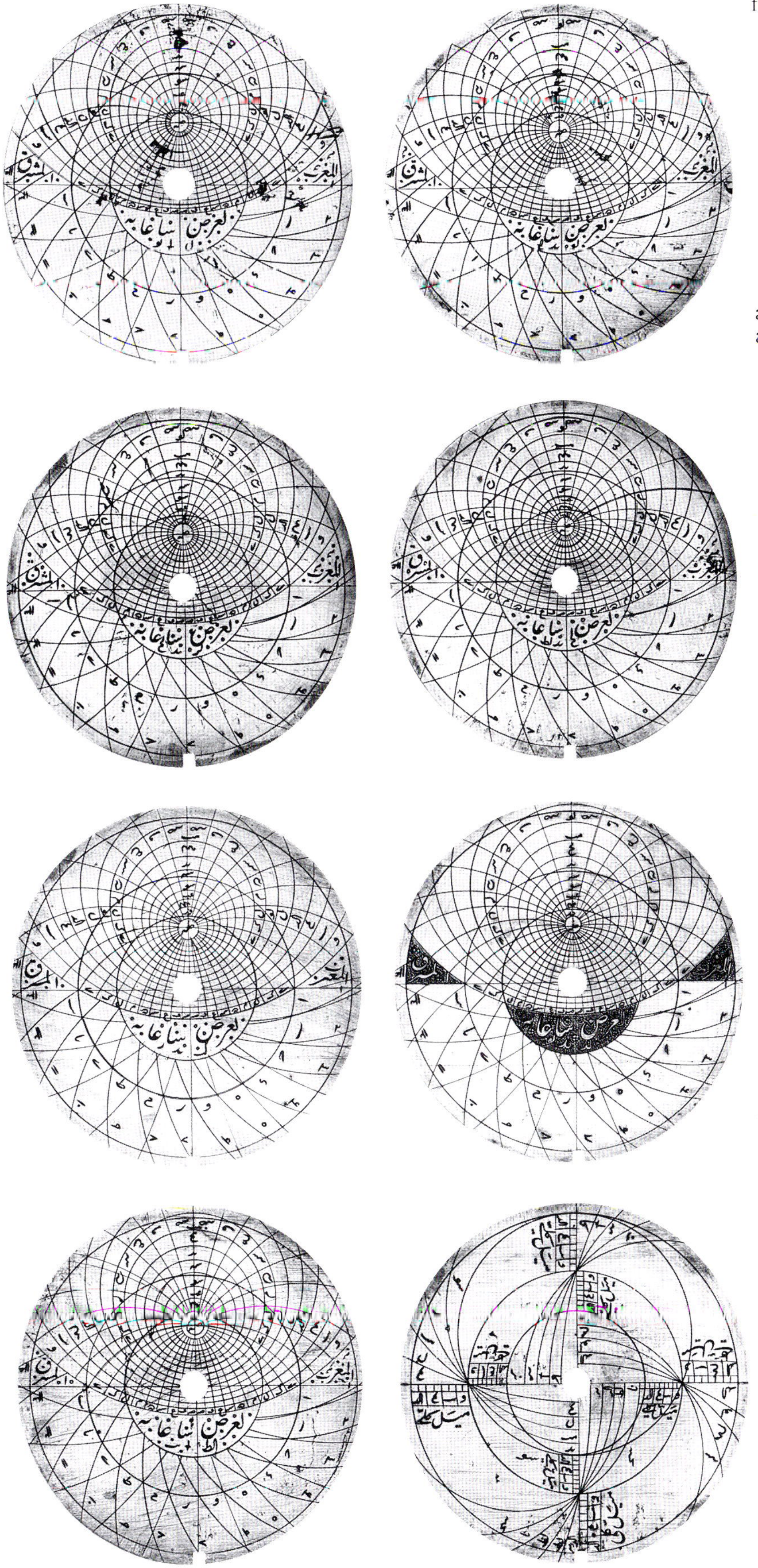

figure 79. Catalogue no. 8. Plates.

a. For latitudes 30° (left) and 36° (right).

b. For latitudes 37° (left) and 38° (right).

c. For latitudes 32° (left) and 34° (right).

d. For latitude 29° (left) and a tablet of horizons (right).

9. Persian Astrolabe

Late 18th century A.D.
Brass
Diameter $2\frac{15}{16}$ *in. (7 mm)*
Not signed
Inventory 3170

The *rete* marks the positions of 19 named stars and is of a light flowing, symmetrical design. The limb is engraved with a degree scale reading to 1°, numbered in groups of five up to 100, when it returns to 0 and recounts up to 100 – which is, however, written 200 – and so to 360°. The *mater* is cast in one piece with the limb and *kursī*, and carries at the centre a table of the latitude, longitude, and *inḥirāf* of 17 places. On the back are scales of degrees (0°–90°–0°), starting from the east-west line) in the upper quadrants surrounding (left) a sine graph with lines drawn at 5° intervals; except at the top where some are omitted, possibly for clarity. This diagram is used in conjunction with the equally divided (5–60) scale on the right-hand half of the beveled edge of the alidade. Across the sine graph, arcs for a scale of unequal hours are drawn. In the right hand quadrant is a diagram of the arcs of the signs of the zodiac crossed from left to right by four arcs of altitude of the sun in the azimuth of the *qibla* at Baghdad, Isfahan, Mashhad, and Herat. From right to left it is crossed by two more vertical arcs of the meridian altitude of the sun at Yazd and Qazwīn. In the lower quadrants is a double shadow square. Along the lower quadrants are a set of co-tangent scales.

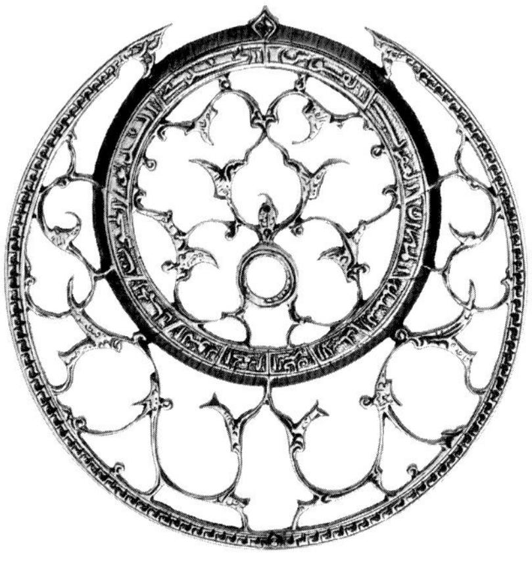

figure 80. Catalogue no. 9. *Rete.*

figure 81. Catalogue no. 9. *Mater* (left) and back without alidade (right).

figure 82. Catalogue no. 9: Persian astrolabe. Front.

figure 83. Catalogue no. 9. Back.

There are four plates, with almucanters drawn for every 6°:

a. for latitudes 32° and 30°

b. for latitudes 36° and 35°

c. for latitudes 33° and 37°

d. for latitude 34° and with a tablet of horizons

The alidade, engraved with declination scales for use with the arcs associated with the unequal-hour scale, the pin, and the horse are original.

Provenance: The collection of Derek J. De Solla Price.

IC/CCA no. 3599.

On the basis of its style, this instrument has been ascribed to the supposed work of ʻAbd al-Ghafūr[257]–with which it does indeed have features in common–although the quality of the engraving is perhaps less good than on that of his finest pieces. Given the great similarity in detail and in decoration among late Persian astrolabes, even by different makers, it is impossible, however, to attribute this astrolabe to him with certainty.

257. See discussion of this problem in Brieux & Maddison.

figure 84. Catalogue no. 9. Plates.

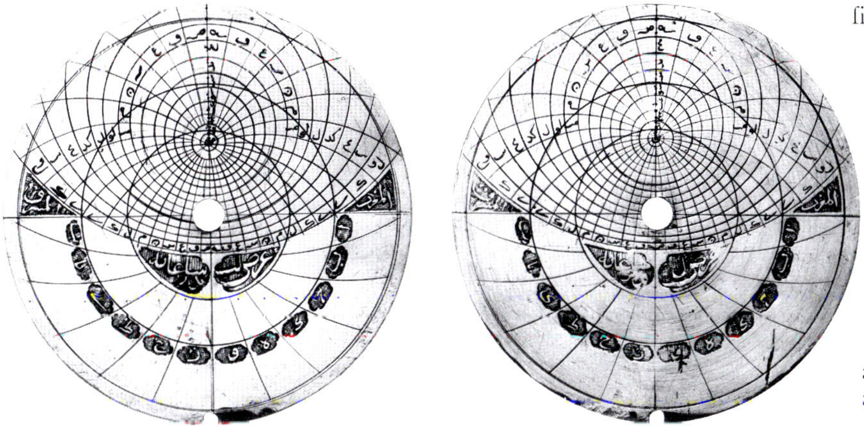

a. For latitudes 32° (left) and 30° (right).

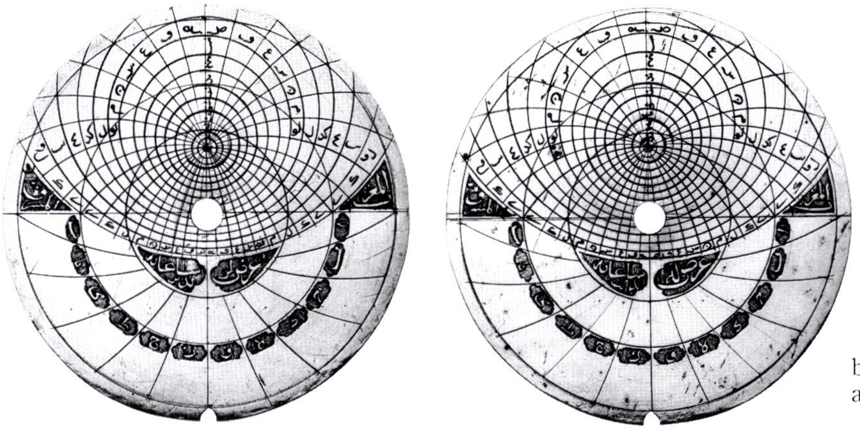

b. For latitudes 36° (left) and 35° (right).

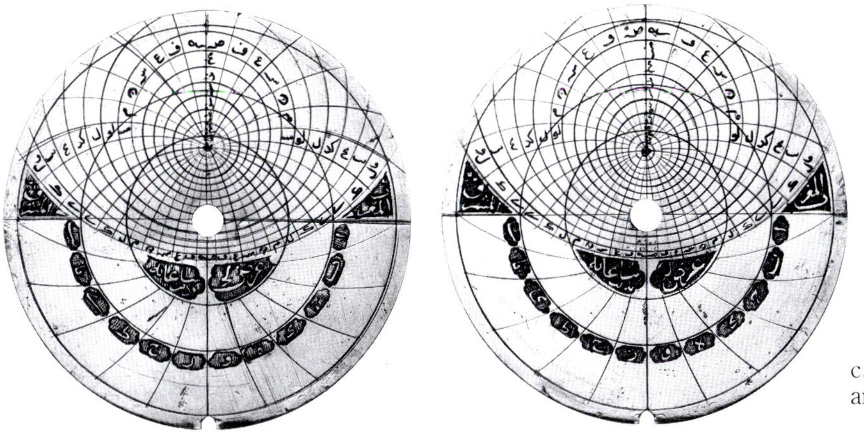

c. For latitudes 33° (left) and 37° (right).

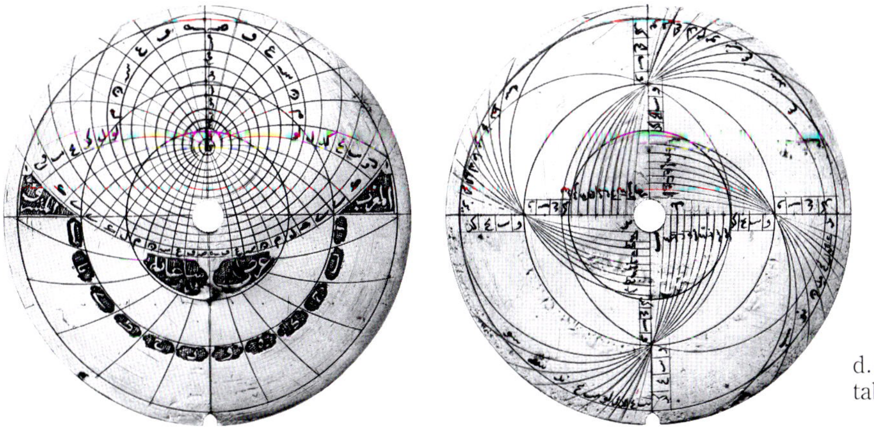

d. For latitude 34° (left) and tablet of horizons (right).

10. Persian Astrolabe

A.H. 1234 (A.D. 1818/9)
Brass
Diameter 8¾ in. (222 mm)
Signed, in a cartouche below the shadow square:
(Made by Muḥammad Akbar 1234)

Inscribed on the front of the kursī:
(Made by order of Muḥammad Mīrzā Qājār Sa'īd)

Inscribed on the back of the kursī:
(He is God the mighty. This was achieved in conformity with the wish which was an order of the very great prince [i.e. the Qājār Shāh Fatḥ 'Alī, A.H. 1211–50 (A.D. 1797–1834)] in the city of Nahāvand [a city of west Iran, 35–40 miles south of Hamadān] the servant of the court Muḥammad Akbar ?Afshār)

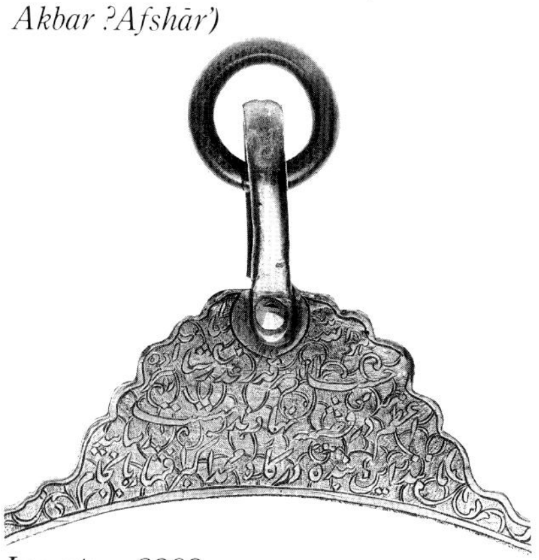

Inventory 2393

The *rete* is symmetrically patterned with thirteen named stars. The *rete* is decorated with heavily cut leaf and scroll patterns, has a counter-changed east-west bar, ecliptic ring with named signs divided to 6°, and one *mudīr.* Neither the equator, nor the tropic of Cancer are shown. The *kursī* is solid, with six lobes and inscribed both back and front; shackle and suspension-ring. The limb is engraved with a 360° scale, each fifth degree indicated by a dot on the division graduation: II•II. The numerals are surrounded by a band of flowing tracery against which they are not easy to read. Engraved on the *umm* is a gazetteer in two rings, four groups each, of 70 places giving their longitude, latitude, and *inḥirāf,* and surrounding a central decorative circle made up of nine leaves and foliage. On the back in the upper-left quadrant is a scale of degrees each numeral being engraved against an arched background and surrounding a sine graph with vertical radial lines. The upper-right quadrant contains a similar degree scale surrounding a diagram of the arcs of the signs of the zodiac. Each sign, which is named in spaces decorated with foliate scrolls at the end of the arcs, is divided into five parts. The arcs crossing the parallels from right to left are the midday arcs for eight latitudes, from 28°–42° at intervals of two degrees. The four arcs going from left to right are those altitudes of the sun in the azimuth of the *qibla* at four different places, Shiraz, Baghdad, Isfahan, and Tus. The lower two quadrants contain a shadow square within which is a table of triplicities. On the limb is a scale of co-tangents and the five scales between these and the shadow square contain astrological tables. The pin and alidade, which is graduated, with a sexagesimal scale and unequal hours, are original. The horse is perhaps a later replacement. There are five plates with the almucantars drawn for every two degrees. The plates are:

a. for latitudes 30° and 34°
b. for latitudes ?32° and 22°
c. for latitudes 36° and 29°
d. for latitudes 37° and 43°
e. for latitude 38° and with a tablet of horizons

A sixth plate, which is unfinished, is marked on one side with the circles for the equator and the tropics of Cancer and Capricorn, and on the other an unfinished trigonometric grid. Plates a) and c) are additionally engraved below the horizon, on the sides for 36° & 34°, with Babylonian hours as is plate e) which on the 38° side also has

figure 85. Catalogue no. 10: Astrolabe by Muḥammad Akbar. Front.

figure 86. Catalogue no. 10. Back.

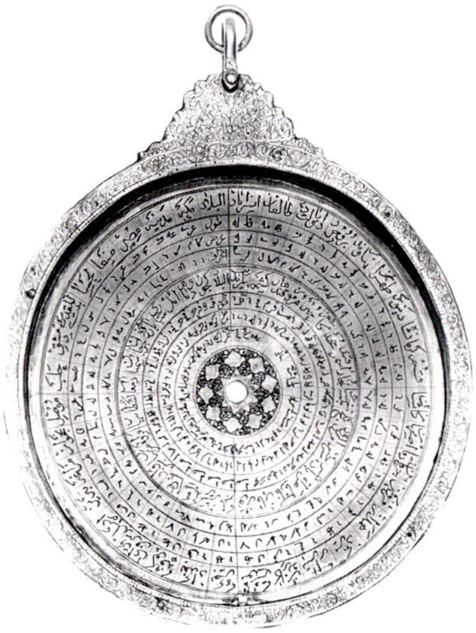

figure 87. Catalogue no. 10. *Mater.*

Italian hours. All the latitude plates are engraved beneath the horizon with an unequal-hour scale, and two of them are also marked with a scale of equal hours.

This is a poorly made instrument, the division of which is somewhat careless and the numerals, especially on the limb, are difficult to read because of the overluxuriant decoration.

Provenance: The collection of Samuel V. Hoffman (no. 11).
New York Historical Society, Acc 1943, 199.
Smithsonian Institution, Washington, D.C., 322459.
Nissen Collection.
IC/CCA no. 58; Brieux-Maddison MHD AKBI.

A second astrolabe by Muḥammad Akbar in the Whipple Museum of the History of Science, Cambridge, is dated A.H. 1236 (A.D. 1820/21).[258]

258. Mayer (I), 62. See also Gunther (V), i, 169–70; Christie, 32, no. 103; Gibbs & Saliba, 100–4.

figure 88. Catalogue no. 10. Plates.

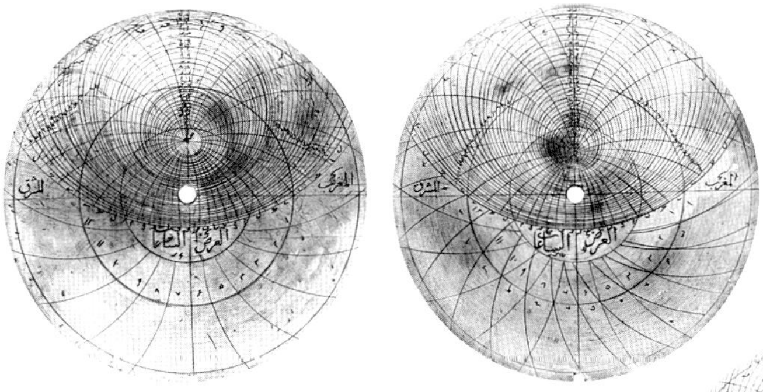

a. For latitudes 30° (left) and 34° (right).

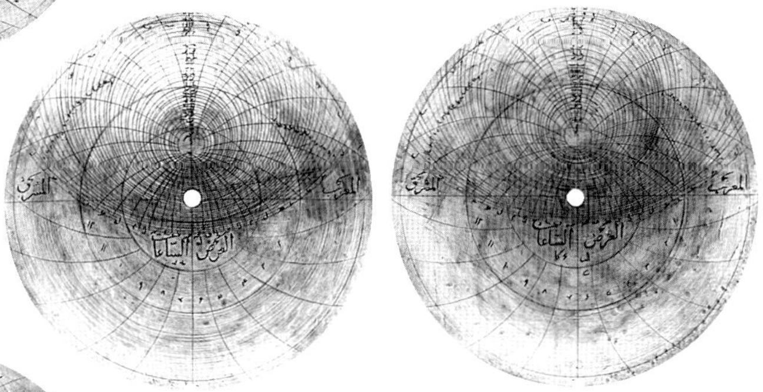

b. For latitudes ?32° (left) and 22° (right).

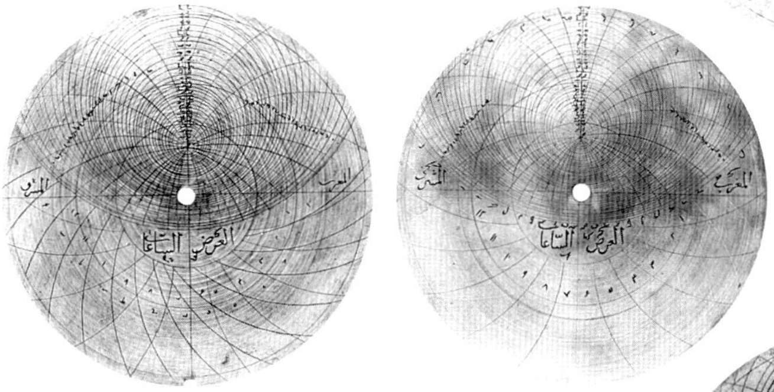

c. For latitudes 36° (left) and 29° (right).

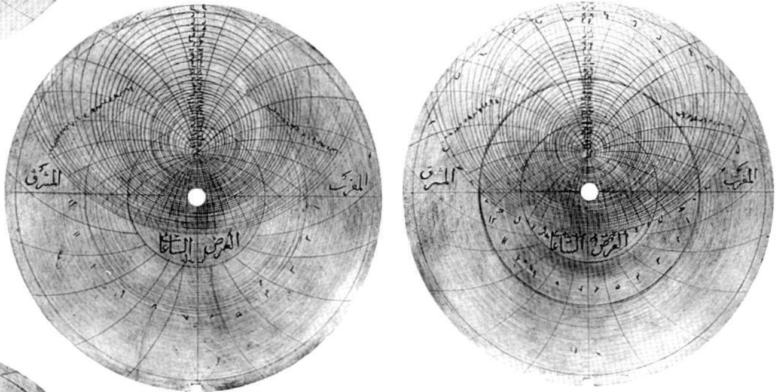

d. For latitudes 37° (left) and 43° (right).

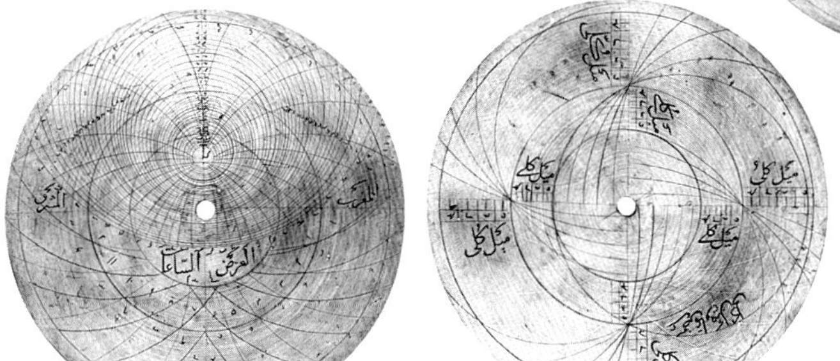

e. For latitudes 30° (left) and a tablet of horizons (right).

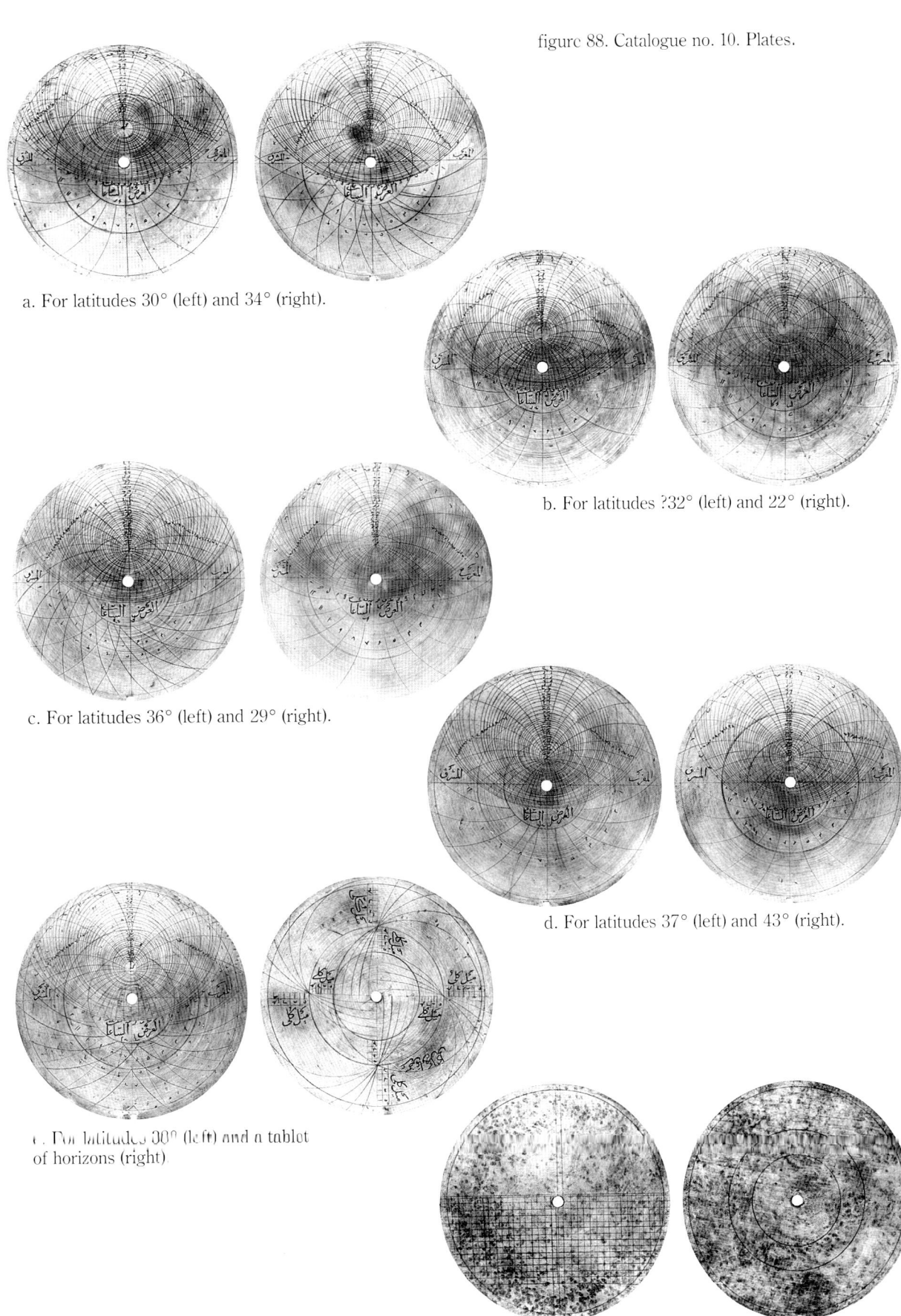

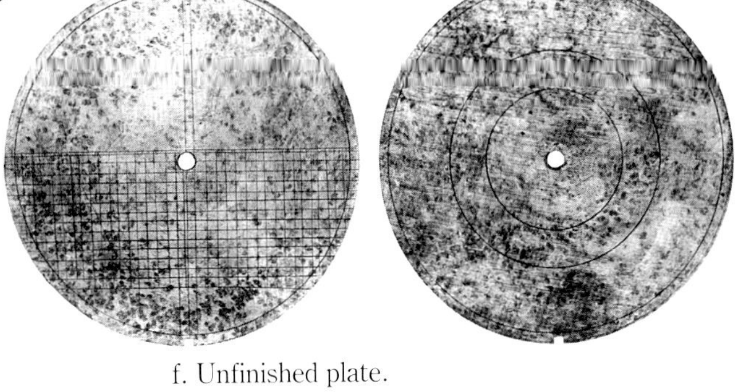

f. Unfinished plate.

11. Maghribi Astrolabe

A.H. 1126 (A.D. 1714/5) [? recte 1261 (A.D. 1845)][259]
Brass, with some rubrication
Diameter $5\frac{1}{8}$ in. (130 mm)
Signed, on the umm:
(By the grace of the Helper, the Virtuous, my Protector, the Arab pilgrim, Muḥammad b. al-Mufaḍḍil b. Aḥmad b. Kīrān made it may God help him in the year 1126 of the Hijra)

Inventory 430

The *rete* marks the positions of nineteen named stars, each indicated by a silver knob and with slender curving pointers springing from a circular base. The *rete* is without tracery and has only small arcs of the equinoctial and Capricorn circles. In the ecliptic circle the signs are named and divided to 5°. There are four silver *mudīrs* and a counterchanged east-west bar. The *kursī* is circular with pointed supports and marked with a horizontal crescent moon, and has a shackle and swivel-pin suspension. The limb, which was made separately and rivetted to the backplate, is marked in rather awkward European-style Hindu-Arabic numerals with a scale of 360° reading to 2° and is rubricated. The *umm* is blank except for the maker's inscription, signature, and date. On the back is a degree scale in four quadrants (90°–0°–90°–0°–90°, from the top) reading to 2°, the divisions in the upper two quadrants being marked in both *kufic* and European-style Hindu-Arabic numerals. Within this is a concentric zodiacal calendar (0° Aries = 6 March), surrounding a double shadow square in units of six, and above this a double unequal-hour scale.[260] The numerals for the divisions of the months and the unequal-hour lines are also in European-style 'Hindu'-Arabic numerals. There are three plates with almucantars drawn for every 5°. The plates are:

a. for latitudes 31°30′ (Marrakesh) and 33° (Salé)
b. for latitudes 33°40′ (Fez) and 34° (Meknes)
c. for latitudes 30° (Sijilmāsa) and tablet of horizons (Western type but with the azimuth lines drawn in)

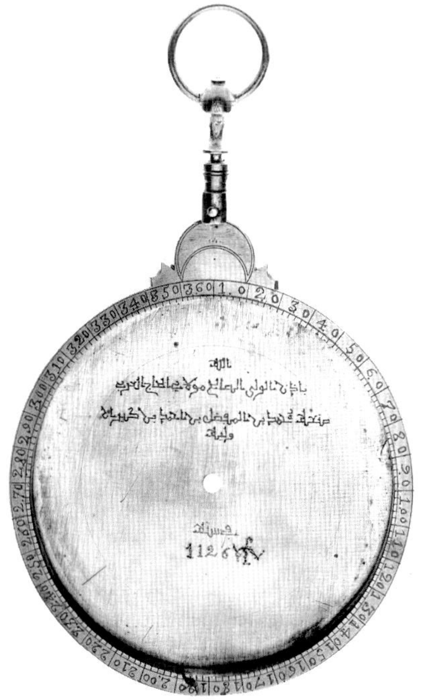
figure 89. Catalogue no. 11. *Mater.*

259. Although clearly marked, this date poses problems since a second astrolabe by the same maker, now in the Adler Planetarium, Chicago, is dated A.H. 1264 (A.D. 1847/8) and two quadrants by him now in private collections in Paris and Fez have the same date. Presumably, if the engraver was unfamiliar with the way of European numerals (as would seem to be the case from his clumsy execution of them and errors in their use, such as including a 32-day month in the calendar) in a moment of distraction he may have confused their positions. The date of 6 March for the vernal equinox confirms this since it is correct for the mid-19th century if one follows, as did Muḥammad b. al-Mufaḍḍil, the Julian calendar.
260. For the redundancy and limited accuracy of this scale, see North (VI), *passim*.

figure 90. Catalogue no. 11: Astrolabe
by Muḥammad b. al-Mufaḍḍil. Front.

figure 91. Catalogue no. 11. Back.

Each plate, except that of the horizons, is engraved below the horizon with an unequal-hour scale and is also marked with the lines of *khaṭṭ aẓ-ẓuhr, khaṭṭ al-'asr* and *khaṭṭ-az zwāl*, together with the crepuscular line.
Provenance: Sotheby's & Co., London, 14 March 1957, lot 153.
IC/CCA no. 3918.[261] Brieux-Maddison MHMD MFDL KIRAN 1.

Of Muḥammad Mufaḍḍil very little is known except that he was a *muwaqqit* at Wazzān, Morocco.

figure 92. Catalogue no. 11. Plates.

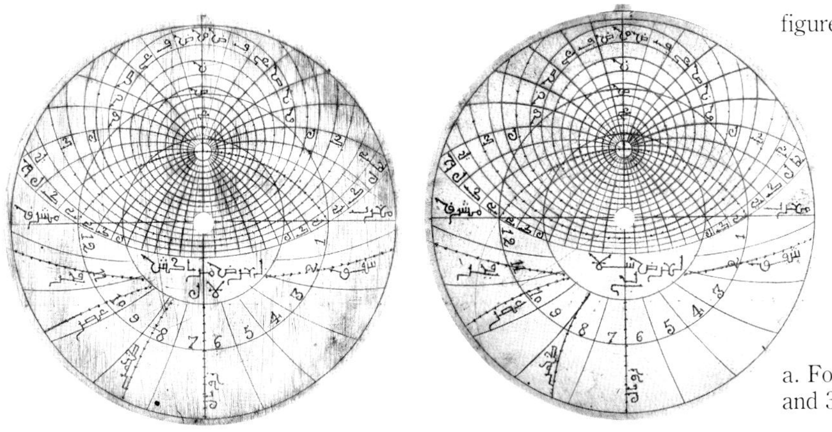

a. For latitudes 31°30′ (left) and 33° (right).

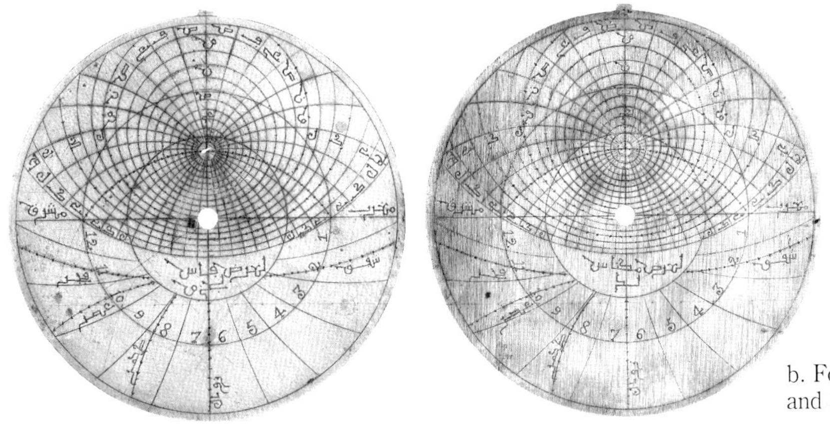

b. For latitudes 33°40′ (left) and 34° (right).

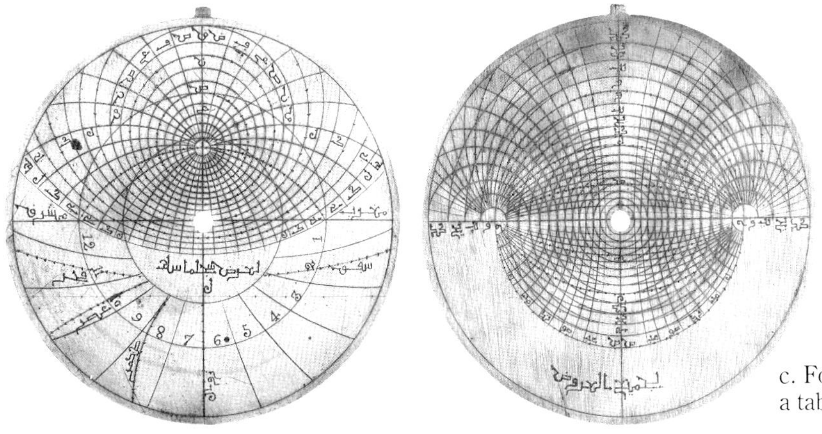

c. For latitude 30° (left) and a tablet of horizons (right).

261. This instrument may also be compared with an unsigned instrument IC/CCA no. 155 described by Gunther (V), i, 300 as 'modern' and dated *c.* 1780. The two instruments resemble each other closely, both being heavily and rather roughly made with European-type numerals, a shackle and swivel ring suspension, and an horizontal, inverted crescent moon engraved on the *kursī*.

12. Indian Astrolabe

?Late 18th century/early 19th century
Brass
Diameter 21¾ in. (552 mm)
Not signed
Inventory 1056

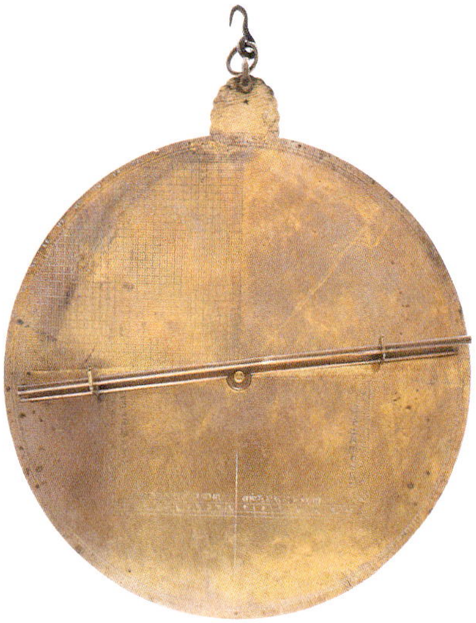

figure 93. Catalogue no. 12. Back.

The *rete* has thirty-five named stars. The position of each star is marked by a triangular or dagger-shaped pointer arising from a narrow rectangular base. The *rete* is of plain design without decorative tracery. The ecliptic is graduated to 1° in groups of five and has the zodiac signs named. The *kursī* is high and lobed, and cut from one sheet of metal with the *mater.* There is no raised limb and there are no plates. On that part of the *mater* extending beyond the *rete* a degree scale is engraved reading to 1° in groups of six numbered sequentially 1 to 60.[262] The back is engraved with a degree scale numbered in the upper two quadrants only. In the upper-left quadrant is a nonagesimal sine-cosine graph, and in the lower two, a double shadow square of twelve and seven divisions. The alidade is fitted with a sighting-tube. The single plate is drawn for latitude 27° (Jaipur) and has almucantars drawn for each degree. The azimuth lines are continued below the horizon[263] along which they are numbered. An unequal-hour scale is also drawn below the horizon.

262. For a similar method of numbering the degrees but in a single sequence up to 60 (= 360°), see cat. no. 13 and the astrolabe formerly in the Plimpton Collection IC/CCA no. 98, Gunther (V), i, 226–7.

263. The continuing of azimuth lines below the horizon occurred occasionally on early astrolabes from east Islam but thereafter is rarely found except on Indian and Indo-Persian astrolabes. See, for example, cat. no. 13 below; Maddison (I), 21–2; Wynter 2–3.

figure 94. Catalogue no. 12:
Indian astrolabe. Front.

13. Indian Astrolabe

Saṃvat 1866 sā va da 5 (=5th day of the kṛṣṇapakṣa of Śrāvaṇa in Vikramasaṃvat 1866 [=A.D. 30 August 1809])[264]
Brass
Diameter 6 in. (152 mm)
Not signed
Inventory 765

The *rete* has nineteen named stars. The star pointers are relatively slender, straight or slightly curved, and arise from square bases. The ecliptic is graduated to 2° in groups of three, with the signs named, and, like catalogue no. 12, the *rete* has a minimum of decorative tracery. There is one *mudīr.* The limb is divided with a 360° scale reading to 1° by groups of six, numbered sequentially 1 to 60, as on catalogue nos. 12 & 14. The *kursī* is high, lobed, and scalloped, carries the date inscription, and is cast in one piece with the *mater* and limb. The *mater* is blank except for a series of five concentric rings and sixteen radii probably intended as the grid for a gazetteer. On the back, the upper-right quadrant is blank except for a degree scale (0°–90°, starting from the east point) reading to 1° by numbered groups of six on the edge. The upper-left quadrant has a similar degree scale (90°–0°, starting from the top) surrounding a 30-division sine-cosine graph marked with an arc. In the lower quadrants are a double shadow square divided into seven and twelve parts, and trigonometrical scales. There are five plates:[265]

a. for latitude, 21° and a quadruple plate for latitudes: 0°, 18°, 42°, and 48°

b. for latitudes 24° and 30°

c. for latitudes 66° and 26°

d. for latitudes 28° and 36°

e. for latitude 27° and with a tablet of horizons

Each latitude plate in addition is inscribed with the length of the noon equinoctial shadow, is engraved with an unequal-hour scale, and has the azimuth lines drawn below the horizon only, except on two plates (latitudes 27° & 66°/26°) where they are drawn both above and below. The alidade is straight and furnished with a sighting tube.

IC/CCA no. 2712.

264. The *Vikrama Saṃvat* era began in 58 B.C. Discrepancies occur because the era year is assumed to begin in March/April in eastern India, and October/November in western India. See Phillips, 80.

265. Indian astrolabes with a series of latitude plates are rare. However, at least two others exist, although neither has been published. Both are in private collections.

figure 95. Catalogue no. 13:
Indian astrolabe. Front.

figure 96. Catalogue no. 13. Back.

figure 97. Catalogue no. 13. Plates.

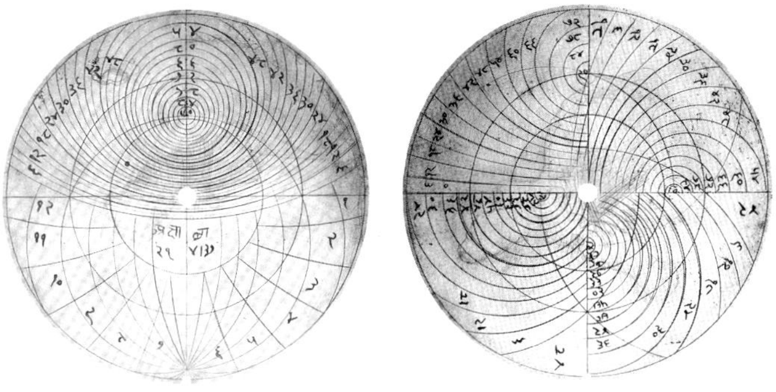

a. For latitudes 21° (left) and for 4 latitudes: 0°, 18°, 42°, and 48° (right).

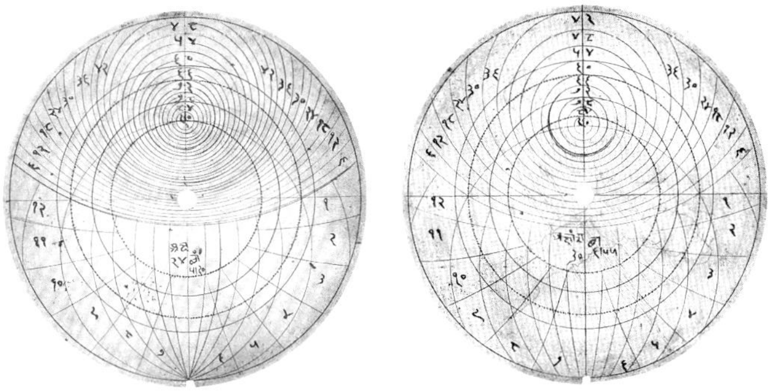

b. For latitudes 24° (left) and 30° (right).

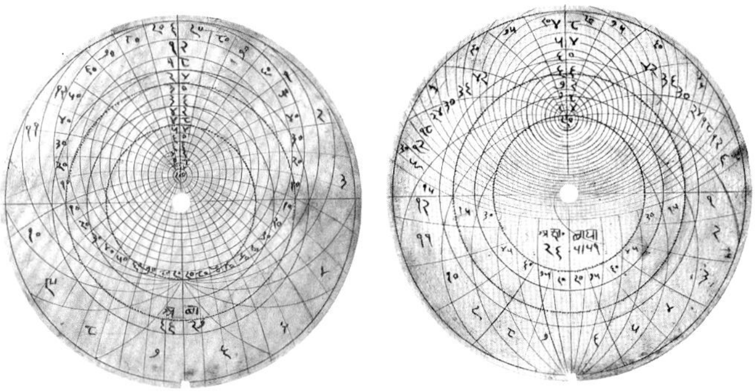

c. For latitudes 66° (left) and 26° (right).

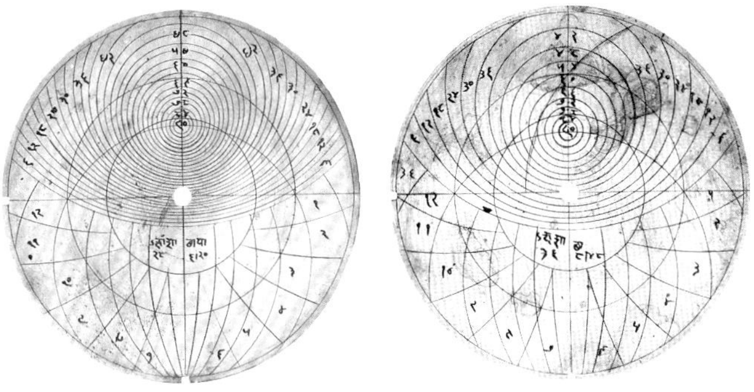

d. For latitudes 28° (left) and 36° (right).

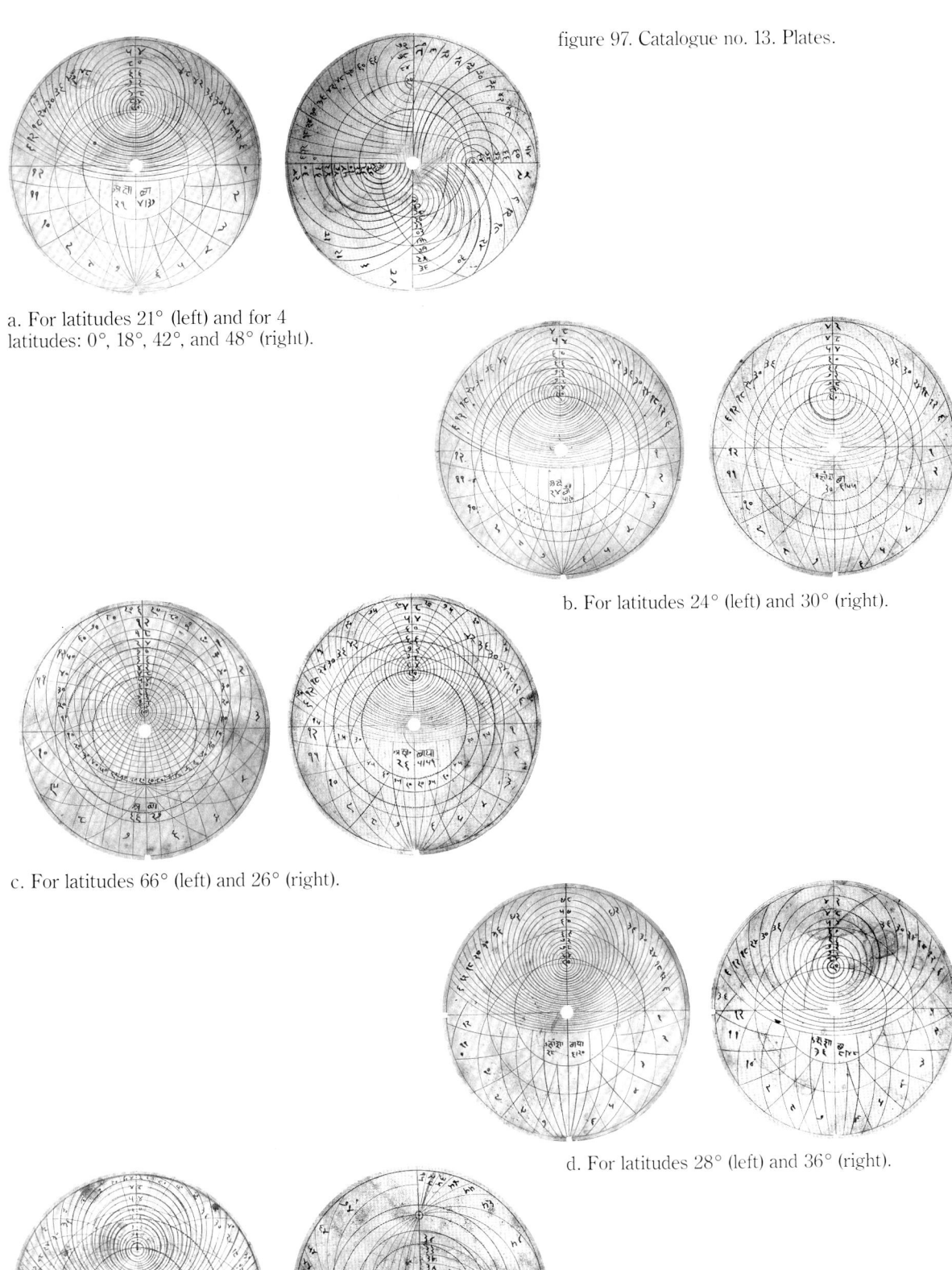

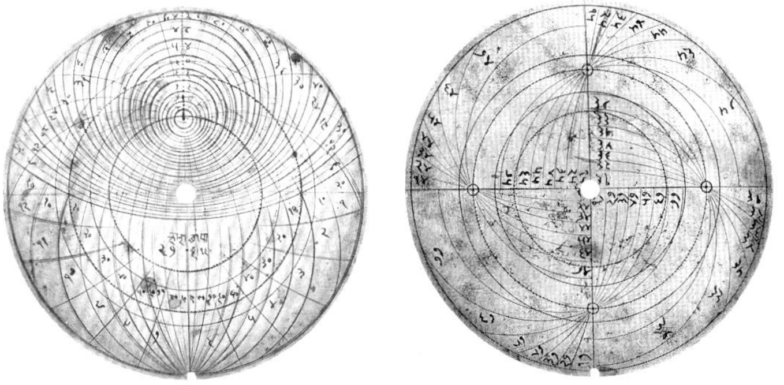

e. For latitude 27° (left) and a tablet of horizons (right).

14. Indian Astrolabe

?Late 19th century
Silver
Diameter 12½ in. (318 mm)
Not signed
Inventory 1752

The *rete* marks the positions of twenty-five named stars. The position of each star is marked by a broad leaf-shaped pointer. These rise either from the circles representing Capricorn, the ecliptic, and the equator, or from the east-west bar which is counter-changed at the west side but not at the east, presumably because of an error by the maker, the *rete* being otherwise symmetrical. No star pointers rise from the limited amount of restrainedly decorative tracery. The ecliptic is graduated to 1° in groups of six and has the zodiac signs named. There is one *mudīr*. The *kursī*, which is high, lobed, and scalloped, is made from one flat sheet with the *mater*. There is no raised limb and there are no plates. On that part of the plate which extends beyond the *rete*, the degree scale is engraved in four quadrants reading to 1° in groups of six, each group being numbered sequentially 1 to 15 from the east point so that 15 = 90°. The back is engraved in the upper-left quadrant with a degree scale surrounding a sine-cosine graph, in the upper-right quadrant with a degree scale surrounding the arcs of the signs of the zodiac. In the lower two quadrants is a double shadow square of six and seven divisions. A graduated rule is mounted at the centre but this should presumably be on the face. The alidade is missing. The plate is engraved for latitude 26° and has the almucantars drawn for every 20°. The plate, as in catalogue no. 12, is unusual in that the azimuths are drawn as full arcs continuing below the horizon, where they are further divided by intermediary half arcs, and where the unequal-hour lines are also drawn.

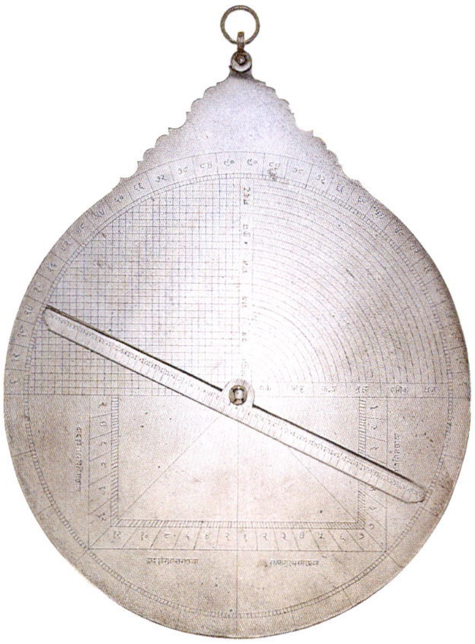

figure 98. Catalogue no. 14. Back.

figure 99. Catalogue no. 14:
Indian astrolabe. Front.

15. Indian Astrolabe

?Early 19th century
Brass
Diameter 10¾ in. (270 mm)
Not signed
Inventory 3173

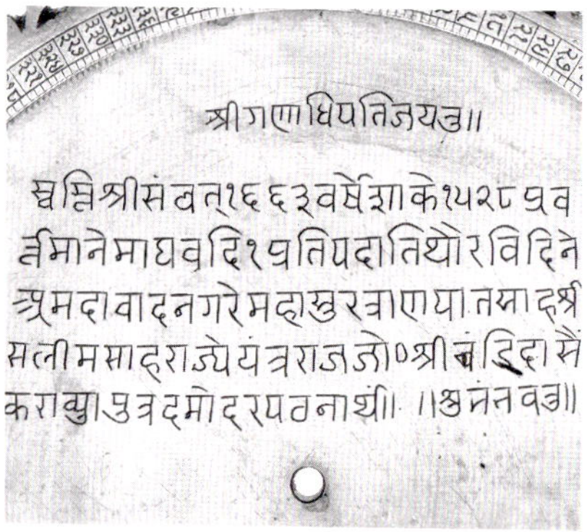

figure 100. Catalogue no. 15. Inscription.

figure 101. Catalogue no. 15. Back.

The *rete* marks the positions of twenty-four named stars indicated by broad, slightly curving pointers of which one is broken and one was originally misplaced. The star names on all of them are poorly engraved. The *mater* is cast in one piece with the high pierced *kursī*, the design of which was clearly influenced by the decorative *kursī* typical of the Indo-Persian astrolabes made in Lahore (catalogue no. 4). At the top of the *kursī*, immediately below the suspension ring, is the inscription *śrīdivyacakṣuṣai namaḥ* (reverence to the divine eye).[266] The limb is engraved with a scale of degrees (0°–360°) numbered in groups of three. The centre of the *mater* is plain except for a long inscription engraved in a formal hand different from that used on the rest of the instrument. This has been translated as follows:

> It is well. In Saṃvat 1663, Śaka 1528 current, on the first tithi of the Kṛṣṇapakṣa of Māgha, on Sunday [1 February 1607], at the city Ahmādabād, during the reign of the Great Sultan, the Pātaśāha, Salīm Shāh [Jahāngīr], the astrolabe for the purpose of the reading of Dāmodara the son of Caṇḍīdāsa [illegible]. Let it be auspicious.

The back of the instrument is engraved round the edge with a degree scale divided to 1° in groups of three, in four quadrants (90°–0°–90°–0°–90°, starting from the top). In the upper-left quadrant is a sexagesimal sine/cosine graph divided by three radii into three sections of 30° each. An arc drawn across the scale from 24° marks the obliquity of the ecliptic. The numbers 1–12 are written, four to each sector, in that portion of the quadrant outside the arc of the obliquity of the ecliptic. In the right-hand quadrant is a scale of unequal hours, in the lower quadrants, a double shadow square of 7 and 12 divisions. There are six plates:

a. for latitudes 18° (Bijapur) and 22°30′ (Ujjain)
b. for latitudes 20°30′ (Burhanapur) and 23° (Ahmadbad)
c. for latitudes 25°56′ (Jodhpur) and 31°50′ (Lahore)
d. for latitudes 27°40′ (Dhaka) and 25°52′ (Benares)
e. for latitudes 26°24′ (Agra) and 28°39′ (Delhi)
f. for latitudes 35°20′ (Kashmir) and 39°40′ (Multan).

In addition, each plate is marked with the value of the *chāyā* (length

266. or '. . . to the divine Sitya' depending whether one takes it as an engraver's error or a badly formed t.

figure 102. Catalogue no. 15.
Indian astrolabe. Front.

of the shadow of a gnomon of 12 units at midday at the equinox), the *Paramadina* (the length of half of the longest day in *ghaṭis*),[268] and the name of a town appropriate to the latitude. Unequal-hour lines are drawn below the horizon.

Provenance: Collection of Roberto Riva. Formerly exhibited in the Museum of Natural History, Houston, Texas.

figure 103. Catalogue no. 15. *Mater.*

This unusual instrument is reminiscent of those made at Lahore but is unlike them in its crude engraving and plain undecorated *rete*. In these respects the instrument resembles other known Hindu instruments (cf. catalogue nos. 12–14). The inscription on the *mater* engraved in a different hand can hardly be admitted as evidence for the date of the instrument. Indeed Pingree[269] has suggested that the inscription was taken from a manuscript treatise (as the phrase 'for the purpose of reading...' would suggest) whence it was inaccurately copied, giving rise to some minor errors and one unreadable word. If this suggestion be accepted, then one might hypothesize that the instrument was made at a relatively late date by a metal-worker who, having some knowledge both of Lahore astrolabes and of astrolabe literature, combined aspects of each with the Hindu tradition to produce this eclectic instrument.[270]

268. This has been omitted on the plate for latitude 35°20′.
269. Private communication in Time Museum files.
270. I am grateful to M. J-P. Verdet for his advice on this astrolabe.

figure 104. Catalogue no. 15. Plates.

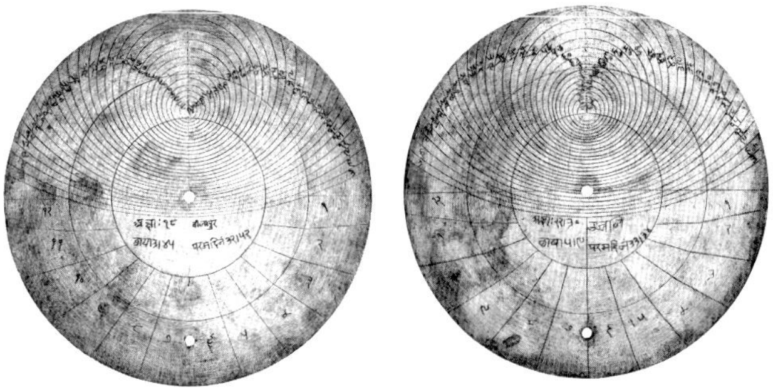

a. For latitudes 18° (left) and 22°30′ (right).

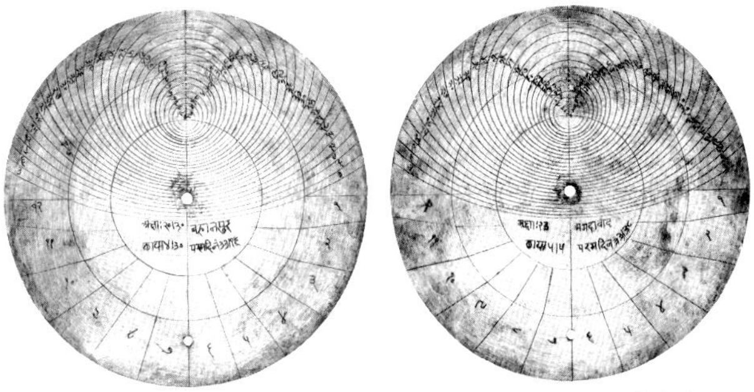

b. For latitudes 20°30′ (left) and 23° (right).

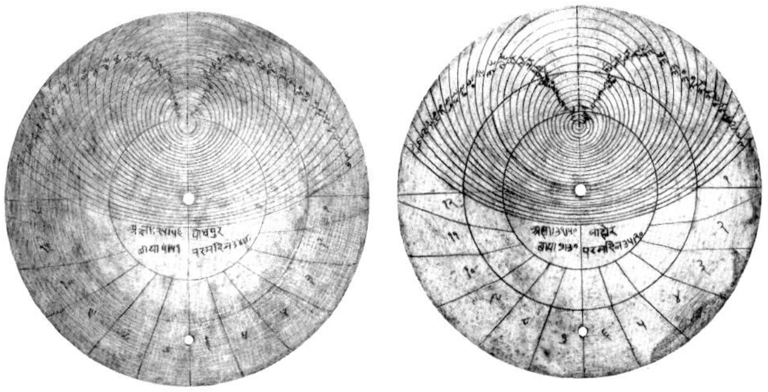

c. For latitudes 25°56′ (left) and 31°50′ (right).

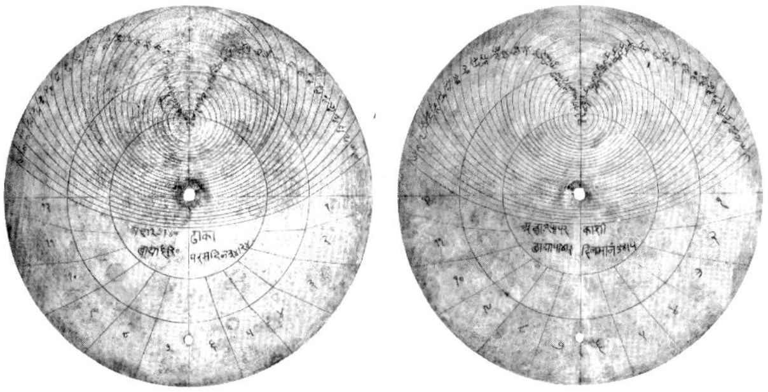

d. For latitudes 27°40′ (left) and 25°52′ (right).

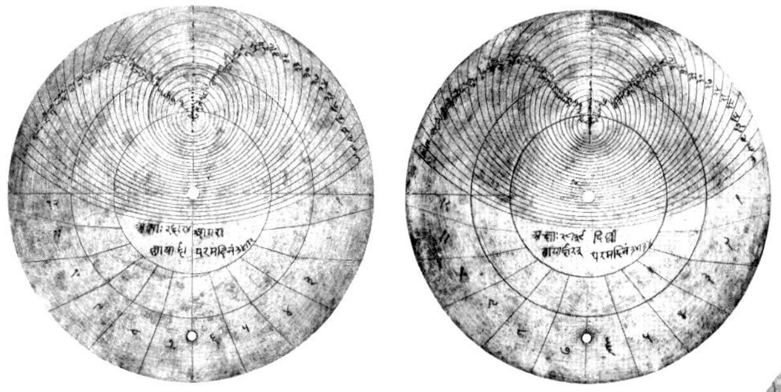

e. For latitudes 26°24′ (left) and 28°39′ (right).

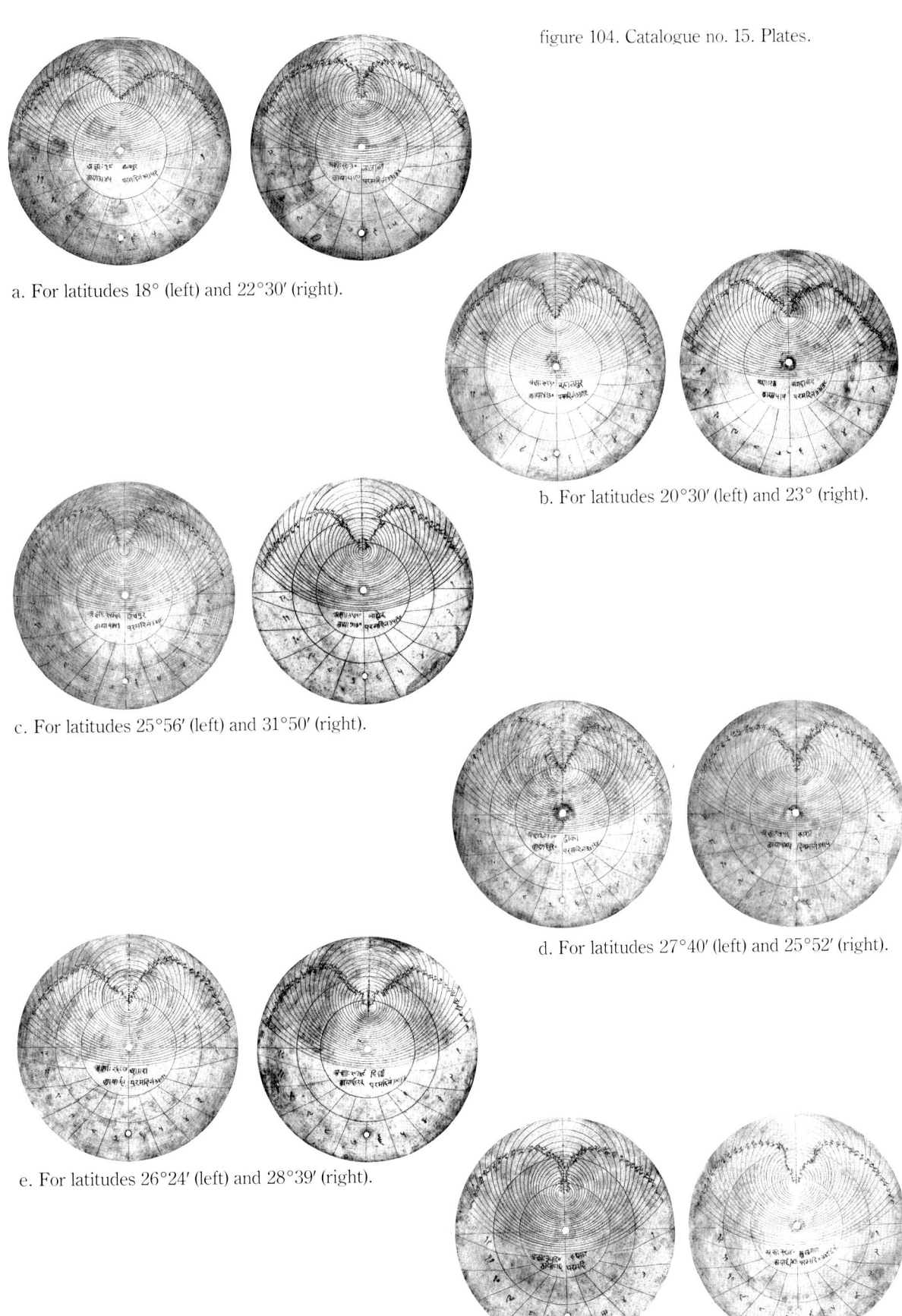

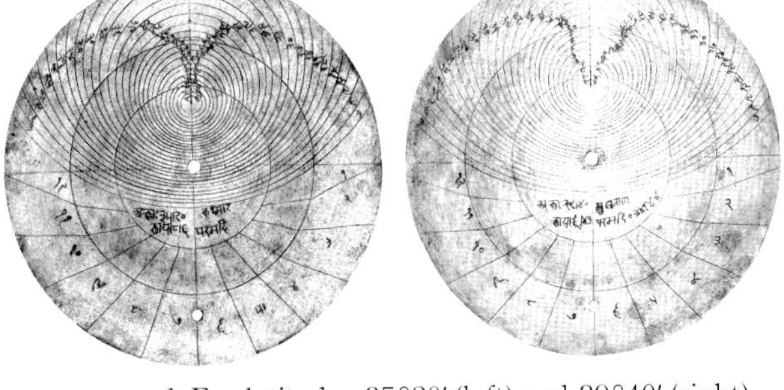

f. For latitudes 35°20′ (left) and 39°40′ (right).

16. French or Italian Astrolabe

Late 14th or early 15th century
Brass
Diameter 3¾ in. (95 mm)
Inventory 3392

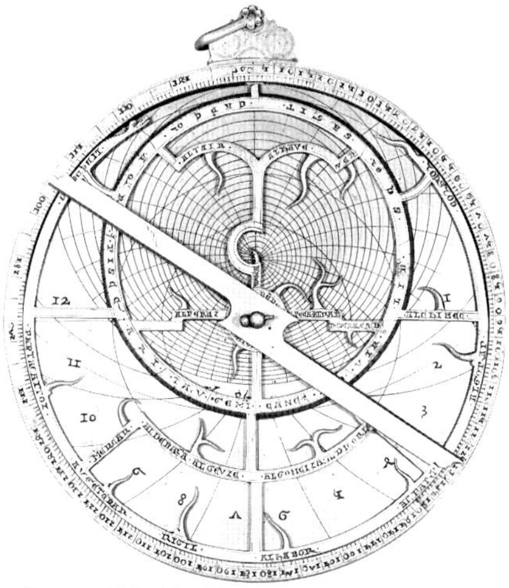

figure 105. Gothic astrolabe, *c.* 1360. Oxford, Museum of the History of Science IC/CCA 168. (Photograph: Museum of the History of Science)

The restrained *rete* carries eighteen star pointers of which fourteen were originally named in the same Gothic script as is used on the rest of the astrolabe, and four have had their names added (or re-engraved) at a later date (? in the late 16th or early 17th century) in a squat italic script. There are elegant, flame-like star pointers and a counter-changed east-west bar. The ecliptic is graduated to 2° in groups of five, with the signs named. The limb and backplate were separately cast and rivetted together to form the *mater*, and there is a squat, rectangular *kursī*. The layout lines are visible on the reverse. The limb is engraved with a degree scale in 2° divisions numbered by groups of 18 and on the edge is a scale of equal hours (1–12, 1–12). The *mater* is plain. On the back, around the edge, is a degree scale in four quadrants (90°–0°–90°–0°–90°, from the top), and within this, a zodiacal calendar (concentric type) with each division representing two days (0°Aries = 12 March). Within these in the upper quadrants is an unequal-hour diagram; below is a shadow square to which the numerals and the words 'Versa' and 'Recta' were added later. There are two plates:

a. for latitudes 40° and 45°
b. for latitudes 49° and 50°

There is space in the *mater*, however, for three plates. Each of the plates has equal-hour lines drawn below the horizon, but on only one of them are they numbered.

The *rete* design of this small astrolabe is typical of instruments of the 'late-Gothic' type (fig. 27d). It is particularly close to that of an unsigned astrolabe in the Lewis Evans Collection (fig. 105), although this is a larger and better-made instrument.[271]

271. IC/CCA no. 168, Gunther, 317. Gunther considered it to be Italian and *c.* 1400. It is now in the Museum of the History of Science, Oxford.

figure 106. Catalogue no. 16:
Gothic astrolabe. Front.

figure 107. Catalogue no. 16. Back.

figure 108. Catalogue no. 16. *Rete.* Front (left) and back (right).

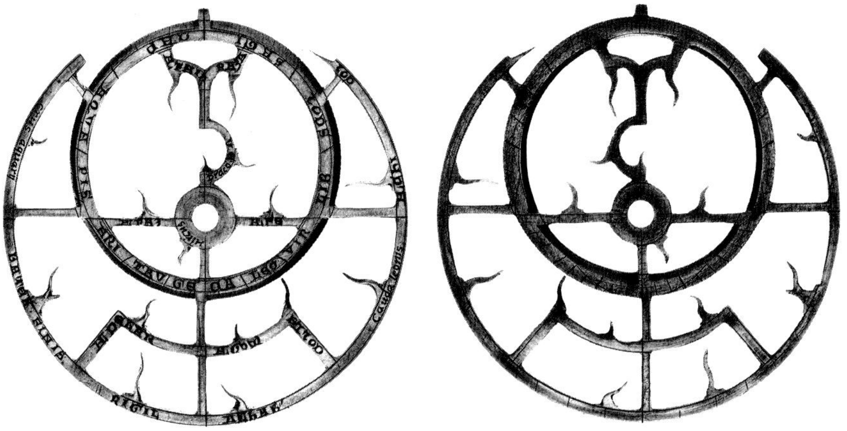

figure 109. Catalogue no. 16. Plates.

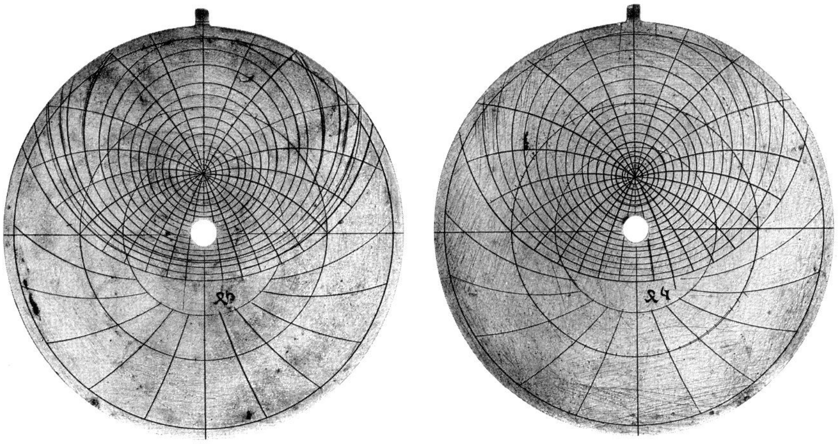

a. For latitudes 40° (left) and 45° (right).

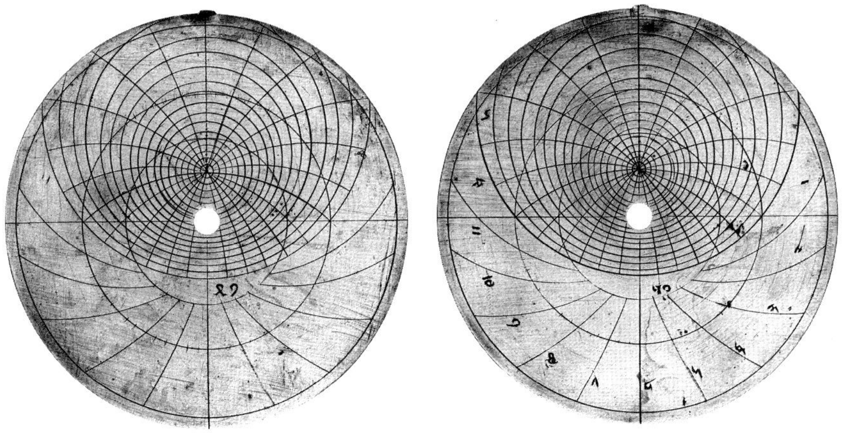

b. For latitudes 49° (left) and 50° (right).

17. German Astrolabe

1532
Brass
Diameter 5⅜ in. (137 mm)
Signed: GEORGIUS HARTMAN NORENBERGE FACIEBAT ANNO MDXXXII'
(George Hartman[n] was making [it] at Nuremberg in the year 1532)

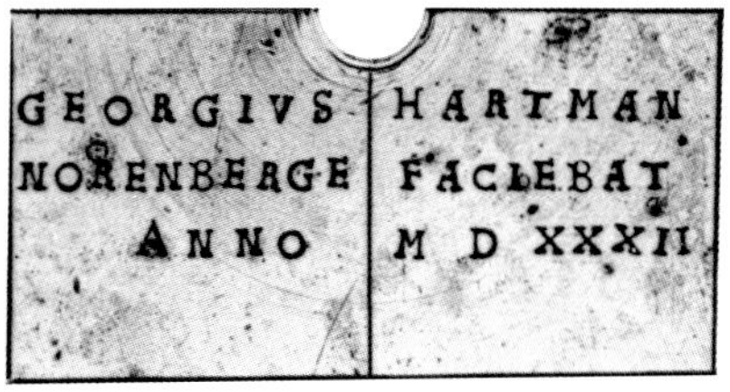

Inventory 3123

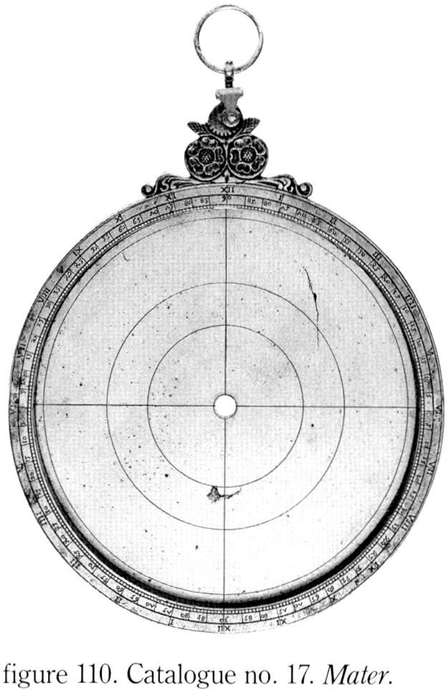

figure 110. Catalogue no. 17. *Mater.*

The *rete* marks the position of twenty-seven named fixed stars. The *mater* consists of a single sheet of metal rivetted to the limb. The body of the suspension mount is cast in one piece with the limb. The suspension mount consists of a low scroll carrying two formal flowers (the resemblance to 'Tudor' roses is probably fortuitous) surmounted by two pine-cones, one on each side of a third flower through which the shackle is pinned. The shackle and suspension ring may be later replacements. This form of suspension mount is characteristic of the work of Georg Hartmann. Engraved on the limb is an equal-hour scale (I–XII times 2) and a degree scale (0°–90°–0°–90°–0°) numbered from the east and west points and reading to 1°. The interior of the *mater* was probably intended to be engraved as a latitude plate but was left incomplete, only the circles of the equator, the tropics, and two meridians having been drawn.

The back of the instrument is engraved around the edge with a degree scale similar to that on the face. Within this is a zodiacal calendar (concentric type, 0° Aries = 11 March), the degrees of the signs of the zodiac being numbered in groups of ten in Roman numerals, and the days of the months in Arabic numerals. Within this are: in the upper-left quadrant, a scale used in conjunction with another scale originally engraved on the alidade for conversions between equal and unequal hours; in the upper-right-hand quadrant, a scale of unequal hours – the 6 o'clock line being marked 'circulus meridie' (meridian circle). Engraved across these hour lines is a semicircle marked 'Opositus Meridiei.' Marked down the vertical line dividing the two quadrants is a scale (0°–60°) marked 'Divisiones diametri.' In the lower quadrants is a double shadow square surrounding the maker's signature. There are three plates:

a. for latitudes 39° and 42°
b. for latitudes 45° and 48°
c. for latitudes 51° and 54°

The latitude of each plate is marked, as are the twelve astrological houses of the heavens and, beneath the horizon only, a set of unequal-hour lines. The alidade, which is not divided and has folding vanes, may be a later replacement. The rule is original.

Four other astrolabes signed by Georg Hartmann and dated 1532 are known. They are:

IC/CCA no. 259 Musée de la Renaissance, Ecouen[272]

272. Previously in the Musée de Cluny & des Thermes and so listed in IC/CCA. Sommerard, 7026, 125–6.

figure 111. Catalogue no. 17: Astrolabe by Georg Hartmann. Front.

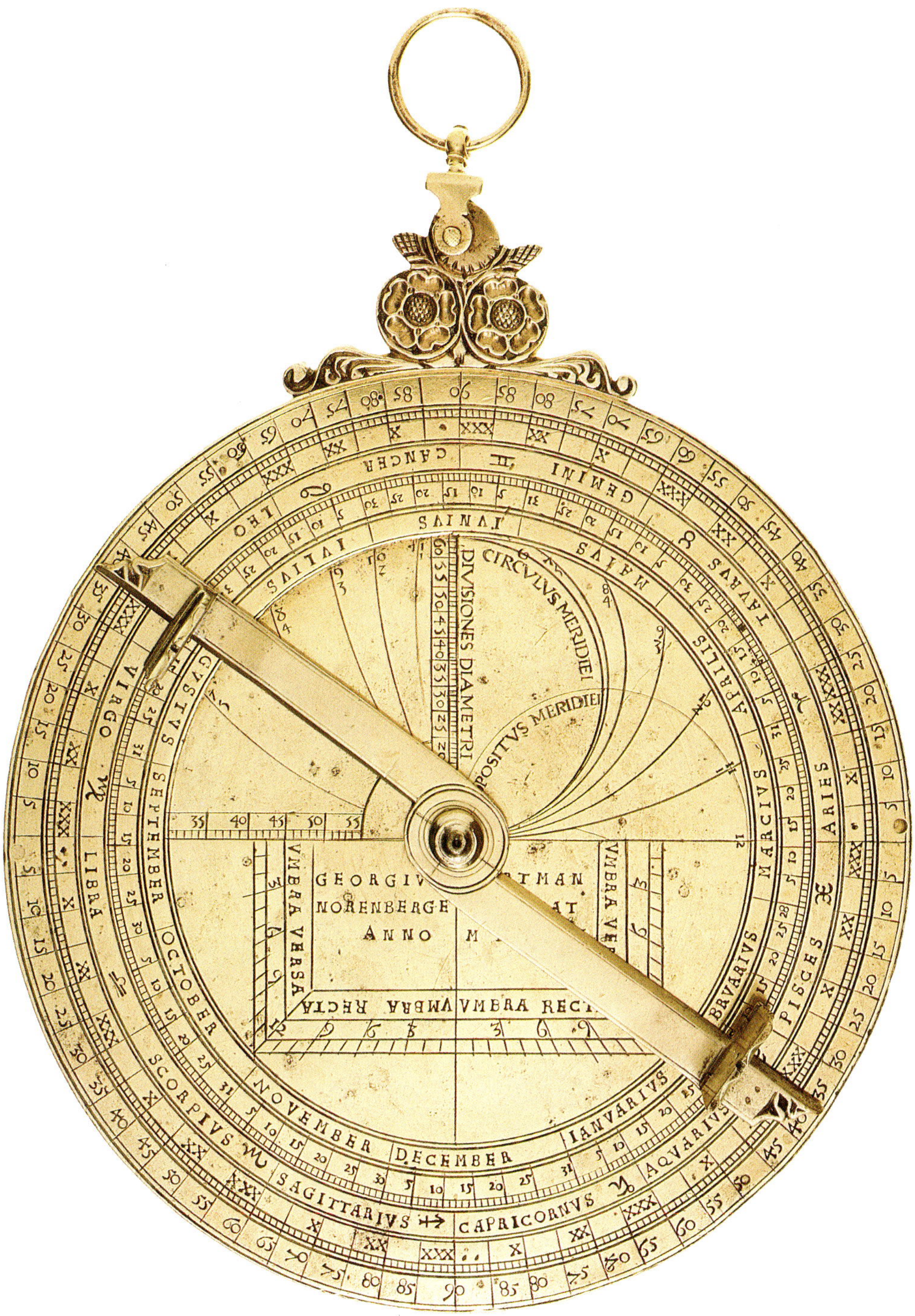

figure 112. Catalogue no. 17. Back.

IC/CCA no. 260 British Museum, London[273]
IC/CCA no. 552 Germanisches National Museum, Nuremberg[274]
IC/CCA no. 565 Národní Technicki Muzeum, Prague[275]

All five instruments are the same size (within a measurement error range of 2 mm), while the similarity of the suspension mounts suggests that they were initially cast from the same mould as were the *retes* which carry the same 27 stars. At least some of the same punches appear to have been used on these astrolabes, although minor differences of positioning and impression occur, as one would expect. Between the five astrolabes there are also some differences in the scales, especially those engraved on the backs.

Provenance: Purchased at Sotheby Parke Bernet, New York, 8 December 1983.[276]

Previously in a private collection in Barcelona, having been earlier at Palma de Mallorca, where it was purchased in *c.* 1923.

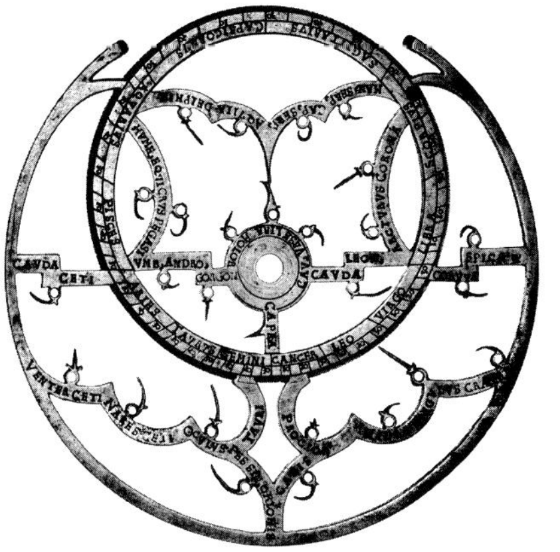

figure 113. Catalogue no. 17. *Rete*.

figure 114. Catalogue no. 17. Plates.

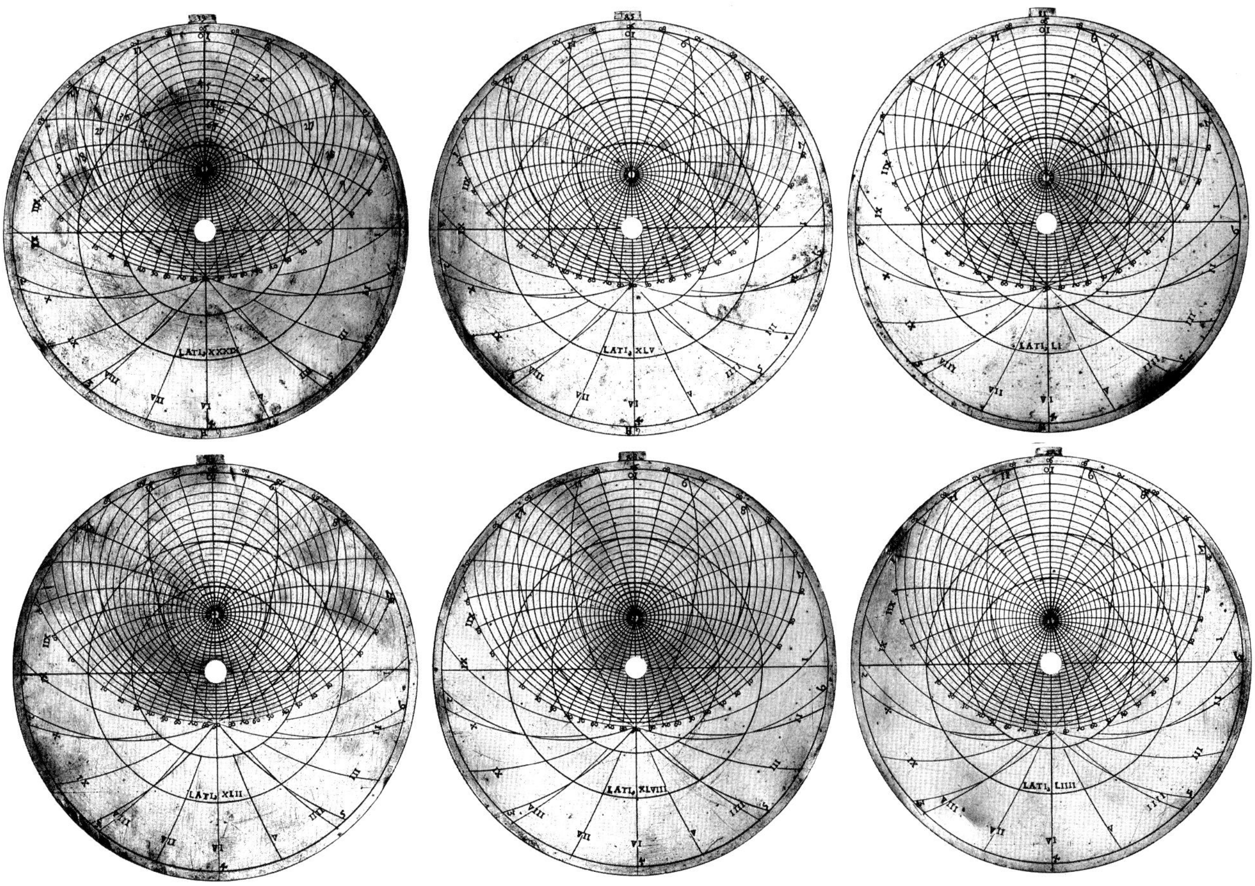

a. For latitudes 39° (top) and 42° (bottom).

b. For latitudes 45° (top) and 48° (bottom).

c. For latitudes 51° (top) and 54° (bottom).

273. Ward, 333.
274. Zinner (II), pl. 25, 3 & 363. Willers & Holtzamer 12, 58–9.
275. Zinner (II), 363.
276. Sotheby (IV), lot 159.

18. Flemish Astrolabe

1568
Brass
Diameter $13\frac{1}{4}$ in. (337 mm)
Signed, on the suspension mount: Gualterus Arsenius Nepos Gemmae Frisij Louvany fecit 1568
(Walter Arsenius, nephew of Gemma Frisius made (it), Louvain 1568)
Inventory 610

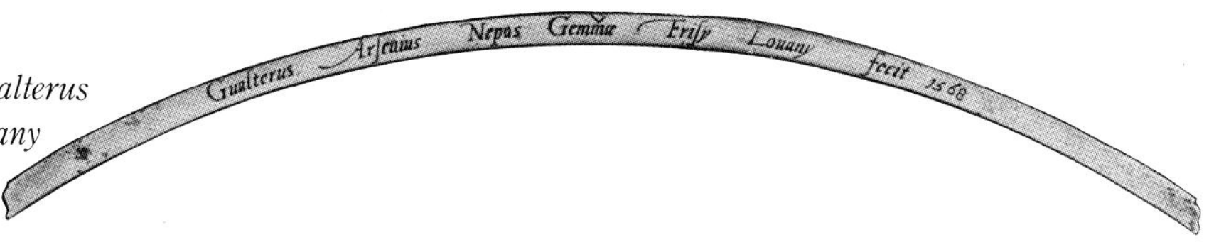

The *rete* marks the positions of fifty-one fixed stars. Since this *rete* is a replacement, probably copied from the astrolabe now at the Adler Planetarium, Chicago, or that at Merton College, Oxford,[277] it will not be described here. The *mater* consists of a single sheet of metal to which one thick and two thin rings were rivetted to provide the limb. The suspension apparatus mount is screwed separately to the limb. The suspension consists of a base carrying two reclining satyrs supporting a central escutcheon with an inset compass (magnetic declination mark approx. $11\frac{1}{2}°$–12° east)[278] surmounted by the shackle and swivel pin for the suspension ring. Except in minor details, this mount is identical with that of the Merton astrolabe dating from three years later. Engraved on the limb is an equal-hour scale (I–XII x 2) reading to four minutes and a degree scale (0°–90°–0°–90°–0°) numbered from the east and west points and reading to $\frac{1}{3}°$ in groups of three. Engraved on the interior of the *mater* is a navigational scale the *Quadratum Nauticum*.

The *Quadratum Nauticum* was probably devised by Gemma Frisius, who first included an account of it in his edition of Peter Apian's *Cosmographia*[279] in 1550 (fig. 117). It is frequently found engraved on the *maters* of astrolabes made in Louvain or by makers influenced by this school such as Thomas Gemini, Humphrey Cole, and Erasmus Habermel.[280] In 1559 William Cuningham gave an account of the device in English, as did Thomas Blundeville in 1597.[281] The purpose of the instrument was to enable a seaman to find, without calculation, the course between two points of which the latitude and longitude were known. It consists of a square divided into four equal quarters by two intersecting straight lines representing the meridian and the equator, and the north, south, east, and west rhumbs of the compass diagram centred on their point of intersection. This compass diagram with thirty-two directions fills the centre of the square, and its circumference represents the horizon. The edges of the square are each divided into scales of ninety degrees,

277. Gunther (V), ii, pl. XCVI facing page 389.
278. Approximately correct for the period. The generally accepted value for 1580, when the declination reached its greatest extension east, was 11°51′.
279. First edition, Anvers, 1529, and numerous other editions and translations thereafter. For bibliographical indications, see van Ortroy, 165–89.
280. For Gemini, see Gunther (VI), fig. 1; for Cole, Gunther (II); for Habermel the astrolabes now in the Museum of the History of Science, Oxford (Josten, no. 19) and the National Maritime Museum, Greenwich (N.M.M. 51 & illus. on front and back covers).
281. Cuningham. *The Cosmographical Glasse*, London, 1559, fol. 162; Blundeville, A 376r–77v. This description did not appear in the first edition of 1594.

figure 115. Catalogue no. 18: Flemish astrolabe by Gualterus Arsenius. Front.

figure 116. Catalogue no. 18. Back

the scales to the left and right of the north/south or meridian line representing degrees of longitude, and those above and below the east/west or equinoctial line representing degrees of latitude. To use it the seaman was to subtract the lesser of his two sets of co-ordinates from the greater, marking the difference of latitude – if negative, below the east-west line (south latitude), if positive, above (north latitude) – with a thread stretched across the latitude scales. The difference of longitude was marked in like manner – to the left of the meridian if negative (east longitude), to the right if positive (west longitude) – and the point of intersection represented the position of the destination, the underlying rhumbs the course to be steered. That the scale was much used may be doubted, not only because it would give an incorrect result (the same difference of longitude was treated as being of equal linear distance at all latitudes, although because of the convergence of the meridians towards the poles, it is not) but also because of the considerable practical difficulties of using it.[282] The compass directions are here indicated in

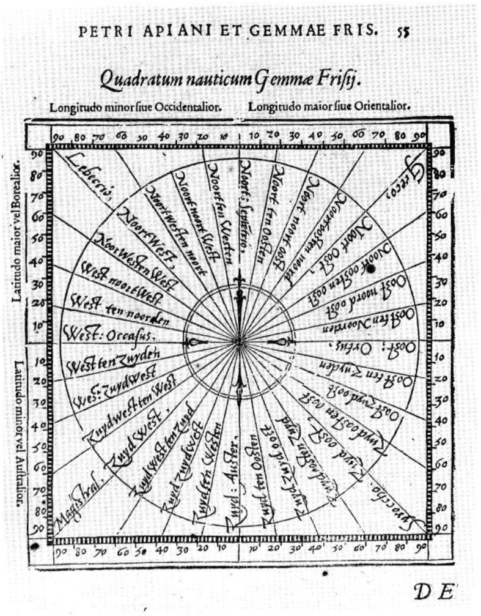

figure 117. The *Quadratum Nauticum* as illustrated in Gemma Frisius' edition of Peter Apian, *Cosmographia…*, Antwerp, 1550, 55.

figure 118. Catalogue no. 18. *Mater* with *Quadratum Nauticum.*

282. Other defects of the device are indicated by Waters, 61.

the classical, Mediterranean way by the names of the winds,[283] whence they were known as 'rhumbs of the winds,' in three bands. Reading north, east, south, west these are:

CENTRE BAND	Tramontana	North	Mediojorno	South
	Grego tramontana	N.N. East	Lebeche mediojorno	S.S. West
	Grego	N.East	Lebeche	S. West
	Grego levante	E.N. East	Poniente lebeche	W.S. West
	Levante	East	Ponie[n]te	West
	Xaloque levante	E.S. East	Poniente maestre	W.N. West
	Xaloque	S. East	Maestral	N. West
	Xaloque mediojorno	S.S. East	Maestre tramontana	N.N. West
MIDDLE BAND	Aparctias	North	Auster	South
	Aquilo	N.E. by North	Libonotus	S.W. by South
	Cecias	N.E. by East	Africus	S.W. by West
	Subsolanus	East	Zephirus	West
	Vulturnus	S.E. by East	Corus	N.W. by West
	Euroauster	S.E. by South	Circius	N.W. by North
OUTER BAND	Greco	N. East	Lebeccio	S. West
	Syroccho	S. East	Magistralis	N. West

The back of the instrument is engraved around the edge with a degree scale reading to 1° in four quadrants, graduated in the upper-left and lower-right quadrants 90°–0° for measuring zenith distance, and in the upper-right and lower-left quadrants 0°–90° (from the east-west points) for measuring altitude. Within this is Gemma Frisius' universal projection (for which see below) with the positions of sixteen stars marked. There are three plates:

a. for latitudes 39° and 42°
b. for latitudes 45° and 48°
c. for latitude 51° and with a tablet of horizons

Each of the latitude plates is engraved with almucantars drawn for each degree, with the unequal-hour lines, and with lines marking the boundaries of the twelve astrological houses. The latitude of each plate is marked, as are the names of the tropics and the equatorial circle. Since the back of the instrument is occupied by the universal projection, the scales usually found there – the shadow square and the graph of the equal hours (used in conjunction with a graduated alidade to convert unequal to equal hours) – are added to the plate of the horizons. In the shadow square both *umbra recta* and *umbra versa* scales are divided 1 to 12 and 10 to 60. Like the *rete*, the alidade, rule, axis, and pin are modern replacements.
IC/CCA no. 3207.

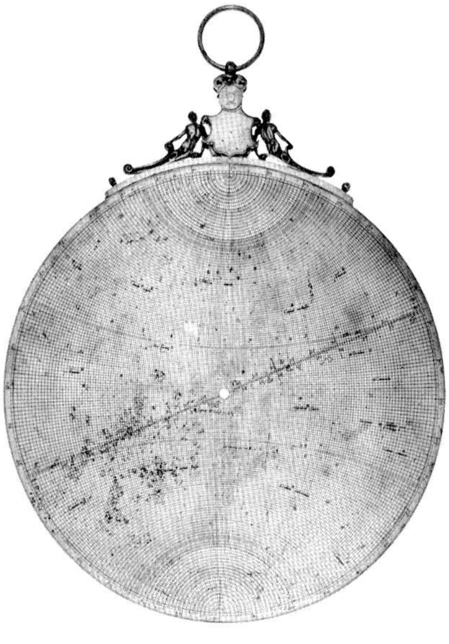

figure 119. Catalogue no. 18.
Back without alidade.

This instrument is an example of the general or 'catholic' astrolabe described by Gemma Frisius in his book *de Astrolabo*[284] [*sic*]

283. For the origins of this practice, see Taylor (I), 352–3.
284. See Ryan, 157, for the suggestion that this is not a printer's error but an attempt at classical purity of Latin diction on the part of Gemma, and the parallel example from George Valla cited by Segonds (II), 24, in support of the same hypothesis.

figure 120. Catalogue no. 18. Plates.

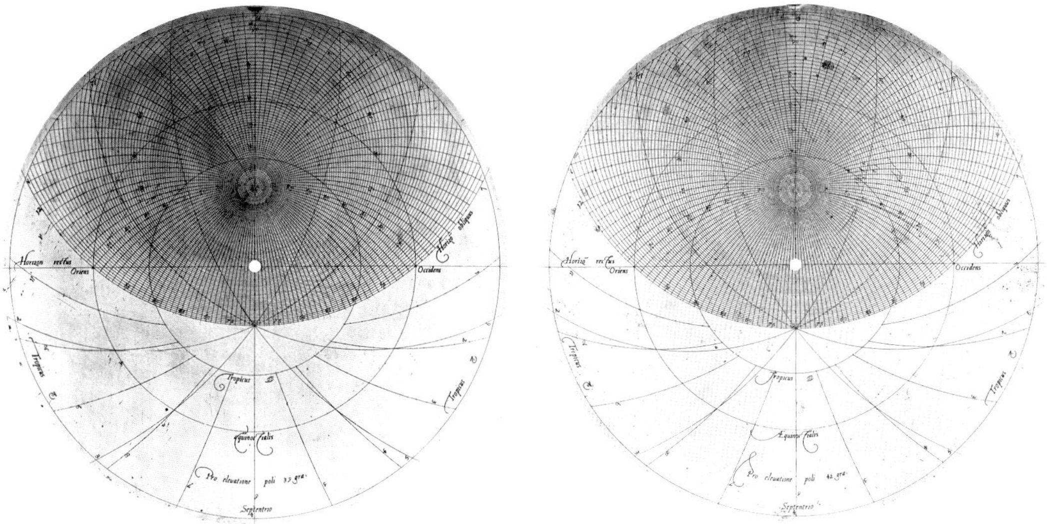

a. For latitudes 39° (left) and 42° (right).

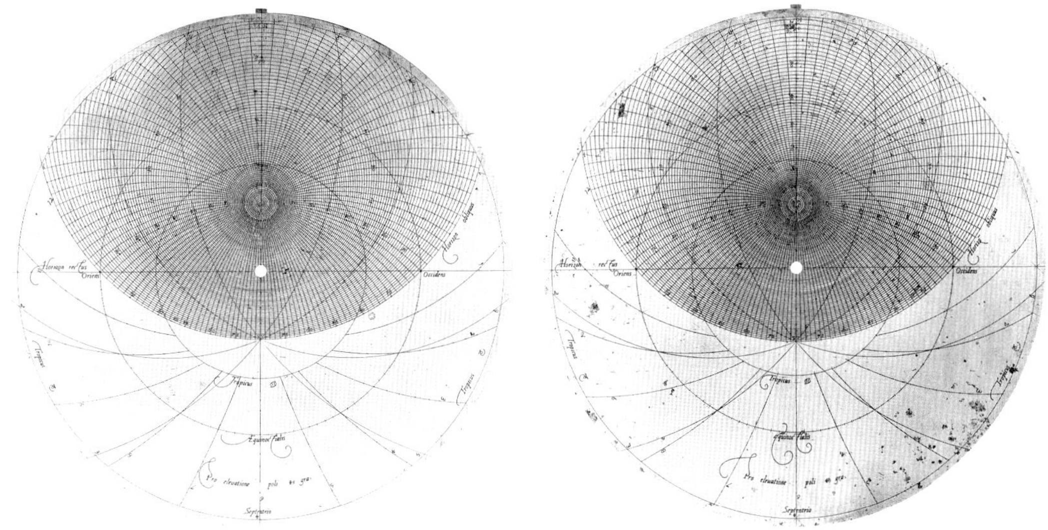

b. For latitudes 45° (left) and 48° (right).

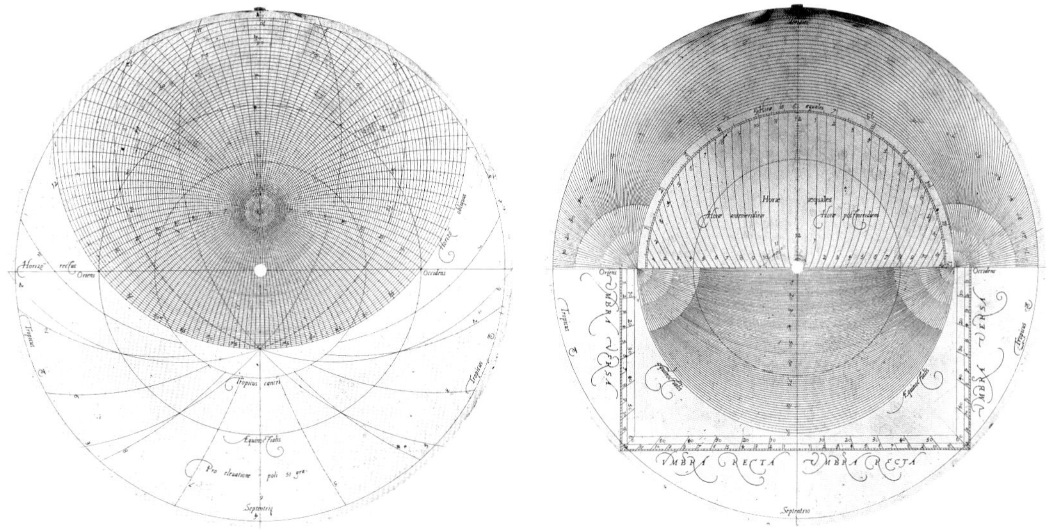

c. For latitude 51° (left) and a tablet of horizons.

Catholico Liber quo latissime patentis Instrumenti multiplex usus explicatur & quicquid vspiam rerum Mathematicarum tradi possit continetur (The Book of the General Astrolabe wherein is explained most amply the manifold use of the Instrument, and the instrument itself described and whatever has been delivered in any matter of Mathematics is contained) published posthumously at Antwerp in 1556. Gemma's intention was to create an instrument that not only contained the fixed-latitude and the universal (see below) forms of astrolabe, but was also equipped for a range of applications in navigation and surveying. It was for the benefit of navigators that he added the *Quadratum Nauticum* and graduated the instrument to read zenith distance as well as altitude, and for the benefit of surveyors that he explained how the instrument could be of use, not only for measuring heights and depth, but also for measuring horizontal angles and establishing the positions of buildings and places. Gemma's instrument, therefore, represents the culmination of efforts to universalize the astrolabe, not only in terms of the number of places in which it could be used, but also in terms of its functions.

Gemma seems to have begun the composition of his treatise on the general astrolabe between 1550 and 1554,[285] although the idea of the instrument itself may go back at least a decade earlier.[286] It was perhaps during this period that he gave instructions for the construction of the instrument to Gualterus Arsenius, whom he mentions in his treatise as making the astrolabes and as being his nephew.[287] An astrolabe which was perhaps the first fruit of this period survives and further documents the situation. It is signed 'Authore Gem[m]a Frisio et exaratu a Gaultero Arsenio Lovanij 1554' (by the originator Gemma Frisius and engraved by Walter Arsenius at Louvain, 1554).[288] The instrument seems to have been a success, and several examples have survived, although Gemma knew nothing of them since he died in 1555.

Of Gualterus Arsenius we know virtually nothing. Presumably it was thanks to his uncle that he was drawn into engraving and the manufacture of astronomical instruments. That he acquired his skills from the two engravers known to have worked with Gemma, Gaspar van der Heyden and Gerard Mercator, seems likely.[289] The calligraphic style of the latter certainly influenced him.[290] From differences of signatures on surviving instruments and in the accounts of the printer Christopher Plantin of Antwerp, who bought several instruments from Arsenius, it was long thought that there were two or even three brothers Arsenius. Surviving instruments by Arse-

285. Van Ortroy, 75.
286. *Ibid.*
287. 'Astrolaba quae nostro instinctu per nepotem nostrum Gualterum Arsenium contructa sunt...' (Astrolabe which at our instigation our nephew Walter Arsenius made...) fol. 14v–15r. Cited in van Ortroy, 100.
288. Fernandez Villar, 6.
289. de Smet (III), 35.
290. Osley, 91.

nius do indeed divide into two groups, one with signatures 'Gualterus Arsenius,' the other with 'Regnerus Arsenius'; the dates on the instruments in the two groups overlap. It is, however, probable that the two forms of signature were both used by Gualterus. Reynier was a family name of Gemma, who matriculated at the University of Louvain on 26 February 1526 as Gemma Reyneri.[291] An early astrolabe by Arsenius is signed 'Nepos Gemme Frisij faciebat Lovanij an[n]o 1557 GAR.' The final initials have been interpreted as G[ualterus] A[rsenius] R[eynerus],[292] giving the translation 'The nephew of Gemma Frisius Walter Arsenius Reyner made [it] at Louvain, 1557.' Gualterus, it is suggested, employed the name Reyner to underline his association with Gemma's family, in much the same way as did Jacob Cools, nephew of the geographer Ortelius, who on emigrating to England signed himself Ortelianus to signify relationship with his more-famous uncle.[293] Study of the calligraphy of instruments signed Gualterus and those signed Regnerus reveals no differences,[294] and the fact that Guiccardini in his *Descrittione . . . di Tutti Paesi Bassi . . .* (1567) commenting on the excellence of Arsenius' instrument, should refer to him as 'Gualteri Rennerio . . .' would seem to confirm that the two men are identical.[295]

The earliest dated instrument by Gualterus is the astrolabe of 1554 mentioned above. From the instruments made thereafter, a series of finely made astrolabes, armillary spheres, a cross-staff, and a few sundials have survived, the latest in date being an armillary sphere of 1575.[296] His instruments were highly esteemed, sold in places as far apart in Europe as London, Paris, and Madrid, and commanded relatively high prices. Large astrolabes in brass cost from 40 to 50 florins, smaller ones from 20 to 25 florins.[297] But of Arsenius himself nothing else is known except that his style influenced the whole succeeding generation of Flemish, Dutch, English, and Spanish instrument-makers.[298]

291. de Smet, (III), 38, n. 16.
292. *Ibid.* The signature is illustrated in Osley, pl. 45.
293. Michel (V), (suppl. III) col. 392.
294. 'The lettering on these instruments . . . is virtually indistinguishable,' Osley 91, n. 1.
295. Cited from van Ortroy, 102. Zinner (II), 237, records in the Landesmuseum, Stuttgart, a brass plate which is signed 'Nepotes Gemmae Frisij Lovanie fecerunt Anno 1579.' This signature perhaps arises from a joint work by Gualterus and Ferdinand Arsenius. Ferdinand, whose precise relationship with Gualterus is not known, signed himself as 'Ferdinandus Arsenius nepos Gemmae Frisij Lovanij fecit anno 1573,' on an astrolabe now in the National Maritime Museum, Greenwich (A54/53-16c). See N.M.M. 49.
296. Zinner (II), 236–9.
297. Van Ortroy, 103.
298. See above, pp. 48–9 and cat. no. 19. For a striking example of Arsenius' influence, see the anonymous manuscript astrolabe in the Settala Collection, Milan. Tomba (III), 636–41.

19. Italian Astrolabe

1573
Brass
Diameter 7 in. (178 mm)
Signed: Adrianus D[e]s[crolière]res facie[bat] Mantue 157[?3].[299]
Inventory 2292

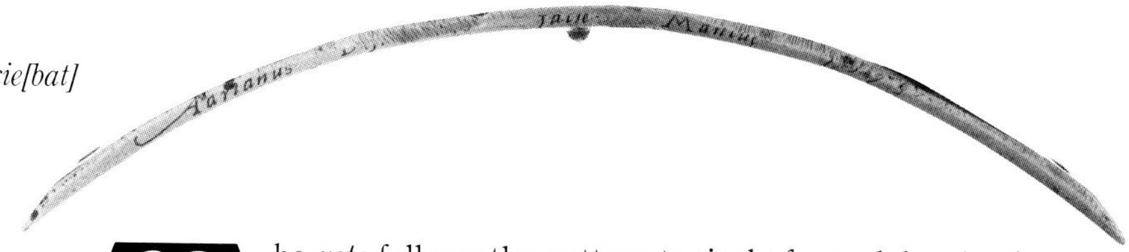

The *rete* follows the pattern typical of astrolabes by Arsenius (catalogue no. 18) with only minor differences. Twenty-two stars are marked with their magnitudes and astrological information. The ecliptic circle contains a calendar as well as the names of the signs of the zodiac (0° Aries = 10 March). The signs are divided to 1° in groups of five. On the reverse of the *rete,* at the base of all but two star pointers, is a number, perhaps indicating the stars' position in a list of astrolabe stars. The *mater* consists of a single sheet of metal to which the limb is rivetted. The suspension apparatus mount is screwed separately to the limb and consists of a base carrying two scrolls which support a central escutcheon originally containing a compass. This is surmounted by a shackle and swivel pin for the suspension ring. Engraved on the limb is an equal-hour scale (I–XII x 2) reading to four minutes, a degree scale (0°–90°–0°–90°–0°, reading from the east point), and sixteen wind names indicating compass directions. These are:

Ponente	Leuante
Ponente ma[estro]	Scirocco leu[ante]
Maestro	Scirocco
Maestro tra[montana]	Ostro scir[occo]
Tramontana	Ostro
Greco tra[montana]	Ostro garbi[no]
Greco	Garbino
Greco leu[ante]	Ponente gar[bino]

Engraved on the interior of the *mater* is a *quadratum nauticum* similar to those engraved on astrolabes by Arsenius (catalogue no. 18), and marking the same 'rhumbs of the wind,' except that *Xaloque* is replaced by *Scirocco, Mediojorno* by *Ostro*, and *Lebeche* by *Garbino*. The back of the instrument is engraved round the edge with a degree scale in four quadrants, graduated in the upper-left and lower-right quadrants 90°–0° for measuring zenith distance, and in the upper-right and lower-left quadrants 0°–90° for measuring altitude. Within this is a universal stereographic projection similar to that of Gemma Frisius and with the positions of fifteen named stars marked. There are five plates:

a. for latitudes 27° and 42°
b. for latitudes 30° and 33°

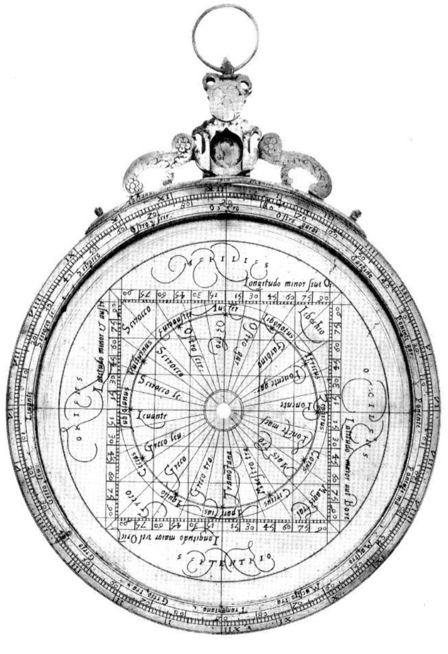

figure 121. Catalogue no. 19. *Mater.*

299. Since only the lower half of this figure survives, it could perhaps be a 5. A slight difference of curvature compared with the first 5 of the date suggests, however, that it is indeed a 3.

figure 122. Catalogue no. 19:
Astrolabe by Adrian Descrolières. Front.

figure 123: Catalogue no. 19. Back.

c. for latitudes 36° and 39°

d. for latitudes 45° and 48°

e. for latitude 51° and with a tablet of horizons, shadow square, and equal-hour diagram

Each of the nine latitude plates is engraved with almucantars for every two degrees and has the twelve astrological houses marked. The alidade is engraved with scales for Italian and Babylonian hours, and with declination scales. The sights have both pin-holes and slits. The rule for use with the universal projection lacks its cursor and *Brachiolus*. It is brazed to a bolt with which the instrument is assembled using a wing nut. On the edge, on either side of the suspension piece, the words 'ASTROLABIUM... RAIAS' are engraved, although for what reason is not clear. Presumably they refer to the universal orthographic projection applied to astrolabes by Juan de Rojas, but this is not the projection used on this instrument.

Provenance: Greppin Collection, Brussels.[300]

Linton Collection, New York.[301]

IC/CCA no. 455.

Adrien Descrolières (*fl.* 1571–80) worked, according to the signatures of his surviving instruments, at Venice, Mantua, Antwerp, and Paris. Little else is known about him except that, as the *rete*, *quadratum nauticum*, universal projection, and Mercatorian calligraphy of the present instrument show, he was heavily influenced by the instruments of Arsenius and was perhaps an apprentice or workman in the Louvain workshops. In addition to four other astrolabes, a dialling instrument and a large astrolabe-quadrant by him are known.[302]

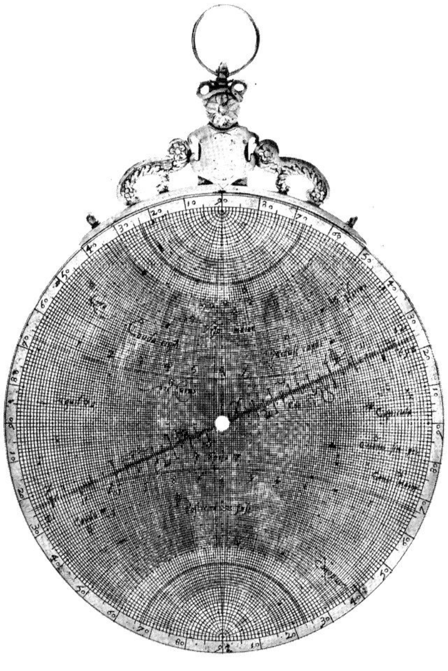

figure 124. Catalogue no. 19. *Saphea Azarchelis.*

300. Jordan, 3133.

301. *Instruments Scientifiques*, 188.

302. The astrolabes and dialling instrument are listed by Webster & Webster, 51; for the astrolabe quadrant, see Guillevic, 150–1; no. 131.

figure 125. Catalogue no. 19. *Rete.* Front (left) and back (right).

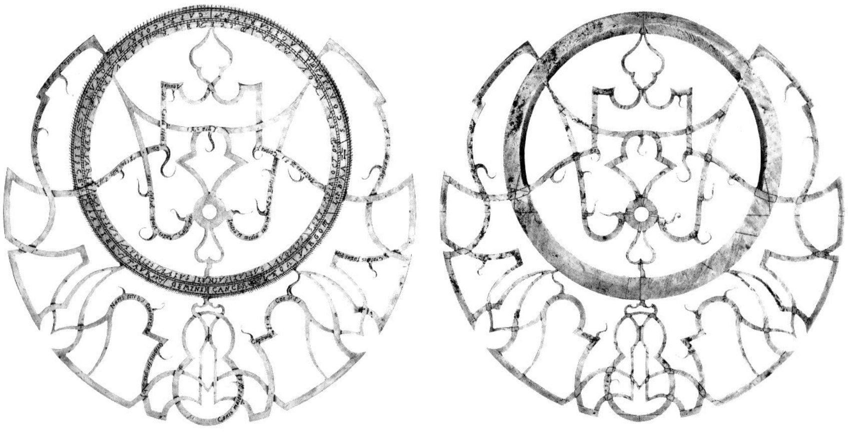

figure 126. Catalogue no. 19. Plates.

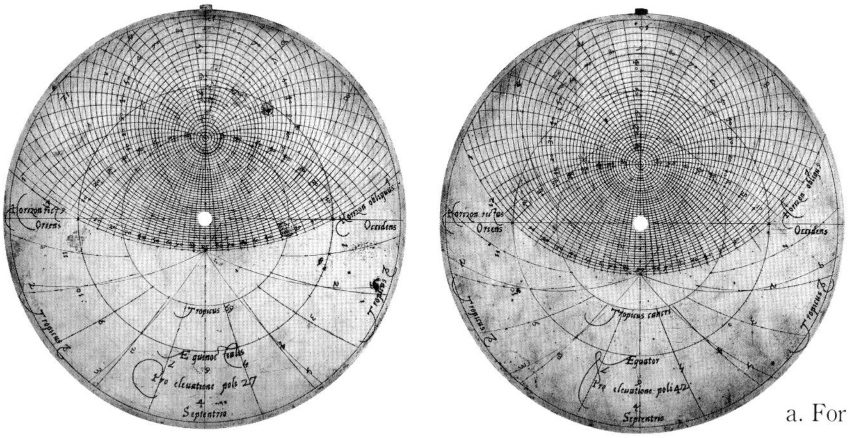

a. For latitudes 27° (left) and 42° (right).

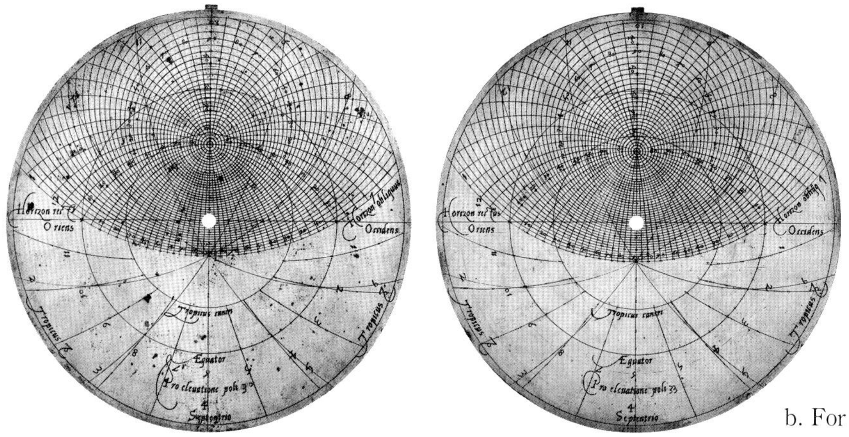

b. For latitudes 30° (left) and 33° (right).

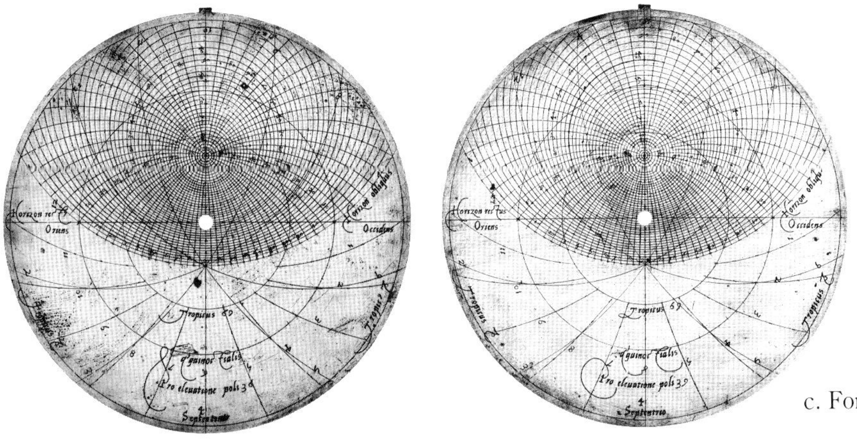

c. For latitudes 36° (left) and 39° (right).

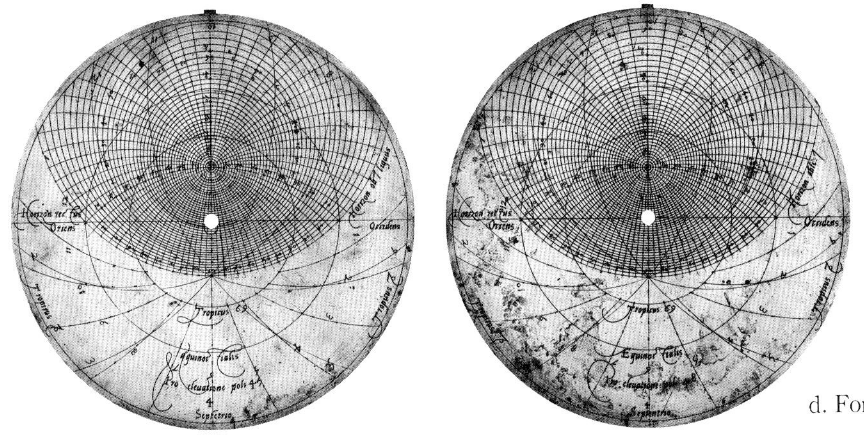

d. For latitudes 45° (left) and 48° (right).

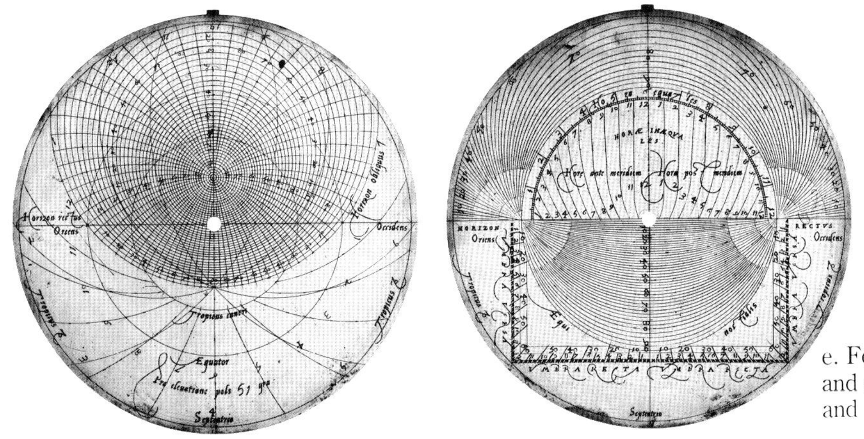

e. For latitude 51° (left)
and a tablet of horizons, shadow square,
and equal-hour diagram (right).

20. ?French Astrolabe

Not dated.
Gilt brass, with case of gold-tooled red morocco
Diameter 3¼ in. (83 mm)
Not signed
Inventory 1438

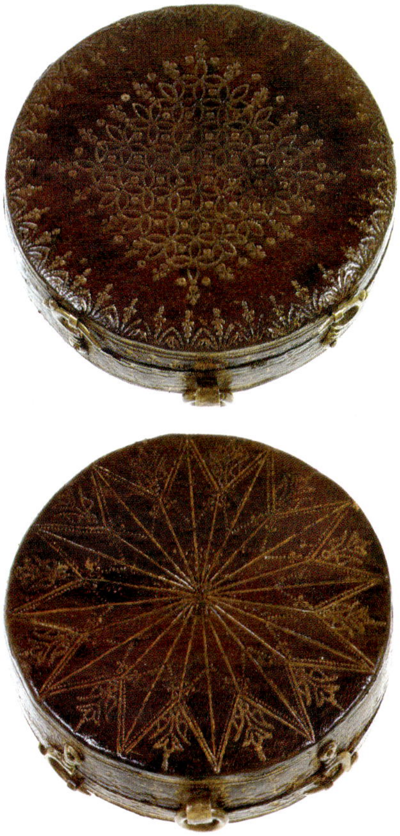

figure 127. Catalogue no. 20. Top and bottom of box.

The *rete* has nine named stars. For the last of these there is no pointer. There is a straight east-west bar, and an ecliptic zodiac band numbered in tens of degrees and marked only by the signs. The *mater* is cast in one piece with a fixed shackle for the suspension piece pinned in. Engraved on the limb is an hour scale (XII–XII x 2) and a degree scale reading to 2° in four quadrants, starting at the east and west points (0°–90°–0°). On the back is a degree scale in four quadrants as on the limb. Within this is a zodiac calendar scale (0° Aries = 21 March), the signs of the zodiac being marked beside the names. Pisces is misspelt 'Pices,' and there are no suspension marks for 'Sagitar,' 'Capri,' 'Aquari' or for abbreviated month names. In the centre is a shadow square and above this an unequal-hour diagram. There is one plate with almucantars drawn for every 5° for latitudes ?45° and ?48/9°. The alidade and rule are plain and fixed in place with a nut and bolt. The red morocco case has brass clasps, hinge and suspension piece, and is decorated with gold-tooled fleur-de-lis and geometrical patterns. It is lined with green velvet.

This is a poorly made instrument of which the authenticity has been questioned.

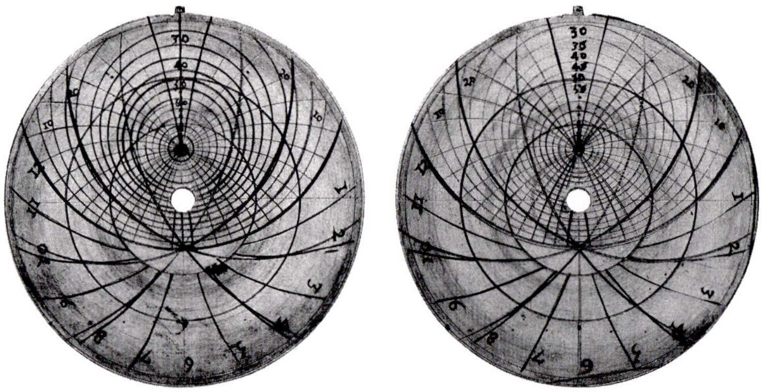

figure 128. Catalogue no. 20. Plate. For latitudes ?45° (left) and ?48° (right).

figure 129. Catalogue no. 20: French astrolabe. Front.

figure 130. Catalogue no. 20. Back.

21. English Astrolabe

?2nd quarter of 20th century
Paper on wood, with brass suspension mount and ring, and cardboard alidade and rule
Diameter 8½ in. (216 mm)
Not signed
Inventory 3171

The *rete* marks the positions of eighteen named stars indicated by squat triangular pointers. The *rete* has no decoration but is unusual in having a full scale of degrees (0°–360°) marked around the circumference (= tropic of Capricorn). The limb is marked with an hour scale (I–XII x 2) and within this is a scale of degrees in four quadrants (0°–90°–0°–90°–0°, from the east point). The back carries a scale of degrees (as on the front) and within this a concentric-type zodiacal calendar (0° Aries = 21 March). The divisions of the ecliptic signs are indicated by Roman numerals, those of the days of the month by Arabic numerals. Within these scales, in the upper centre, is an unequal-hour diagram, and in the lower centre, a double shadow square. There are no plates but the interior of the *mater* is drawn as an astrolabe for latitude 51° or 52°. The rule has a scale of declinations marked on it. The alidade, marked with a scale 1–6, lacks both sights.

This astrolabe is eclectic in style and was clearly drawn by an *amateur* inspired by earlier instruments. The design of the *rete* recalls that of the printed-paper astrolabes of John Prujean (*fl. c.* 1646–1692), who was probably an apprentice of Elias Allen[303] and who worked in Oxford from 1664 onwards.[304] On Prujean's astrolabes one finds similar squat triangular star pointers and the marking of a degree scale around the circumference of the *rete*. In other respects, however, the *rete* of the present example is greatly simplified in comparison with those of Prujean.[305] The back of the instrument is very closely modelled on the astrolabes of Georg Hartmann (catalogue no.17).[306] Particularly striking is the similarity of the suspension piece, where the maker has not only drawn in two circles such as are found on Hartmann's astrolabes (although filled with floral decoration), but also a decorative foot (?preparatory to cutting round it) which is virtually identical with that on Hartmann's instruments.

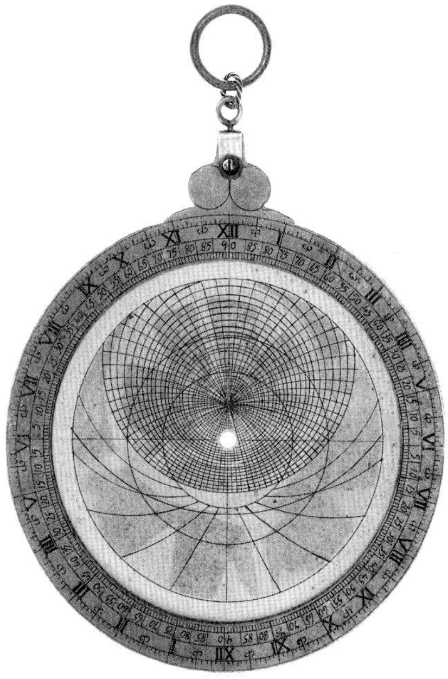

figure 131. Catalogue no. 21. Plate. For latitudes 51°.

303. Reasons for this suggestion are given in Turner (III), 120 n. 86.
304. Gunther (V), ii 519–20.
305. For an astrolabe by Prujean, see Gunther (I), facing 220.
306. For several other astrolabes by Hartmann, see Gunther (V), ii 436–40. Pl. CVIII shows an example which shares the peculiarity of both the words 'Umbra' and 'Versa' on the shadow square reading from the left.

figure 132. Catalogue no. 21:
English astrolabe. Front.

figure 133. Catalogue no. 21.
Back.

UNIVERSAL ASTROLABES

The planispheric astrolabe, as it was transmitted from Islam to Christian Europe from the 10th century A.D. onward, was an instrument which, although perfected in theory, suffered from several practical disadvantages. Chief among them was the need for a different plate for each latitude; this meant that it was impossible for the instrument to cover more than a limited range of places without becoming unmanageably heavy. Such a multiplication of plates also meant that instruments became expensive. But astrolabes needed to be large for ease in dividing and for legibility, so the point when an astrolabe became unmanageable and too expensive was soon reached. Thus, users of the astrolabe were forced to make an unsatisfactory compromise between ease of use and portability. These problems, combining perhaps with a natural inclination of mathematical scholars to explore all the possibilities of stereographic projection, led in the A.D. 11th century to the appearance of universal forms of astrolabe which greatly simplified the physical form, if complicating in some cases its manner of use.

The necessity of different plates for different latitudes can be overcome if the centre of projection is moved from the celestial pole to the vernal equinox and the plane which passes through the solstices perpendicular to the equator is taken for the plane of the projection. The resulting projection (fig. 134) gives a vertical section, from pole to pole, of the celestial sphere, rather than the horizontal section given by ordinary polar stereographic projection. The poles

figure 134. Universal stereographic projection.

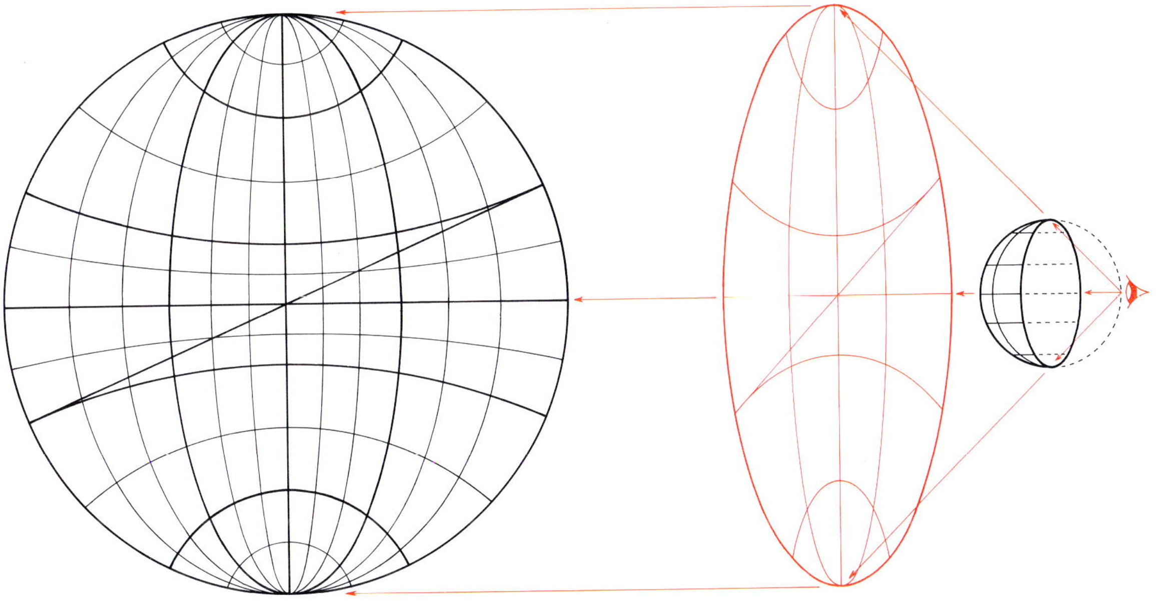

are situated at the top and bottom of the instrument and are joined by arcs between them which serve either as meridians or as hour lines. The equator is projected as a straight diameter across the centre of the instrument, being flanked on each side by arcs of circles representing parallels of latitude which cut the meridians at right angles. The ecliptic is represented by a diametral straight line running between the extremities of the two tropics and cutting the equator at an angle equal to that of the obliquity of the ecliptic.

Of the exact origin of universal stereographic projection in this form it is not yet possible to be certain. That it may have originated in the course of investigations of the horizon plate is suggested by the existence of an astrolabe plate engraved on its two sides with nine quarters of astrolabe plates similar in arrangement to the tablet of horizons used in eastern Islam;[307] by the fact that studies of the tablet of horizons were written in both the A.D. 9th and 10th centuries;[308] and by the name 'universal horizon' given to one of the forms of universal astrolabe developed during the 11th century. At Toledo two forms of the instrument were developed independently by Abū-l-Hasan 'Alī b. Khalaf b. Aḥmar and by Ibn az-Zarqellu. On the relationship between the two men, who were contemporaries, there is no information, and 'Alī b. Khalaf is very little known. Az-Zarqellu, by contrast, was revered even in his own lifetime as the leading astronomer of his day, being described by Ṣā 'id al-Andalusī as '. . . the man of our century most versed in astronomical observations, in the knowledge of the nature of the celestial spheres, and in the calculation of stellar motions, he who understands best astronomical tables and the methods of constructing astronomical instruments.'[309]

In the treatise he wrote on his universal astrolabe, 'Alī b. Khalaf records that while he was considering how to avoid the drudgery of making several plates for an astrolabe, '. . . it happened that I understood how might be made an instrument for the whole world, that would not have in it more than one plate and one *rete*. And I gave it the name of the *universal horizon*, and put it up for my lord the king Ma'mūn [of Toledo, Yaḥyā-Ma'mūn b. Ismā'īl, reigned A.H. 429–67 (A.D. 1037–74)] and made this book. . . .'[310] The instrument that he there describes consisted of a single plate engraved with a universal projection of the sphere (as described above), over which rotated a *rete* divided into two halves. One half was a star map of conventional form but bounded by the equator (not the tropic of Capricorn) and with the ecliptic represented by an arc subtended by a diameter of the *rete*. The other half of the *rete* was made up of a net of meridians

307. Now in the Berlin Staatsbibliotek. See Maddison & Turner, 115 & no. 54 and *cf.* 148 n. 150.
308. By Habash at Damascus (not extant) and by as-Sijzī (Ms Damascus Zahirya 9255). See Morley (I), 7, n. 12 & King (IV), 255.
309. Maddison & Turner, 125. For az-Zarqellu's life and work, see Millás (III), and more briefly J. Vernet in *D.S.B.*, xiv, 592–5.
310. The treatise is extant only in the Castilian translation included in the *Libros del Saber*, *c.* 1276–7. Cited from Maddison & Turner, 124.

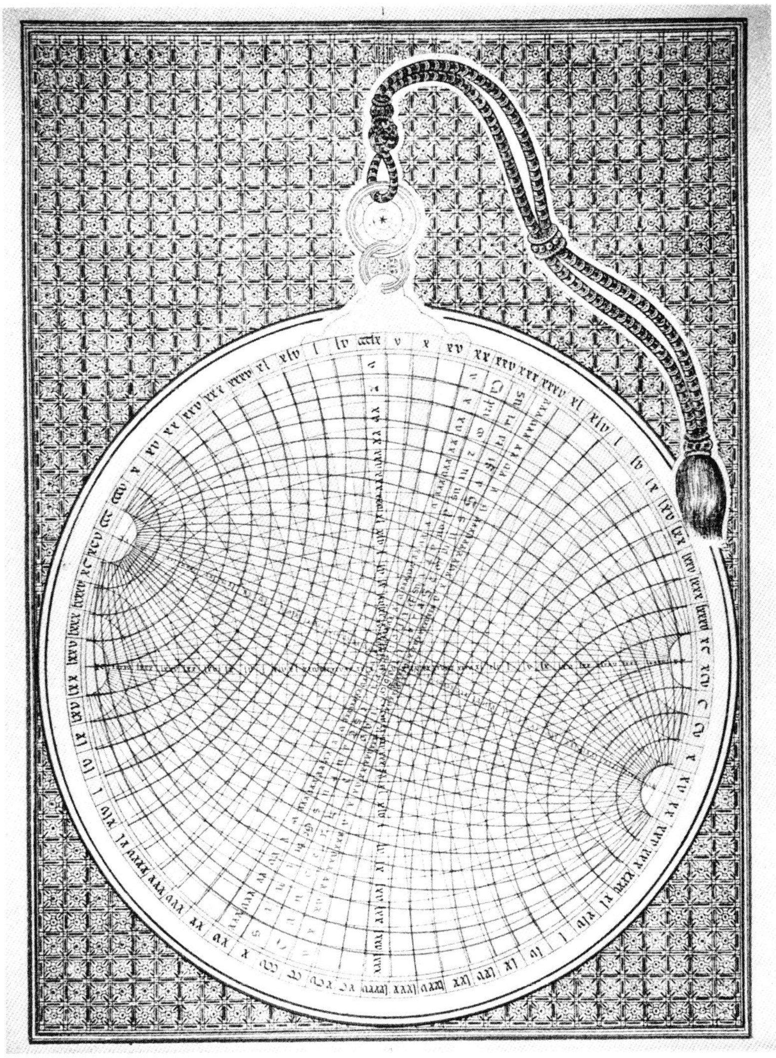

figure 135. *Saphea Azarchelis* from the *Libros del Saber*. 19th-century drawing after the manuscript. From Rico y Sinobas, iii, facing 148. (Photograph: Museum of the History of Science, Oxford)

and parallels similar to those of the universal projection on the plate beneath (figs. 135 & 138). In the hands of az-Zarqellu, however, simplification went even further, the *rete* being abandoned and the instrument reduced to a single plate. This was done by engraving on the plate the celestial co-ordinates for both the ecliptic and the equatorial pole axes and marking the positions of a number of stars. On the reverse, the degree scales, zodiacal calendar, sine diagram, and shadow square usually placed on the back of astrolabes were engraved, together with a number of diagrams peculiar to az-Zarqellu himself.[311] In addition, the instrument was fitted with both a diametral ostensor or rule (on the back) and (on the front) an alidade with cursor – that is, a short straight-edge which could slide along the rule making a right angle with it.

Although 'Alī b. Khalaf's form of the universal astrolabe (conventionally referred to as *lamina universal*) has usually been considered

311. Maddison & Turner, 124; Maddison, 24–6; Millás (I), *passim*.

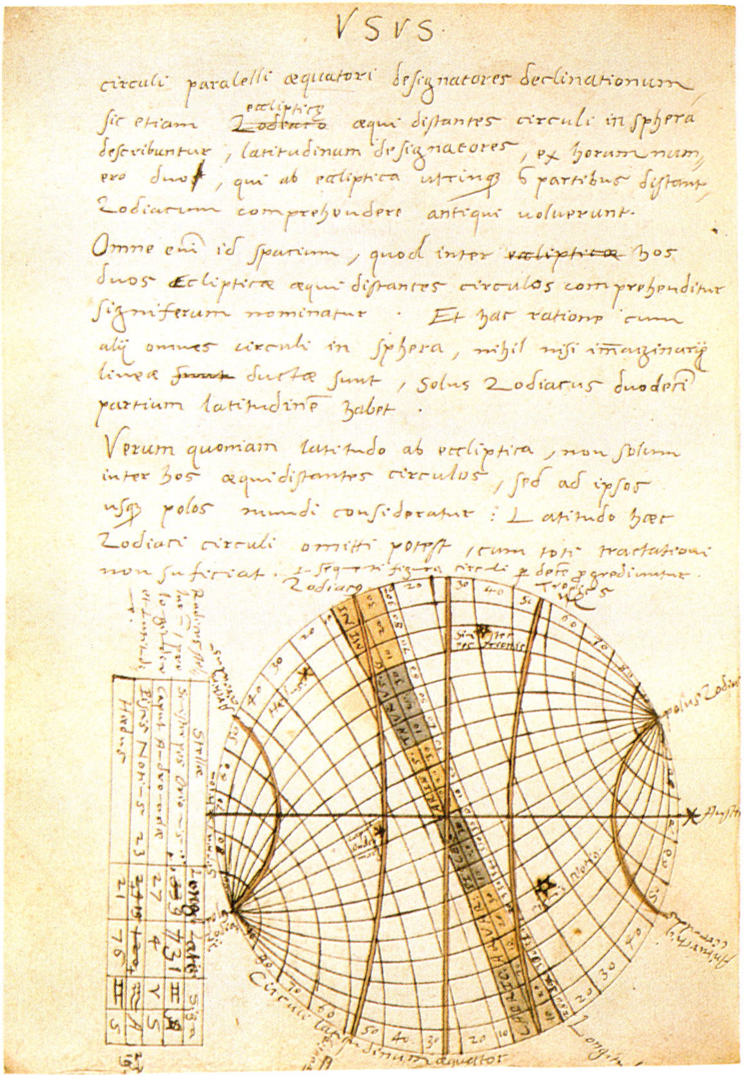

figure 136. A *saphea Azarchelis*, probably drawn by Sebastian Fabritius in 1551 to illustrate his copy of the lectures given in 1550 and 1551 by Josias Simler at Zurich, on the making and use of a planisphere. London, British Library Add. Ms. 24010 fol. 37v. (Photograph: British Library)

the earlier of the two, remarks in some recently examined manuscripts suggest that the development of the two forms may have been virtually simultaneous. Thus, the anonymous Egyptian author of a treatise on time-keeping written in the 14th century and citing Ṣā'id al-Andalusī states that az-Zarqellu wrote a treatise in one-hundred chapters on the instrument called the *zarqālliya,* which he invented *c.* A.H. 440 (A.D. 1048/9), while an instrument called *al-aṣṭurlāb al-Ma-mūnī* 'which had a universal horizon' was presented to al-Ma'mūn of Toledo by 'Alī b. Khalaf in A.H. 464 (A.D. 1071/2).[312] If these remarks are accurately reported, then it may have been az-Zarqellu who first developed a form of universal astrolabe which was further developed by 'Alī b. Khalaf before being given its final single-plate form by az-Zarqellu.[313] Since, however,

312. Ms Leiden Universiteitsbibliotheek 468 fol. 90v, King (IV), 250–2.
313. Although the passage quoted by King (IV), 248 from Ms Escorial ar. 962, ff 81v–82r may confirm this, as there translated it is not without ambiguity.

there is no need to suppose that the instrument presented to al-Ma'mūn by 'Alī b. Khalaf was newly invented, the traditional sequence of events may well be equally correct.[314]

The typical form of az-Zarqellu's instrument – usually known by its Latin name, *saphea Azarchelis* – is shown in figs. 135 & 136. Apart from not clearly demonstrating the rotation of the celestial spheres around the pole, the capabilities of the instrument were similar to those of the fixed-latitude instrument, augmented, thanks to the two sets of co-ordinates marked upon it, by a much greater facility for converting from ecliptic to equatorial co-ordinates and vice versa. Thus to convert from an equatorial to an ecliptic co-ordinate it was only necessary to set the rule along the equator and move the cursor to mark the point of intersection of the lines representing the given declination and right ascension. Rule and cursor were then rotated until the rule lay along the ecliptic; the cursor would then mark the corresponding position in ecliptic co-ordinates, i.e. celestial latitude and longitude.[315]

A further idea of how the instrument was used is given by the procedure for finding the hour. First the observer measured the altitude of the sun (or at night one of the stars marked on the astrolabe) by means of the alidade and the degree scale engraved on the back of the instrument. Turning the astrolabe over, he set the rule to the latitude of the place of observation against the degree scale on the edge. When thus set, the position of the rule represented the horizon of the place of observation. Having found, from the zodiacal calendar on the back, the sun's position in the ecliptic for the day of observation, the cursor was slid along the rule until the graduation on the cursor corresponding to the observed altitude of the sun met the parallel representing the sun's position in the ecliptic. The meridian of the equatorial pole that passed through this point (and which is equivalent to an hour arc) then indicated the time. If a star had been observed, the cursor was moved until the graduation corresponding to the star's observed altitude met one parallel of the star as marked on the instrument, the time being read off from the meridian passing through this point.

Although never rivalling the fixed-latitude instrument in the number of treatises devoted to it, the *saphea Azarchelis*, seems to have been reasonably popular. A few examples of the instrument have been preserved from later medieval Islam,[316] and az-Zarqellu himself wrote three accounts of it (works which are distinguished by their number of chapters: sixty, eighty, and one hundred).[317] Of

314. North (IV), ii, 190, discussing a universal astrolabe by 'Alī b. Ibrahīm al-Harrār A.H. 728 (A.D. 1327/8) now in the Museum of the History of Science, Oxford, (Mayer (II), ii, 194), shows that although it uses the same universal projection, it is not the same as either az-Zarqellu's or 'Alī b. Khalaf's astrolabes. North further suggests that the instrument, which uses a *rete* which, if folded, gives the *rete* of 'Alī b. Khalaf's instrument, may be of an earlier type. This, however, in the light of 'Alī's account of how he invented his device, seems unlikely.
315. For a more detailed account, see Poulle (VIII), 10 and King (I).
316. These are listed below, pp. 177–8.
317. Millás (IV), 114; King (IV), 253.

these, the versions in sixty and one-hundred chapters are well-known through later translations. That in eighty chapters, which may have been presented to al-Ma'mūn, seems to have had very little circulation. The shortest treatise, in sixty chapters, provided the basis for a Latin work on the *saphea* by William the Englishman, probably working with Yehuda bar Mosé, in 1230–31,[318] and this work may have been known to Pierre de Maricourt (Petrus Peregrinus), who draws a comparison between Ptolemy's and az-Zarqellu's projections in his treatise on the astrolabe.[319] The whole of az-Zarqellu's work was jointly translated by Ibn Tibbon (Prophatius Judaeus) and John of Brescia in 1267.[320] Meanwhile, the version in one-hundred chapters was translated into Castilian by Bernardo al-Aráuigo and Abraham su Alfaquī[321] for inclusion in the *Libros del Saber*. Since the *Libros del Saber* were translated into Italian in 1341,[322] two vernacular versions of az-Zarqellu's treatise were thus available by the mid-14th century. The *saphea* was equally not forgotten in Islam. The 13th-century Abū-l-Ḥasan al-Marrākushī, writing at Cairo, included a description of it in his account of contemporary astronomical instruments,[323] and there were several later treatises in Persian and Turkish as well as Arabic.[324] In Christian Europe the device also continued to attract attention, in particular from John of Lignères,[325] who in the early 14th century devised a form of the instrument with *rete*, but quite different from that described by 'Alī b. Khalaf.[326] That John could have known of 'Alī b. Khalaf's instrument is possible, for he was a leading figure in the small group of scholars at Paris who introduced the Alphonsine tables to the Latin West and was, therefore, more likely than most European scholars to have seen the account of the *lamina universal*, as 'Alī's instrument became known, contained in the Alphonsine corpus.[327] Both the form of the *saphea* described by az-Zarqellu, with a simple rule, and that described by John of Lignères, with a sort of *rete (circulus mobilis)* and a rule, may have continued to be known in Europe, the term *saphea* being applied indifferently to each. A treatise on the *saphea* was ascribed by Leland to the little-known William Batecumbe,[328] and a *saphea* diagram is engraved on

318. Bibl. Nat., Paris, Ms Lat 7195 f. 89; Sédillot, 185–190.
319. Poulle (X), 496, who discusses William's instrument in some detail, showing its differences from other forms of the instrument.
320. Sédillot, 190. It is edited by Millás (II), who adds a translation of Prophatius' Hebrew version. For Prophatius, see below.
321. Millás (IV), 114; Rico y Sinobas, iii, 135.
322. Maddison (III), 23 & n. 41; Knecht, *passim*.
323. Sédillot, 183–91.
324. King (IV), 246 & n. 7.
325. *Fl.c.* 1320–35, still alive in 1350, dead by 1355. For what is known of his life and work, see Emmanuel Poulle in *D.S.B.*, vii, 122–8.
326. Bibl. Nat., Paris, Ms Lat 7295. Sédillot, 188 n. 1, Poulle (X), 499.
327. Poulle (X), 499–502, who describes John's instrument in detail. It is interesting to note that at the same time the *lamina universal* was being studied in Aleppo by Aḥmad b. as-Sarrāj, who in A.H. 729 (A.D. 1328–9) made an example, which has survived, for Muḥammad b. Muḥammad aṭ-Ṭanūkhi. It is now in the Benaki Museum, Athens. Maddison & Turner no. 61, and for details, King (I) and (IV), 126. Mayer (I), 34.
328. North (V), 279–80, & n. 52.

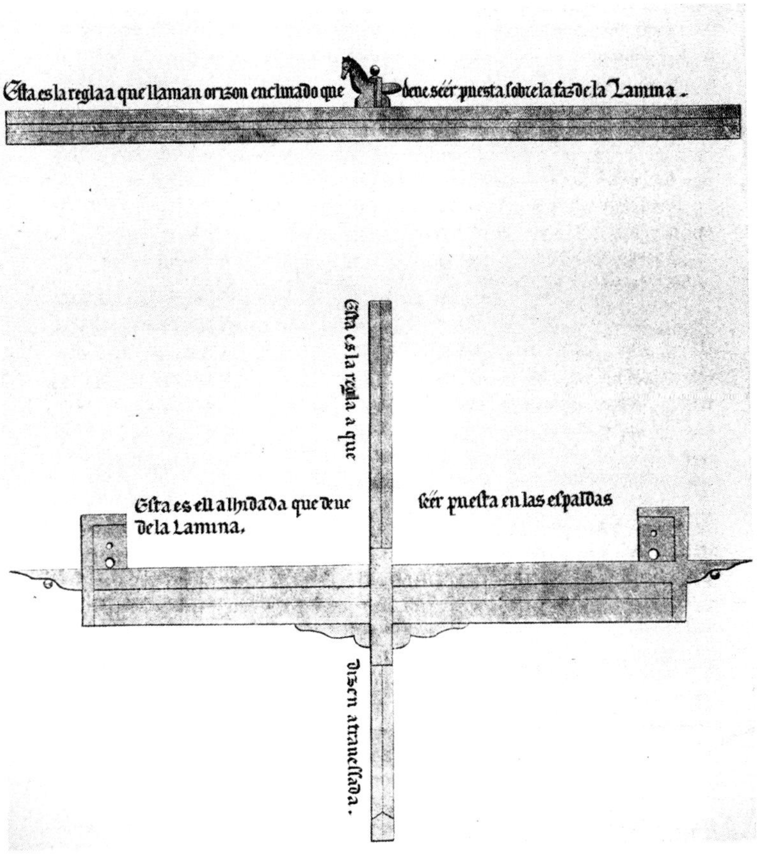

figure 137. Rule and alidade for a *saphea Azarchelis*. 19th-century drawing after the manuscript. From Rico y Sinobas, iii, 145. (Photograph: Museum of the History of Science, Oxford)

the *mater* of a composite late-Gothic astrolabe of *c.* 1350.[329] Evidently such instruments were used, and it is found listed, for example, among the instruments for instruction in the college of Maître Gervais, founded at Paris by Charles V.[330]

Influential, perhaps, in disseminating knowledge of the instrument was the inclusion of a *saphea* by Richard of Wallingford in his composite astronomical instrument the Albion.[331] The description he gave of it in his treatise on this instrument was very brief, however, and both Simon Tunsted (d. 1369) and John of Gmunden (*c.* 1380/84–1442) felt obliged to expand it.[332] We may therefore assume that it was thanks to the latter's lectures on the astrolabe and the Albion given at Vienna that knowledge of the universal projection was disseminated in central Europe. Certainly the projection was well-known there,[333] and the anonymous author of a treatise sometimes attributed to Regiomontanus was perhaps the first to describe the use of an articulated arm *(Brachiolus)* that could be attached to the *cursor* to facilitate the marking of a position on the curved meridians and parallels. He also suggested the use of a *rete*

329. In the Museum of the History of Science, Oxford. See Maddison (I), 175a, 32. A *saphea* plate is also included in a late Gothic astrolabe in the Civici Musei d'Arte Antica, Milan (see Tomba (I), 303), although it is probably a later addition (Poulle (IX), 162 and *cf.* Tomba (II)).
330. Wickersheimer, 4; Poulle (X), 502.
331. North (IV), i, 330–3. Poulle (X), 498–9, who points out an echo of William the Englishman's treatise in Richard's work.
332. North (IV), ii, 191.
333. North (I), 59, n. 9.

of transparent horn marked, like the *rete* of the *lamina universal*, with a co-ordinate grid similar to that engraved on the plate below.[334] Two manuscripts about the *saphea* belonged to Regiomontanus; a supplement to one of them described John of Lignères form of the instrument. [335] It was also known to Regiomontanus' pupil, Hans Dorn, who included a *saphea* in the astrolabe that he made in 1486 for Martin Bylica of Olkusz.[336] The following decades saw considerable attention paid to the projection. Johannes Volmar wrote a description of the projection and owned a manuscript that described a *rete* similar to that of John of Lignères.[337] In 1504, Jacob Ziegler imagined a form of the instrument using threads instead of a rule or a *rete*.[338] Johan Werner (1468–1522) gave a detailed account of the theory of the projection and applied it to the special use of graphically solving problems involving spherical triangles. His unpublished work on this was copied by Peter Apian in his 'Meteoroscope.'[339] In 1525, Johann Schöner published a description of the instrument[340] while, perhaps a little earlier, Andreas Stibonius (Stöberl, *fl.* 1497–1514) had written a comprehensive survey of the different forms of the instrument drawing on the work of his predecessors, whom he lists. They were az-Zarqellu, Profatius, John of Lignères, an anonymous author, Johannes de Machlinea, and George Peurbach. No works on the *saphea* by the last two have yet been found.[341] In 1532, Oronce Fine described an application of the diagram to cartography,[342] and in 1550, Josias Simler included a *saphea* in a 'sphere' he had designed.[343] It is clear, therefore, that Gemma Frisius was not original when he included a *saphea* in the composite instrument which he called the *astrolabum* [*sic*] *catholicon* in the treatise he began to write about it probably *c.* 1552–4 and which was published posthumously by his son, in 1556.[344]

Thereafter, the projection was widely used on the astrolabes made by Gemma's nephew, Gualterus Arsenius, and by the makers inspired by him.[345] Of these, Adrian Zeelst[346] and Gerard Stempel produced a treatise on the astrolabe which included a detailed description of the universal instrument and which passed through four editions between 1602 and 1629, while an interesting development was made by the English *amateur* John Blagrave.

334. Poulle (X), 504–5.
335. *Ibid.*, 506.
336. Copernic, no. 71; Zinner (II), 296.
337. Poulle (X), 505.
338. *Ibid.*
339. North (I), *passim*.
340. Zinner (I), no. 1304, 165. It was reissued in 1534. *Ibid*, no. 1577, 183. I have not been able to see a copy of either edition.
341. Poulle (X), 507. One might wonder, however, if one or both of the anonymous manuscripts that belonged to Regiomontanus may have been the treatise of Peurbach.
342. Fine, fols. 87r–v.
343. B.M. Add Ms 24010 ff 24–56. Poulle (X), 504, n. 27.
344. Van Ortroy, nos. 129, 130 and see above cat. nos. 18 & 19. The Islamic antecedents of Gemma's instrument were, however, pointed out by Egnatio Danti in 1569. *Ibid.*, 77, n. 3.
345. See above, pp.134 & 136.
346. *Utriusque Astrolabii tam particularis quam universalis fabrica et Usus sine ullius retis aut dorsi adminiculo*. . . , Leiden, 1602. An astrolabe signed by Zeelst in 1569 which includes Gemma's universal projection is in the Museo Arqueológico, Madrid. See García, 207–9.

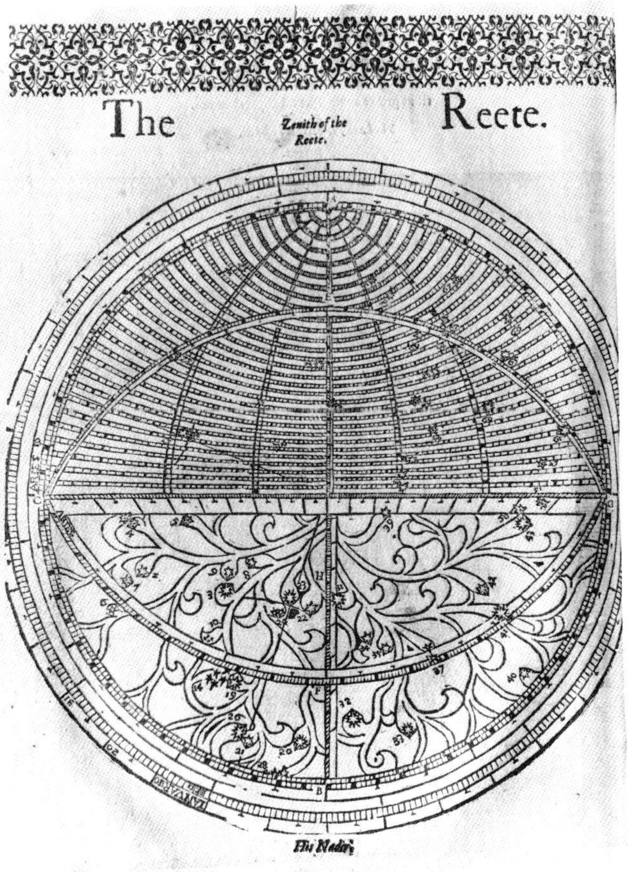

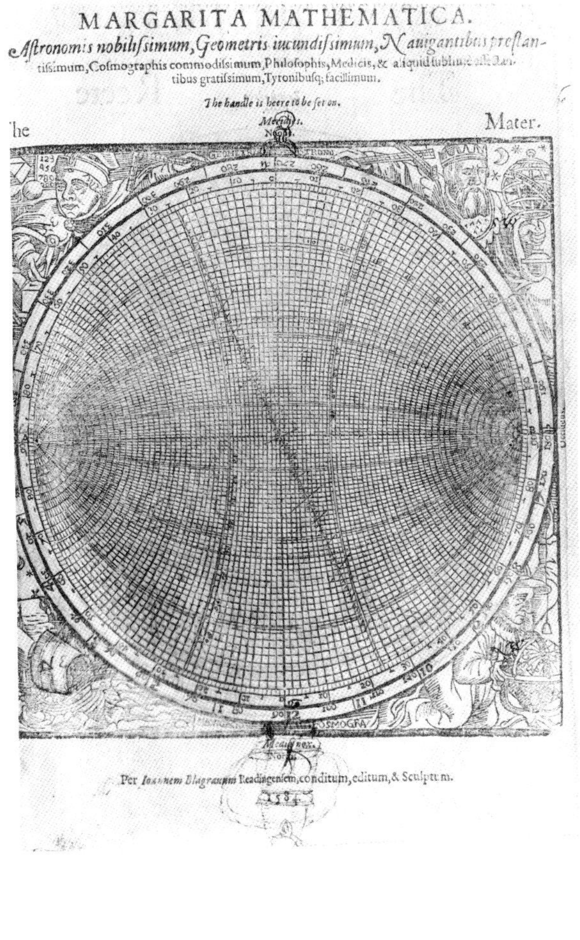

figure 138. *Rete* (left) and plate (bottom left) for a universal astrolabe. From John Blagrave, *The Mathematical Iewel...*, London, 1585.

figure 139 (opposite). Universal astrolabe. From Juan de Rojas, *Commentarium in Astrolabium quod planispherium vocant*... (2nd edit.), Paris, 1551, 278.

From an attentive reading of Stöffler and Gemma, both of whom he constantly acknowledges, praises, and criticizes,[347] Blagrave virtually reinvented the *lamina universal* of 'Alī b. Khalaf (fig. 138). Khalaf's instrument he called the 'Mathematical Iewel,' and it and his book describing it enjoyed some popularity in England. Three examples of the instrument have survived.[348] It was described at length by Thomas Blundeville in the fifth book of his *Exercises*... (1594). It was given a further lease of life in 1658 when the scientific publisher and cartographer Joseph Moxon published a new account of the device written by John Palmer, presumably to accompany a 17-inch version of the instrument printed on paper that Moxon had published.[349] Palmer's account was far briefer and a great deal clearer than Blagrave's own description, and it led John Aubrey to recommend the instrument as suitable for schoolboys.[350] The Palmer account remains even today perhaps the best account in English of the universal astrolabe. It was, however, also the last, and like the astrolabe for single latitudes, the stereographic universal astrolabe quietly disappeared from use during the final decades of the 17th century.

Although forms of universal astrolabe based on stereographic projection were the most widely used during the Middle Ages and Renaissance, they were not the only possibility. If the point from

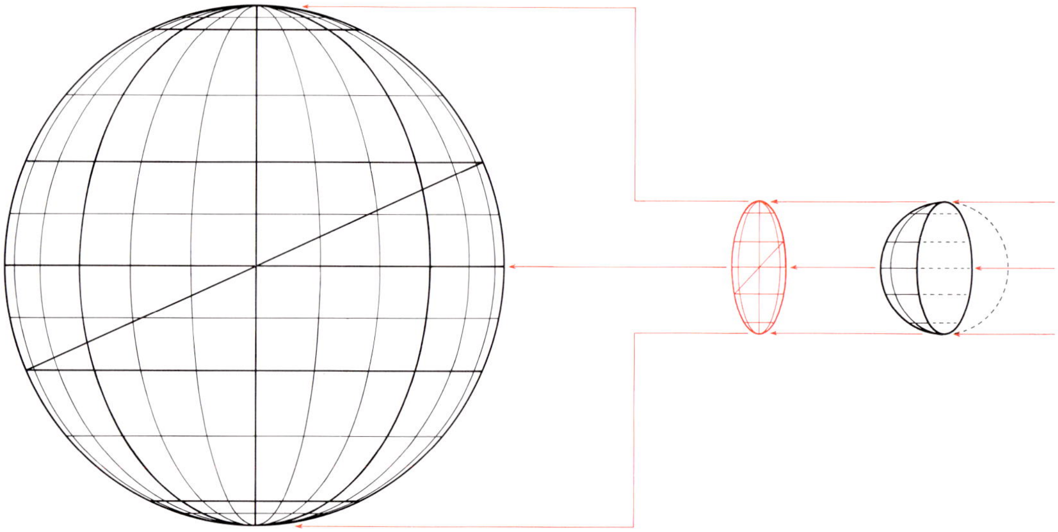

figure 140. Universal orthographic projection.

347. For his life and other inventions, see Gunther (IV). Throughout his work Blagrave gives cross references to Gemma whenever they are dealing with the same subjects. He does not, however, hesitate to criticize him. Thus, describing how to find the time in any part of the world he says that the problem '. . . is one of the chiefest that any young beginner can delight in, and by my *Iewel* is one of the easiest whereas in Gemma F. instrument it was one of the hardest. . . .' Blagrave, 26.

348. These are an unsigned instrument at the Adler Planetarium, Chicago (see Gunther (V), ii 492–501); an unsigned instrument dated 1598 in the British Museum, probably by James Kynfin (see Turner (III), Appendix A1); and an astrolabe by Charles Whitwell dated 1595 in the Museo di Storia della Scienza, Florence (see Bonelli, no. 87, 162, and for an illustration Bonelli & Settle, 66). That so few examples have survived may be explained by the fact that Blagrave strongly recommended that the instrument be made of paper and in his book supplied a printed example which could be cut out and mounted on pasteboard for use. Similarly in 1658, Moxon issued a printed-paper version. (Gunther (V), ii 500).

349. *The Catholique Planisphaer which Mr. Blagrave calleth the Mathematical Iewel*..., London, 1658. It will be noted that Palmer's title recalls that of Gemma.

350. Turner (I), 73–4.

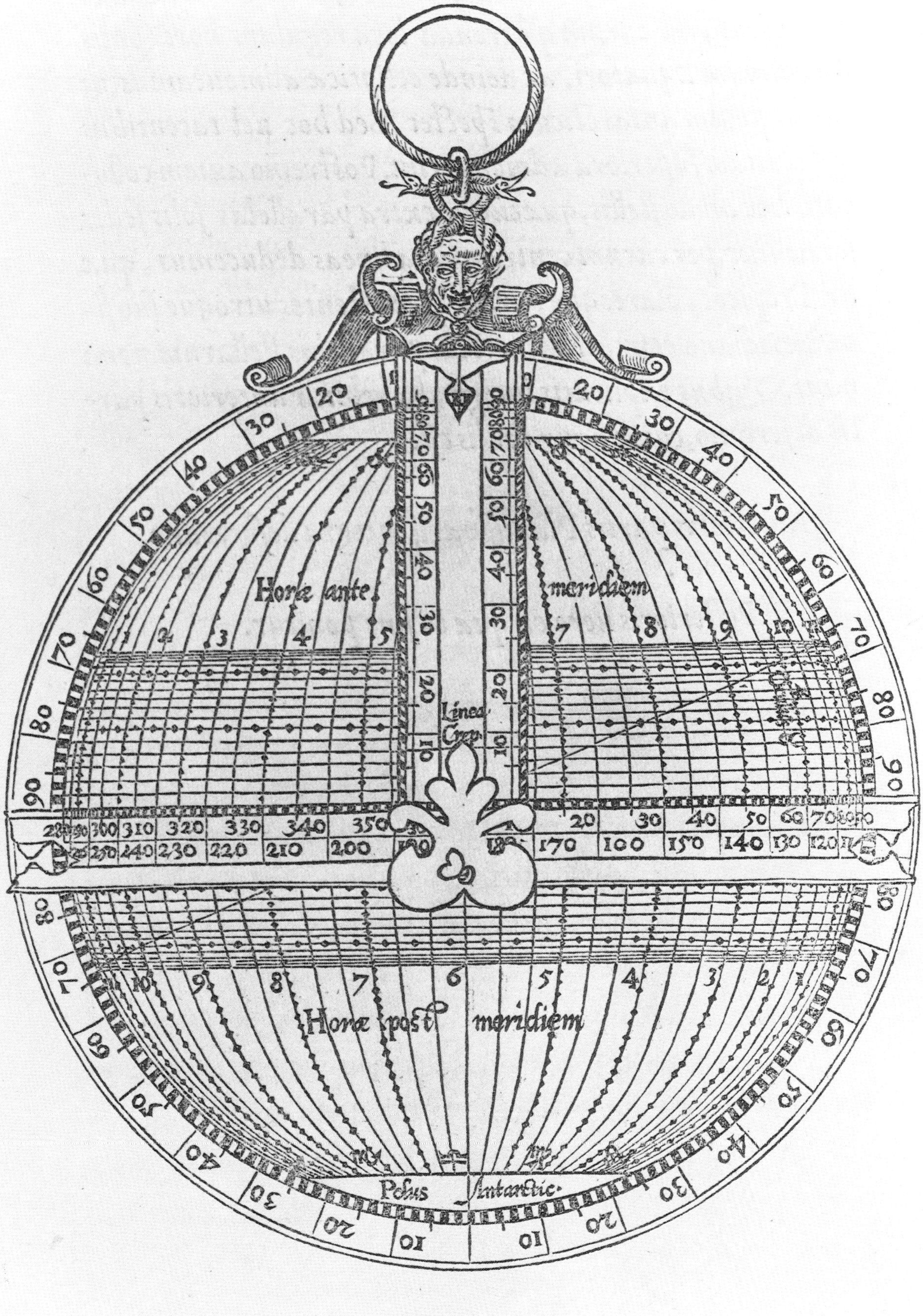
Horæ ante meridiem
Linea Crep.
Horæ post meridiem
Polus Antarctic.

which the projection of the sphere is made is taken back from the surface of the (imaginary) celestial vault (where in stereographic projection it is supposed to be) to a point at infinity so that the projecting rays are parallel, the projection becomes orthogonal or orthographic – that is, the meridians project as semi-ellipses and the parallels as straight lines (fig. 139). It was this form of projection that was employed in a variant form of universal astrolabe usually known as the Rojas astrolabe after Juan de Rojas. In 1550, he published in his *Commentarium in Astrolabium quod Planisphaerium vocant, libri sex nunc primum in lucem edidit* so comprehensive an account of this form of instrument that his name became identified with it (fig. 140).[351] As de Rojas was aware, however, it was not original to him[352] and the source of this projection, like that of stereographic projection, is in fact to be found in antiquity. Exactly when, where, and by whom orthographic projection was first investigated in antiquity is unknown, but it is discussed briefly in the context of sundials by Vitruvius in the 1st century B.C., by Hero in the 1st century A.D., and was the subject of a complete treatise by Ptolemy, of which only parts have survived.[353] That Ptolemy's treatise on the *Analemma* survived long enough to be known in medieval Islam, however, is clear from the fact that in the 13th century William of Moerbeke produced a Latin transation of it from a now lost Arabic original.[354] That it was very widely known in Islam is uncertain. Al-Bīrūnī, mentioning in passing[355] that he had discussed the projection (which he called cylindrical) in his treatise on astrolabes, also states that he had found no mention of it by earlier mathematicians. However there must have been some diffusion, particularly in Muslim Spain, where in the 13th century a trigonometric diagram, which was perhaps drawn with the help of orthographic projection, is found on a number of astrolabes by Muḥammad b. Fattūḥ al-Khamā'irī (*fl.* A.H. 609–34 [A.D. 1212/3–1236/7]), on one by Muḥammad b. Muḥammad b. Hudayl (A.H. 650 [A.D. 1252/3]), and in the 'libro del Açafeha' in the *Libros del Saber* (1276/7).[356]

In all of these cases, however, with the possible exception of al-Bīrūnī, we find evidence only for knowledge of orthographic projection, not for the existence of orthographically projected astrolabes. This is perhaps not surprising. The orthographic projection in both Ptolemy and Vitruvius was primarily a sundial projection, and it was as such that it would have been disseminated. In *Protomathesis* (1532), Oronce Fine gave a brief description with illustration of a sundial using the projection,[357] and when Federico Commandino

351. For description, construction, and use of the Rojas astrolabe, see Michel (III), 20, 103–9.
352. Pereira da Silva, 337.
353. Biblioteca Ambrosiana, Milan. Cod Gr. L99 sup, now 491, which dates from the 7th century or earlier. That orthographic projection may go back further, perhaps to Eudoxus (*fl. c.* 370 B.C.), has been suggested by Maula, but no copy of his paper on this subject has been available to me.
354. Maddison (III), 21.
355. Sachau, 357–8. The passage is given in Maddison (III), 21–2.
356. For detailed discussion of these, see Sauvaire & Rey Pailhade, 93–6; Maddison (III), 25–7.
357. Fine, fol. 194 r.

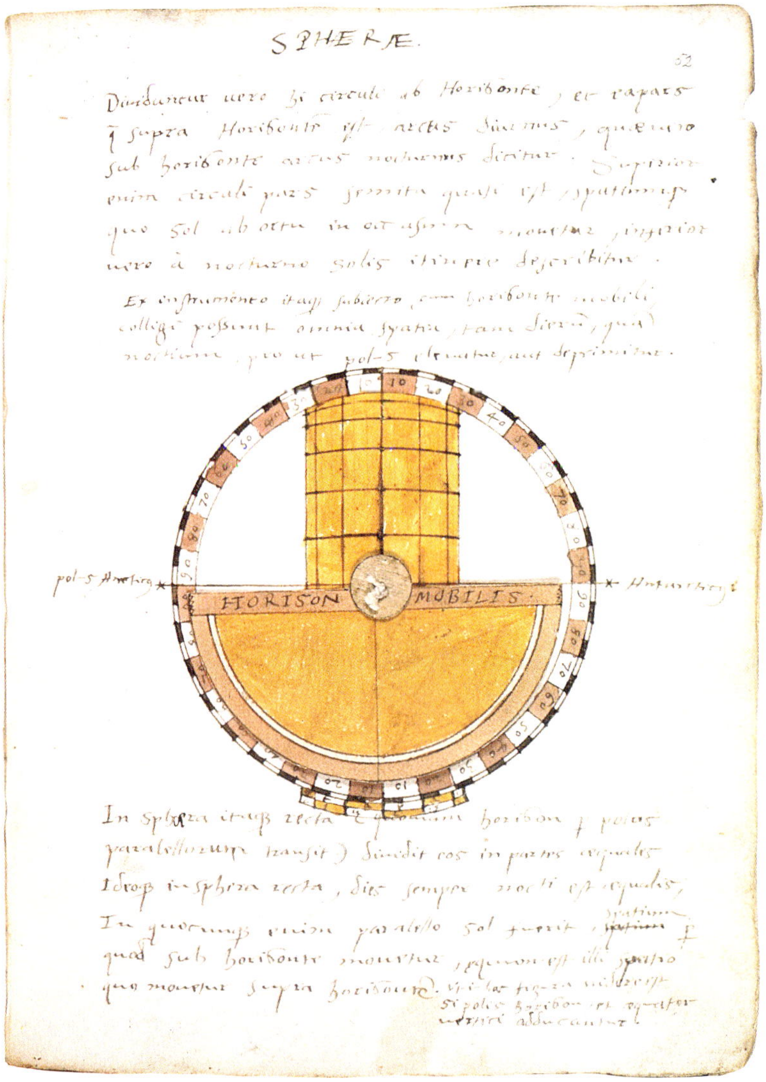

figure 141. Orthographic projection of the sphere, probably drawn by Sebastian Fabritius in 1551 to illustrate his copy of lectures given in 1550 and 1551 by Josias Simler at Zurich, on the making and use of a planisphere. London, British Library Add. Ms. 24010 fol. 52r. (Photograph: British Library)

edited Ptolemy's *Analemma* in 1562 he added a commentary of his own on its dialling uses. These aspects of it were further discussed in 1581 by Christopher Clavius.[358] Although we know as yet very little with certainty of the diffusion and influence of Vitruvius in the Latin Middle Ages, it was clearly of some importance,[359] and thus would also have tended to attract attention to the gnomonic aspects of the *Analemma*. This emphasis may therefore help to explain the use of the orthographic projection in a number of 16th- and 17th-century dials and compendia,[360] just as it relates them to the older tradition. No doubt it was this same tradition that supplied the basis for application of the orthographic projection to the astrolabe but it perhaps needed to be inspired by the similarity of the *saphea Azarchelis*. Certainly it was contemplation of the universal stereographic astrolabe in the context of orthographic projection that seems to

358. For a modern treatment of the *analemma* and dialling, see Gibbs, 105–17.
359. See, for example, the interesting case study by Harrison.
360. For a basic list, see Maddison (III), Appendix.

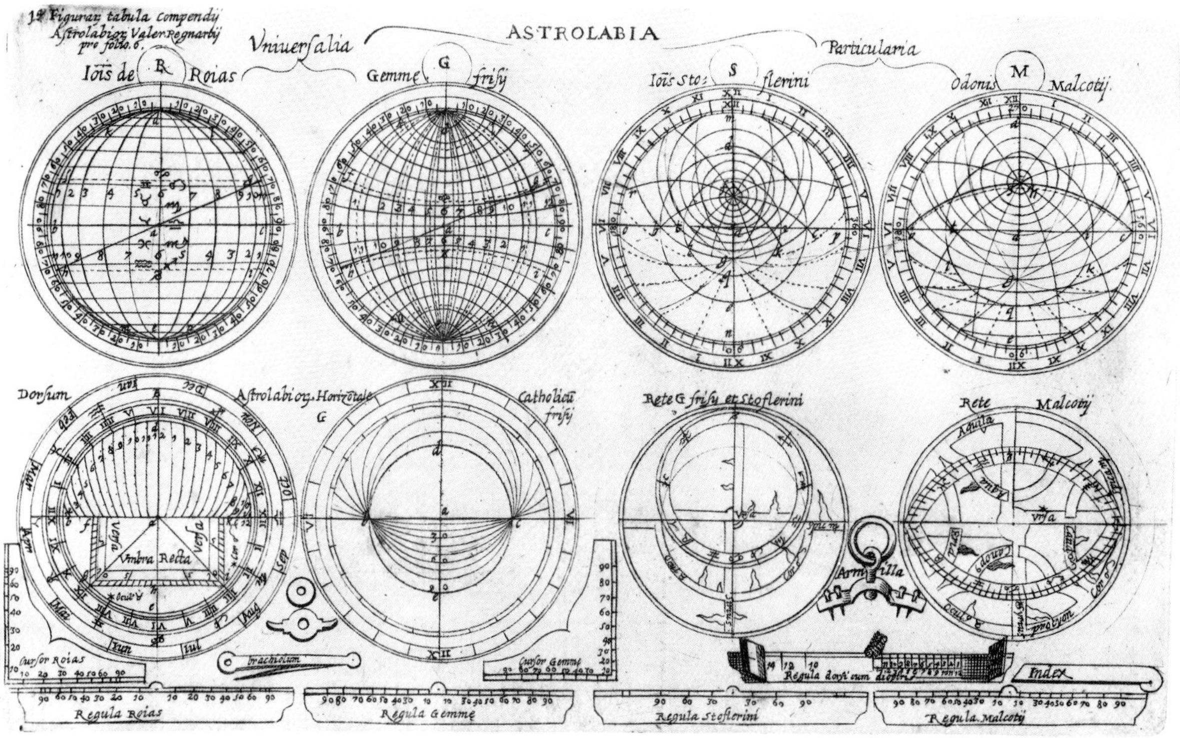

figure 142. The single latitude astrolabes of Stöffler and Odo Malcote, and the universal astrolabes of Gemma Frisius and Juan de Rojas. From Valerianus Regnartius, *Astrolabiorum seu utriusque Planispherii universalis et particularis Vsus*, Rome, 1610.

have stimulated de Rojas. The combination of orthographic projection and astrolabe, however, had taken place at least a century earlier, perhaps in Central Europe, where it is found in two instruments of 1480 and 1483. Of these, the earlier is an astrolabe which surmounts the globe made for Martin Bylica of Olkusz by Hans Dorn. The second, an astrolabe now in the Museo di Storia della Scienza, Florence, although not signed, may also be attributed to Hans Dorn.[361]

By the late 15th century it is clear that neither orthographic projection nor its use for astrolabes was a mystery, and the availability of several editions of Vitruvius *(editio princeps* 1486*)* made knowledge of it readily accessible. Even so, the lack of a manuscript tradition concerning the projection, and of many examples of its use,[362] suggest that it did not attain any great popularity before the appearance of de Rojas' book in 1550. De Rojas had been a pupil of Gemma Frisius, who included orthographic projection in his mathematical lectures. Although de Rojas apparently wrote his book in Spain, he had there the assistance of another Frisian pupil of Gemma's, Hugo Helt, who was acknowledged sole author of the sixth book, describing the actual construction of the instrument.[363] On its appearance the work, which is beautifully printed, must have

361. IC/CCA 492; Bonelli, no. 88, 162. The projection is also found in a 15th-century text entitled 'Instrumentum universale ad inveniendum horas in quocunque climate fueris fabricare' (*Incipit:* Prima in materia aut ligni solidi . . .'). Maddison (III), addenda 1.
362. For examples of the projection before 1550, see *Ibid.*, 29–30.
363. *Ibid.*, 37–8.

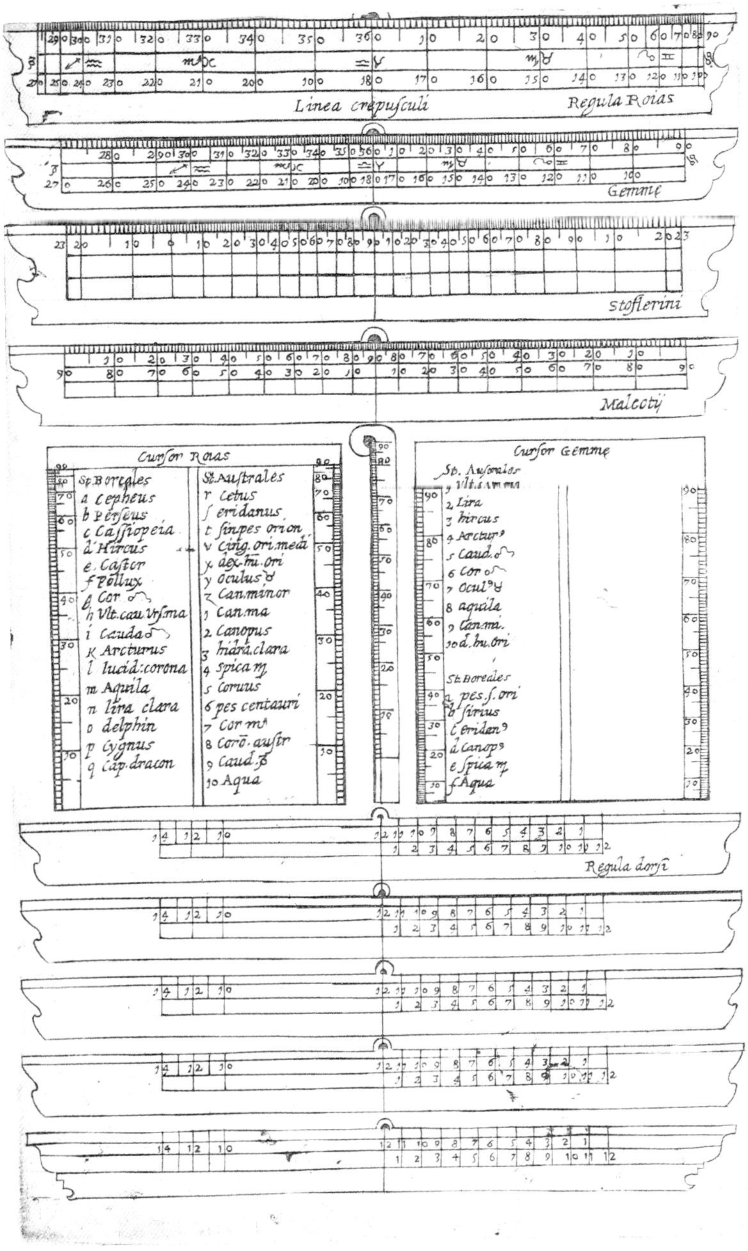

figure 143. Rules for the astrolabes of Stöffler, Malcote, Gemma Frisius, and Rojas. From Valerius Regnartius, *Astrolabiorum seu utriusque Planispherii universalis et particularis,* Rome, 1610.

seemed novel and appealing. Although in 1568 Egnatio Danti pointed out the orthographic diagram in the *Libros del Saber* which could have provided de Rojas with a source,[364] it may not be incorrect to ascribe to de Rojas' book the increased interest in the projection that may be detected in the following decades.[365]

While successfully solving the problem of the multiplicity of plates in the ordinary fixed-latitude astrolabe, both the *saphea* and

364. *Ibid.*, 23. A further possible source are orthographically projected volvelles, included in some editions of Sacrobosco's *de Sphaera*, e.g. those of Creutzer (Wittenberg) in 1545 or Cavellat (Paris) in 1552, and in Peter Apian's *Cosmographicus Liber*. . . , first edition, 1524, and numerous later issues.
365. See the list of examples given in Maddison (III), Appendix.

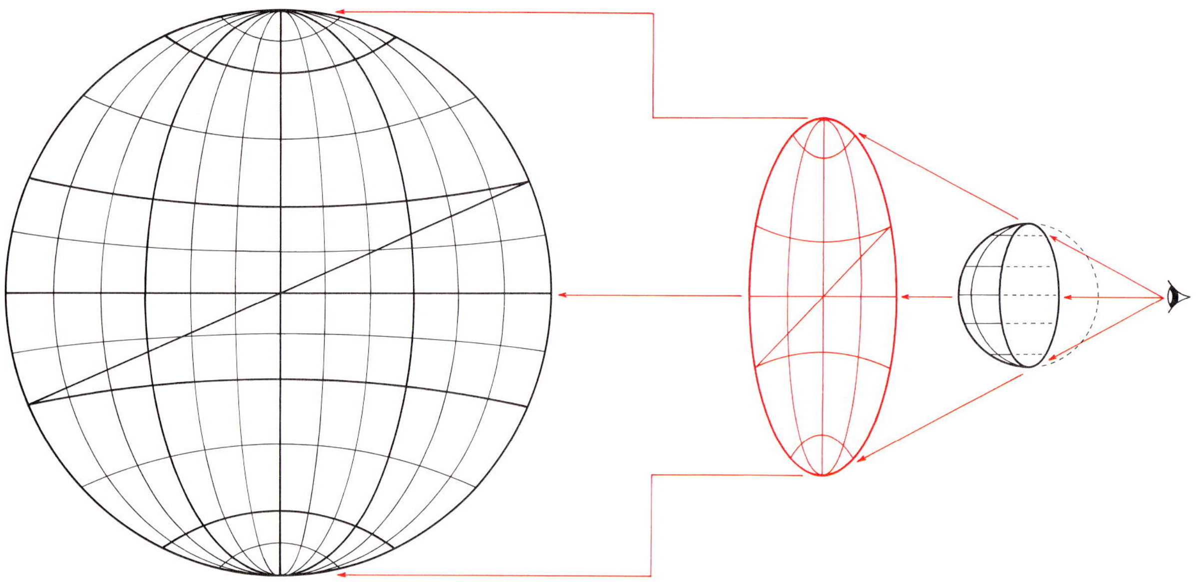

figure 144. Universal projection of Philippe de la Hire.

the Rojas astrolabe posed some problems of their own. Their chief inconvenience arose from the increasing closeness of the meridians and parallels, making them difficult to read on large instruments and impossible on small ones. To overcome this, Philippe de la Hire (1640–1718) displaced the centre of the projection to a point somewhere between infinity and the vernal point, chosen so that the projected parallels and meridians (which thus become arcs of ellipses) were more or less equally spaced on the instrument (fig. 144). This simple solution was enunciated by la Hire in his lectures at the 'College Royal' and was enthusiastically advocated by the noted Parisian instrument-maker and writer, Nicholas Bion (1652–1733),[366] who incorporated it into a number of astrolabes he produced of printed paper and pasteboard.[367] Several years later he described the projection in his *l'Usage des astrolabes, tant universels que particuliers*. . ., Paris, 1702. Bion, however, seems to have been alone in using la Hire's new projection and it was probably quickly forgotten as the astrolabe fell out of use. It was, however, reinvented in the early 20th century by E.A. Reeves (1862–1945) as part of an instrument with which desert travellers could find both time and direction by a single observation.[368]

366. The projection was presumably devised before 1685 since a vignette on page 1 of la Hire's, *Sectiones conicae* published in that year includes an example. Bion, 21–9, 73–140; Michel (III), 21, 111–3; Maddison (III), 11.

367. Only one example of Bion's pasteboard astrolabe has so far been recorded. See above n. 194.

368. Reeves was Map Curator and instructor in practical astronomy and surveying to the Royal Geographical Society, London. For a brief notice, see *Who Was Who*, 961. His instrument was manufactured by Stanford's. For a description, see Reeves *British Patent Specification*, No. 14008, 1908. Examples are preserved in the Museum of the History of Science, Oxford, and The Science Museum, London.

Astrolabe Catalogue Universal Astrolabes

22. Syro-Egyptian *Saphea Azarchelis* and Astrolabe Plate

Mid-late 8th/14th century or 9th/15th century
Brass with inlaid silver
Diameter $6\frac{11}{16}$ in. (170 mm)
Signed:[369]
(Work of 'Alī al-Wadā'ī)

Inventory 3529

The *rete* of the instrument is missing. The remaining part of the instrument is a single plate cast in one piece with the *kursī*. This is pierced with the maker's signature in a cursive script in such a way that it can be read from both sides of the instrument. There is no raised limb, which was probably considered unnecessary since there are no loose plates. Reading from the outer edge inwards, in the upper two quadrants and lower-right quadrant is the first scale (0°–90°, reading from the east and west points); that in the lower-left quadrant is a scale for finding the shadow length (on the standard medieval base of 12), equivalent to an altitude read from the scale on the opposite quadrant. It is thus equivalent to the shadow square on a conventional instrument. Within these scales is a complete scale (0°–360°, reading from the bottom). This scale and the outer scales in the upper two quadrants are engraved on inlaid silver, presumably for ease of reading. Within the degree scale is a zodiacal calendar made up of a scale of months in the Coptic calendar (12 x 30 days + 5), a scale of months in the Syrian calendar, and a scale of zodiac signs (0° Aries = approximately 16 Taremhotep in the Coptic calendar and 12 Ādār [March] in the Syrian calendar). Within these scales is a double plate on which almucantars for latitudes 30° and 36° are drawn. Although they are not specified as such, these are the latitudes of Cairo and Aleppo. The inscription on each half reads 'For each place of latitude 30°/36°.' This rare form of plate, usually called 'ogival' from the pattern produced by the almucantars, was useful for the economy of space and metal and weight that it offered, but had disadvantages in use. In each of the lower quadrants two small holes have been carefully drilled through the plate on the outer scale 30° below the east-west line.

The obverse of the instrument is engraved around the edge with degree scales (0°–90°–0°–90°–0°, reading from the top and bottom). Within this a *saphea Azarchelis* is drawn on which the positions of twenty-two stars are marked by inlaid silver points (three missing). The stars are not named, but the meridians for every 30° (i.e. for each pair of zodiac signs) are emphasized by being made with a series of small cuts of the burin and not by a continuously engraved line. A similar technique is used for the centre

369. For certain peculiarities of this inscription, see the discussion in Brieux & Maddison. The reading 'amal 'Alā al-Wadā'i has been proposed by Dr. E. Savage-Smith.

figure 145. Catalogue no. 22:
Syro-Egyptian *saphea Azarchelis* and astrolabe plate. *Saphea Azarchelis.*

figure 146. Catalogue no. 22. Plate.

line of the ecliptic and for some almucantars on the reverse. There is a shackle and ring suspension system. The rule and cursor, used with the *saphea Azarchelis*, and the *rete* are missing. An alidade was perhaps mounted above the *rete* although it would have been difficult to read accurately without a raised limb. An alternate possibility is that two sights were mounted on the *rete* itself, as is known to have occurred on other Islamic astrolabes when a conventional alidade was difficult or impossible to use.[370]

Although this astrolabe displays a certain eclecticism, the inclusion of the Coptic and Syrian calendars suggests strongly that it should be ascribed to the eastern Islam and was probably made in either Egypt or Syria. The latitudes of 30° and 36°, which correspond with those of Cairo and Aleppo, tend to confirm this, and the use of silver inlay is also characteristic of this region. The scales on the instrument are marked out in eastern *abjad* numerals in which 60 is represented by س (s) and not ص (ṣ) as in western *abjad*. The script of the instrument, however, is western *kufic*, and the use of the triangular-dashed line is also typical of Maghribi instruments.[371] Possibly, therefore, the maker, although working in the Syro-Egyptian region, originated from or learnt his craft in the Maghrib. Similarly contradictory is the evidence which can be derived directly from the instrument for its date. The marker for *Regulus* is at an approximate ecliptic longitude of 138½°, corresponding to a date of about A.D. 1200, but the position of 0° Aries at 12 March suggests a date no earlier than the mid-14th century. Since it is more likely that an instrument-maker would use out-of-date star tables than that he would mark out a zodiac in advance of his own period, a later date is probably to be preferred.

In the 3rd/9th and 4th/10th centuries, Muslim astronomers, working mainly in Iraq and Iran, introduced three innovations to the astrolabe.[373] Firstly, they noted that the *rete* on a standard astrolabe, though symmetrical with respect to the line representing the solstitial colure, was not symmetrical with respect to the line joining the equinoxes. They therefore devised various kinds of *retes* symmetrical with respect to both.[374] Secondly, they noted that the markings on a standard astrolabe plate enjoyed a symmetry with respect to the meridian but not with respect to the east-west line. They therefore devised plates using just one-half of the markings for one

370. The double astrolabe, cat.no.4 above, is so fitted, as is the geared astrolabe by Muḥammad b. Abī Bakr b. Muḥammad ar-Rāshidī al-Ibarī al-Iṣfahānī A.H. 618 (A.D. 1221–2) in the Museum of the History of Science Oxford. See Maddison & Turner, no. 52, 112; Brieux & Maddison, MHD ABI BAKR 1.

371. Another instrument from eastern Islam showing Maghribi influences is the *saphea Azarchelis* made by 'Abd [?Allāh] b. Yūsuf in Damascus, A.H. 695 (A.D. 1295/6), now in the Victoria & Albert Museum, London. See Maddison & Turner, no. 64, 128; Brieux & Maddison, ABDALLH YUSF 1 and *cf. Ibid.* 2.

372. On the Syro-Egyptian tradition of astronomy, see King (VI).

373. The following notes prepared by D.A. King are supplementary to the account in the introduction, the present astrolabe having been added to the collection after the rest of the text had been completed.

374. The only published material on these is in Sédillot, 181–2 & pls. 25–8; and in Frank.

latitude. In this way, markings for two latitudes could be represented on one side of a single plate.[375] Thirdly, in the mid-9th century, the astronomer Ḥabash al-Ḥāsib devised the plate of horizons using which certain problems involving the local horizon could be solved for any latitude.[376]

Further development took place in Andalusia in the 11th century, probably inspired by earlier developments in the Islamic East. The Toledo astronomer 'Alī b. Khalaf, known as ash-Shajjār or ash-Shakkāz, devised a universal astrolabe[377] which can be seen as the culmination of earlier eastern developments. It can be used to solve any problem of spherical astronomy for any latitude. The appropriate setting of the upper grid upon the universal plate enabled the user to convert from any one set of orthogonal co-ordinates (ecliptic, equatorial, or horizon) to any other.[378]

Ash-Shakkāz's contemporary, az-Zarqellu, simplified this universal astrolabe to produce two different plates.[379] In one of these a special alidade, along which could slide a perpendicular cursor, replaced the *rete*. A given celestial position in one co-ordinate system could then be represented on the alidade *cum* cursor, and the corresponding position in the other co-ordinate system read off from the *Shakkāzīya* markings below. This instrument, later called *aṣ-ṣafīḥa ash-Shakkāzīya,* could thus be used for effecting any co-ordinate conversion.

Az-Zarqellu's second plate consisted of two sets of *Shakkāzīya* markings superimposed with their axes inclined at an angle equal to the obliquity of the ecliptic (fig. 135). With the special alidade and cursor, this device, called *aṣ-ṣafīḥa az-zarqellīya,* could be used for converting ecliptic and equatorial co-ordinates, but the plate was difficult to use for any other conversion because of the profusion of the markings.

These developments were known in Ayyubid Syria. The special *retes* and plates were made known by al-Bīrūnī's book on astrolabes,[380] and the universal plates of az-Zarqellu (but not the universal astrolabe of ash-Shakkāz) were introduced into Syria, probably during the 7th/13th century. The Rockford astrolabe presents a combination of the two traditions.

Although not common, enough examples of the *saphea Azarchelis* have survived to show that it had a relatively widespread use.[381] The ogival plate, *maftūḥ* (open) in medieval Arabic astronomical termi-

375. Indeed, by folding over the markings south of the east-west line, one could represent the markings necessary for one latitude in a single quadrant of the plate. Four latitudes could thus be represented on one side of a plate. *Cf.* an unsigned, undated astrolabe, perhaps of the 8th/14th century, now in the Staatsbibliotek, Preussicher Kulturbesitz, Orientalabteilung Berlin (Sprenger no.2049), which uses similar techniques to draw 9-quarter astrolabe plates on the 2 sides of a single plate. See Maddison & Turner, no. 54, 115.
376. See Morley (I), 7, n. 12 (for Morley's 'Ḥanash' read 'Ḥabash').
377. See King (IV), *passim* and above, to which the present notes are supplementary.
378. For details of the various operations, see King (I), *passim*.
379. See King (IV).
380. Sezgin, vi, 268.
381. See list below.

nology, is rare presumably because its disadvantages outweighed the saving of space and weight that it offered. The basic problem of the plate was that it required a fairly competent astronomer to use it. Since no azimuth circles are shown, the user needed to know enough about the mathematical equivalence of the problems of determining the hour angle and the azimuth from an observed celestial altitude to determine azimuths using the almucantar curves. With the plate, moreover, the basic function of the astrolabe – namely, to display the instantaneous configuration of the heavens relative to the local horizon – is lost; with these markings, there is only a mathematical abstraction. The markings for each latitude are discontinuous at the meridian; caution is needed when using them in operations which involve moving from one quadrant to another. The amount of rotation of the *rete* from one position to another, which is a measure of the passage of time, must take into consideration these discontinuities.[382]
Published in Sotheby (III), lot 46.

382. For further developments of unusual astrolabes and double plates in the 8th/14th century, see King (VI).

23. Maghribi Astrolabe and *Saphea Azarchelis*

9th/15th or 10th/16th century
Brass
Diameter 7⅝ in. (194 mm)
Not signed
Inventory 1935

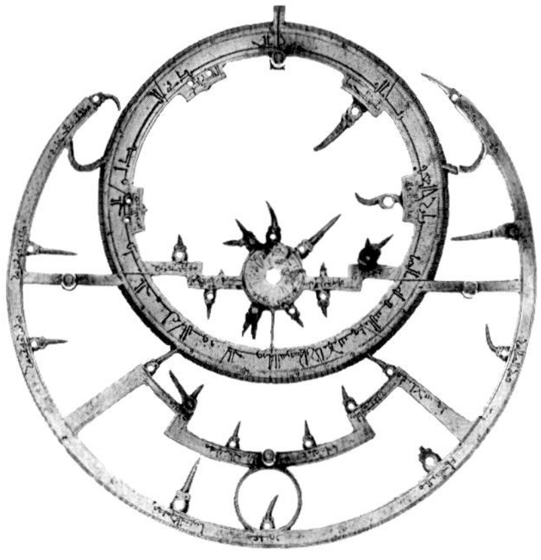

figure 147. Catalogue no. 23. *Rete.*

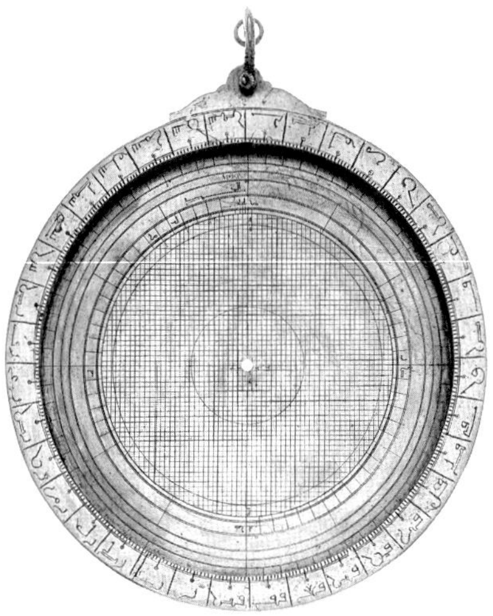

figure 148. Catalogue no. 23. *Mater.*

The *rete* marks the positions of thirty fixed stars indicated by straight or slightly curved and slender pointers arising from bases formed of a large circle surmounted by two smaller ones. Originally, silver knobs were almost certainly set in each of the larger cavities, although all are now missing.[383] There is considerable wear to the *rete*, the astrolabe having evidently been much used; several star pointers are broken, and four have been rather crudely repaired. The *rete* has a double counter-changed east-west bar and four *mudīrs*, one of which is a replacement. The zodiacal signs are named on the ecliptic circle, which is divided to 3°, and the 15° position or mid-point of each sign is also marked by a dot within a small circle. The *mater* of the instrument is composed of a backplate cast in one piece with the *kursī*, and two rings are rivetted to it to make up the limb. Possibly the instrument was originally made in the usual manner with the *mater* and one ring for the limb, the third piece having been added later to accommodate an extra plate since lost. That this was the case is suggested by the fact that, as now assembled, the *rete* rests below the level of the limb with which it is usually flush.[384] The limb is engraved with a scale of degrees (0°–360°) reading to 1° by tens. The 5° point of each group is indicated by a dot in a circle. The *kursī* is low and wide with three shallow lobes. The shackle, suspension-ring, and *'ilāqa* are all probably later.

The back of the instrument is engraved on its outer edge in the upper two quadrants with a degree scale (0°–90°–0°) reading to 1°, and in the lower two quadrants with a scale equivalent to a shadow square[385] since this could not be marked in its usual rectangular form on the instrument. Within this is a zodiacal calendar (eccentric-type; 0° Aries = approximately 13½ March) marked with the Western names for the months. The centre of the instrument is filled with a *saphea Azarchelis* on which are marked, by means of a dot in a small circle, the positions of thirty-six named stars. On the equatorial projection the tropics, equator, polar circles and certain meridians (every 30°) are marked by arrowed lines

383. For an impression of how the astrolabe would originally have appeared, see the instrument by Muḥammad b. Fattūh, A.H. 621, (A.D. 1224/5) now in the Museum of the History of Science, Oxford, illustrated in Gunther (V), i, 176–7, pl. LXI. This astrolabe, the *rete* of which is similar to that of the instrument described here, has retained most of its silver studs.
384. Alternatively, one may wonder if a new limb was not engraved and added to the instrument in consequence of the original having been broken or become illegible.
385. For a description of an equivalent scale on an astrolabe by Muḥammad b. Fattūh, A.H. 609 (A.D. 1212/3), see Sauvaire & Rey Pailhade, 86–8.

figure 149. Catalogue no. 23:
Maghribi astrolabe. Front.

figure 150. Catalogue no. 23.
Back, with *saphea Azarchelis*.

thus, ⥤⥤⥤⥤, and equatorial meridians and parallels are drawn for every 5°. For the ecliptic projection, however, only the ecliptic itself and some meridians are drawn, presumably in order not to overcrowd the diagram. There are three plates (? of four) held in the *mater* by a tab which enters a small hole at the south point of the limb. They are marked for latitudes, as follows:

a. 33° Fez, Salé, Tunis, Damascus, Baghdad & Ascalon
34°30′ Tlemcen, Tripoli, Mahdia, Aleppo, Samarra
b. 21°40° Mecca, aṭ-Ṭā'if, al-Yamāma, Jeddah, &c.
25° 'Medina, city of the Prophet may the blessing of God be upon Him and peace'
c. 31° Marrakesh, Alexandria, Kufa, Basra and all [? places of that latitude]
36°30′ Almería, Algeciras, Bougie, Isfahan, Harran and all places of that latitude

The plates have the almucantars drawn for every 3°. Both the azimuths and the almucantars are numbered, as are the unequal-hour lines. The azimuths are drawn above the horizon only. The crepuscular line is marked, as are the *khaṭṭ az-ẓuhr*, the *khaṭṭ al-'asr*, and the *khaṭṭ az-zwāl*. The missing plate probably included a tablet of horizons. The interior of the *mater* is incomplete. It was intended to be engraved with a degree scale surrounding a zodiacal calendar (concentric type), the Arabic name for Capricorn having been begun in one of the divisions, but this was not finished. The centre is completely filled with a rectangular grid (complete) as if for a sine/cosine graph. The alidade is of the straight-bar type and has a single pin-hole sight. The horse and pin are both missing.

This is a well-made instrument, clearly in the mainstream of Andalusian and Maghribi astrolabe-making. The conservatism of this tradition can be seen in the resemblance of the *rete* of this astrolabe to those of astrolabes by Muḥammad b. aṣ-Ṣaffār[386] in the 5th/11th century, and by Muḥammad b. Fattūḥ in the early 7th/13th century.[387] The instrument may be further compared with the small number of other *saphea Azarchelis* known to survive. These (in chronological order) are by:

Muḥammad b. Fattūḥ al-Khamā'irī, A.H. 613 (A.D. 1216/7)[388]
Muḥammad b. Fattūḥ al-Khamā'irī, A.H. 615 (A.D. 1217/8)[389]
Muḥammad b. Muḥammad b. Hudayl, A.H. 650 (A.D. 1252/3)[390]
Ibrāhīm ad-Dimashqī, A.H. 669 (A.D. 1270/1)[391]
'Abd (?Allah) b. Yūsuf, A.H. 695 (A.D. 1295/6)[392]
Abu-l-Ḥasan 'Alā 'ad-Dīn 'Alī b. Ibrāhīm b. Muḥammad b.

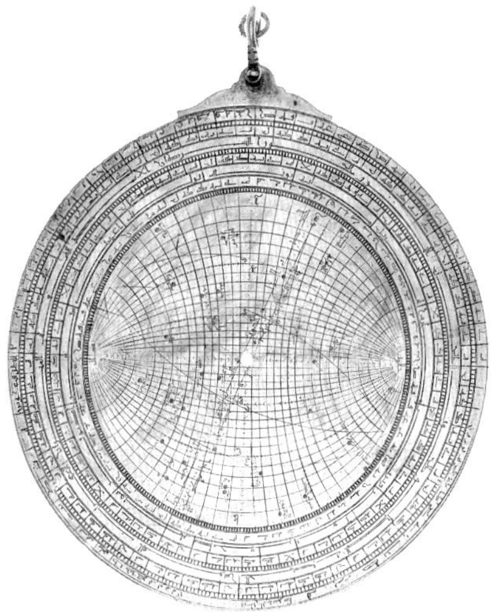

figure 151. Catalogue no. 23. Back without alidade.

386. Gunther (V), i, 116, 251 & pl. LIX; Plenderleith; Mayer (I), 75.
387. Gunther (V), i, no. 130, 276 & pl. LXI–II; Mayer (I) 64–6.
388. Mayer (I), 65, II. Gunther (V), i, 270–3, now in the Oservatorio Astronomico, Rome.
389. Mayer (I), 65, III; Gunther (V), i, 274. Now in the Bibl. Nat., Paris.
390. Mayer (I), 73; Garcia, 313–19; Millás (IV), *passim*, now in the Observatorio Fabra, Barcelona.
391. Mayer (I), 48; Gunther (V), i, 238, now in the British Museum, London.
392. Maddison & Turner, no. 64, 128. Now in the Victoria & Albert Museum, London.

figure 152. Catalogue no. 23. Plates.

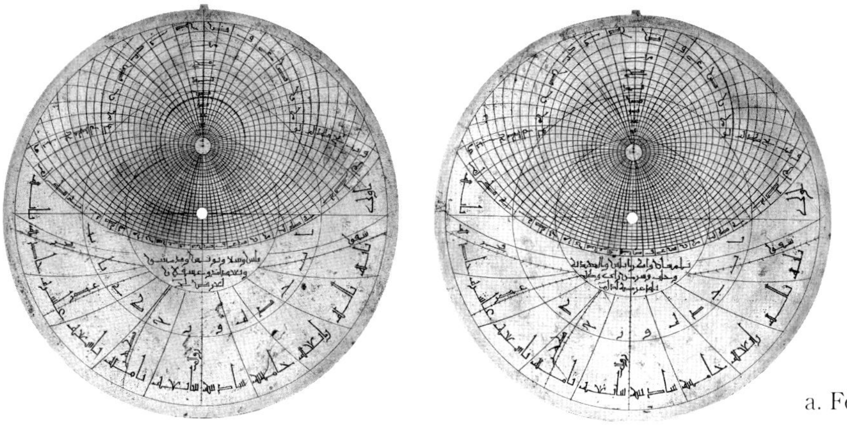

a. For latitudes 33° (left) and 34°30′ (right).

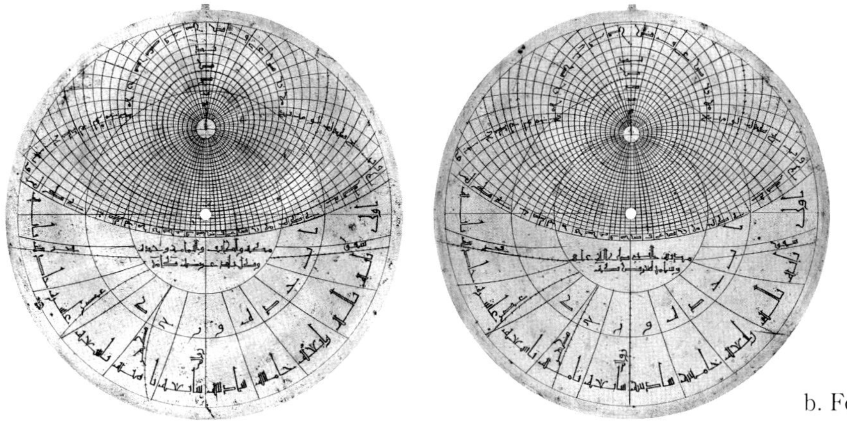

b. For latitudes 21°40′ (left) and 25° (right).

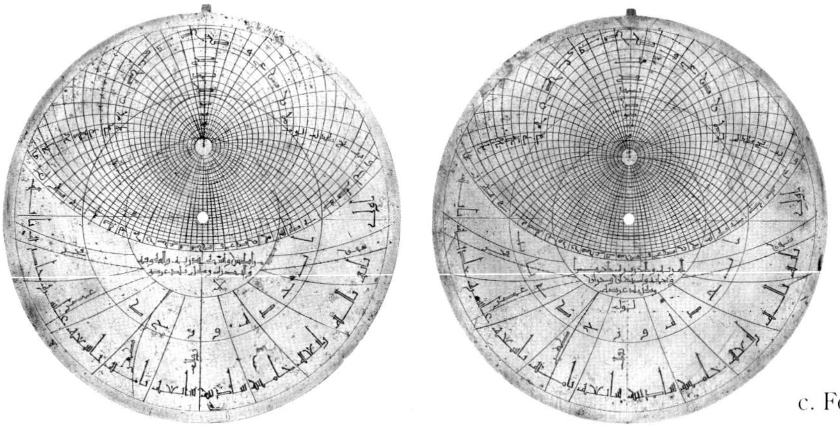

c. For latitudes 31° (left) and 36°30′ (right).

al-Himām al-Anṣārī, known as Ibn ash-Shāṭir, A.H.738 (A.D.1337/8)[393]

'Alī al-Wadā'ī, not dated, probably 8th/14th or 9th/15th century[394]

Muḥammad b. Aḥmad b. al-Ḥasan al-Baṭṭūṭī, A.H. 1141 (A.D. 1728/9)[395]

Not signed or dated[396]

393. Mayer (I), 41, II; Gunther (V), ii, 287. Now in the Bibl. Nat., Paris.
394. Sotheby (III), lot 46. See cat. no. 22.
395. Renaud, 22, now in the Batha Museum, Fez (Inv. 20993).
396. Gunther (V), i, 284–5. Now in the Museum of the History of Science, Oxford.

Of these, the present example resembles most closely the unsigned and undated example (now in the Museum of the History of Science, Oxford, fig. 153), in which the *saphea Azarchelis* is engraved on the back of a single-latitude astrolabe rather than, as in all the other cases except the last, being engraved on the face of a single plate, the reverse of which was engraved like an ordinary astrolabe back. The *rete* of the Oxford astrolabe also resembles that of the present example very closely and has the same number of stars. The *kursī*, although with rather more lobes, is similarly squat and without inscription or decoration.

figure 153. Astrolabe and *saphea Azarchelis*. Not signed or dated. ? 13th century A.D. Front (left) and back (right). IC/CCA no. 139. See Gunther (IV) i, 84-5. Oxford, Museum of the History of Science. (Photograph: Museum of the History of Science)

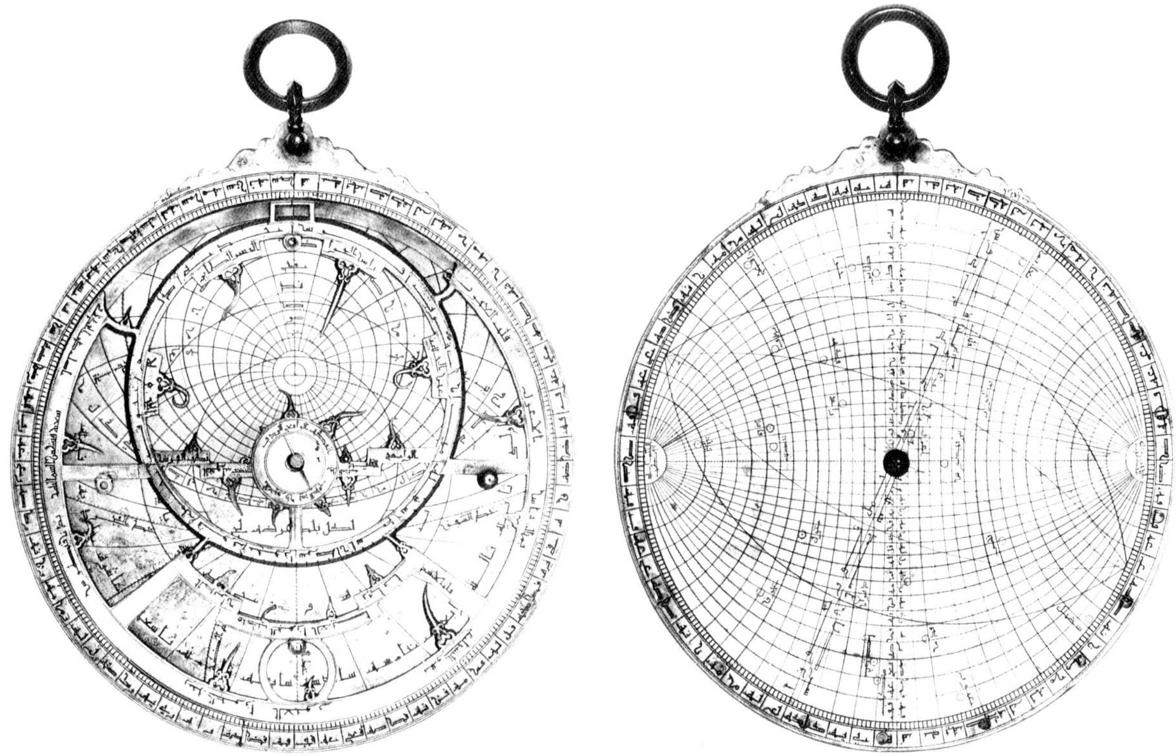

24. ? Flemish Astrolabe *Mater*

Not dated, late 16th/early 17th century
Brass
Diameter 7¾ in. (197 mm)
Not signed
Inventory 2505

The limb of the instrument, which was made separately and rivetted to the plate, is engraved with scales for degrees (0°–90°–0°–90°–0°, starting from the east point) and for equal hours (1–12 x 2). The *mater* is blank. It has not been smoothed and polished preparatory to engraving, and there is no hole in the inner edge of the limb to retain a latitude-plate lug. The swivel ring is mounted on a casting separately attached to the limb. On the reverse a degree scale is marked in four quadrants (0°–90°–0°–90°–0°, starting from the top) around the edge. Within this is a Rojas universal projection with eighteen named stars. Of these stars, the positions of four (Canopus, Caput medusae, Hircus, Lyra) are not indicated. On the edge of the instrument, starting at the north and south points and reading respectively to east and west, are two numerical scales: 0–25 and 1–10. The purpose of these scales is unclear, unless they represent an unsuccessful attempt by the maker to supply an equivalent to the shadow square he could not place on the back of the instrument.

The combination of the unsmoothed *mater*, the lack of a hole for the lug of a latitude plate, the unmarked positions of four stars, and the scales on the edge suggest that this instrument was abandoned by its maker and was never completed.

figure 154. Catalogue no. 24. *Mater*. Front.

figure 155. Catalogue no. 24: Flemish astrolabe. Back of the *mater* with Rojas universal projection.

25. English Universal Astrolabe

Not dated [c. 1970/5]
Brass
Diameter 5 ⅞ in. (149 mm)
Not signed
Inventory 1025

The single plate is engraved on one side with a degree scale in four quadrants (90°–0°–90°–0°–90°, reading from the east point), surrounding a Rojas universal projection without stars. On the reverse is a degree scale (0°–90°–0°–90°–0°, reading from the east point), surrounding an hour scale and a zodiacal calendar (0° Aries = 11 March). Within this is a four-sided shadow square. Mounted on the reverse is an alidade, with double pinnule sights, attached by a bolt which also carries the graduated rule and cursor for use with the Rojas projection on the obverse. The suspension ring is attached to a gimbal mounting carried in a cradle screwed to the rim of the instrument.

This is a well-made modern imitation of a Rojas universal instrument.

figure 156. Catalogue no. 25. Reverse.

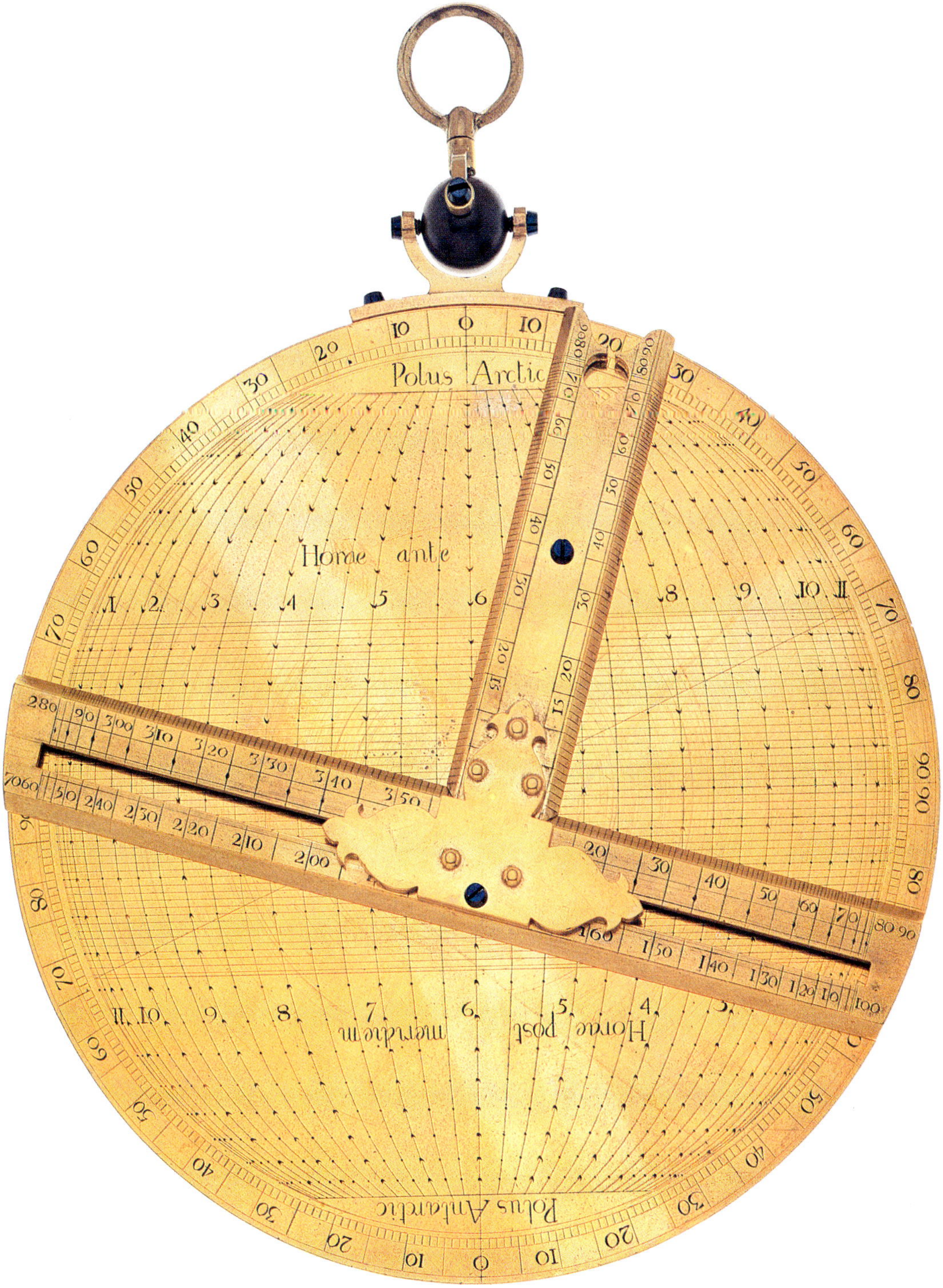

figure 157. Catalogue no. 25: English universal astrolabe. Back with Rojas universal projection.

LINEAR ASTROLABES

The desire to reduce the bulk of the astrolabe, which helped to generate the invention of universal astrolabes, also gave rise to a simplified form of single-latitude instrument in which the instrument was reduced to a simple stick. This, the 'linear astrolabe,' is an obscure device that appears to have been very little used in Islam. No examples of the instrument are known to have survived,[397] and it does not appear to have been transmitted to the Christian West. All that is known of the device, which was invented by the Persian astronomer and mathematician Sharaf ad-Dīn al-Muẓaffar b. Muḥammad b. al-Muẓaffar aṭ-Ṭūsī (d. *c.* A.H. 610 [A.D. 1213/4]), derives from the description of it given a few decades later by Abū 'Alī al-Ḥasan b. 'Alī 'Umar al-Marrākushī,[398] which, however, throws no light at all on its history.

figure 158. The principle of the linear astrolabe.

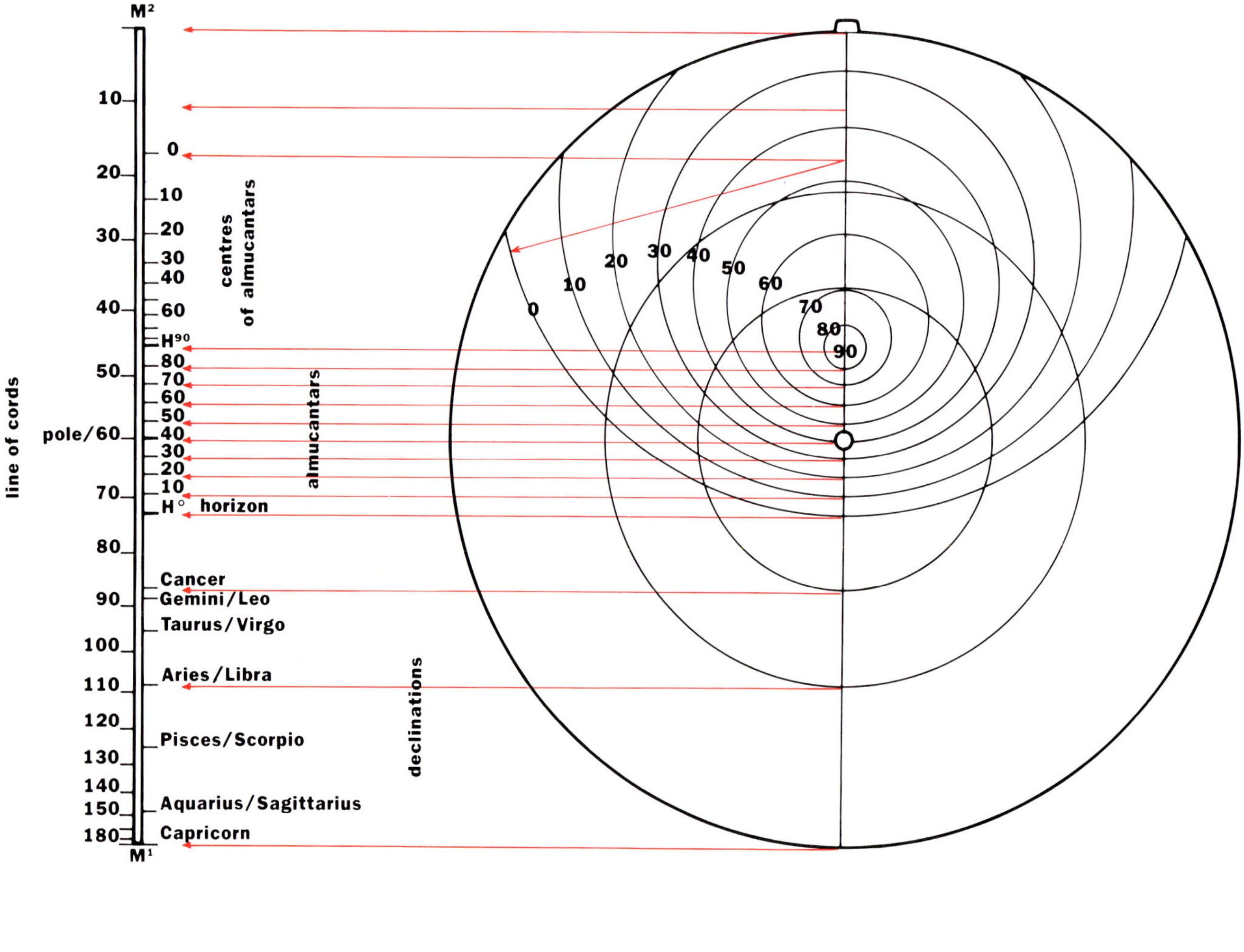

397. A number of modern reconstructions of the device have been made in ivory by M. Henri Michel. One example, signed 'H.M. 1943,' is in the Museum of the History of Science, Oxford (Josten, 13, no. 15). A second is in a private collection, Paris.
398. Edited and translated by Carra de Vaux.

The linear astrolabe consists of a graduated rod equipped with a sight at each end, a plumb-line fixed at the centre, a fixed thread at one end of the instrument, and a third movable thread. The rod itself represents the meridian line of an astrolabe (MM^1 in fig. 158) and on it are marked the points of intersection of the horizon (H°) and the almucantars (H°–H^{90}), usually marked on the astrolabe plate, and the divisions of the ecliptic, usually found on the *rete*, as they intersect the meridian in the course of one year's complete rotation. The point P represents the projection of the pole, and also the centre of the equator and of the two tropics. On the opposite face of the rod, a scale of chords is marked, used for angular measurement. Although no doubt relatively easy and inexpensive to make, the linear astrolabe is difficult to use and very limited in what it can do.[399] These facts are probably sufficient to explain why it disappeared from history almost immediately after it was invented.

SPHERICAL ASTROLABES

Although all the astrolabes so far discussed have employed some form of projection to produce a flat depiction of the heavens, there is in principle no reason why the instrument should not retain the spherical form of its original. This the spherical astrolabe does. It is composed of a globe on which are marked the horizon, the almucantars, the meridians and, in the lower portion of the globe below the horizon, the unequal-hour lines. Enclosing this globe and rotatable around it is a *rete* made up of the ecliptic, the equator, a number of fixed stars, and a quadrant of altitude. A series of holes is pierced down the northern portion of the meridian, and the instrument can be adjusted for latitude by means of an axis which passes through a hole at the northern equatorial pole of the *rete* and then into the appropriate hole of the meridian. The instrument was completed by some arrangement (equivalent to an alidade) for making solar and stellar observations.

The spherical astrolabe is an exceedingly rare instrument of which two examples only are known to have survived. One signed by Mūsà in A.H. 885 (A.D. 1480/1) is from eastern Islam (fig. 159),[400] the other – represented by the globe only – is Maghribi and probably from the early 10th/16th century. It is not signed or dated.[401] Despite the late date of these two specimens, the history of the instrument goes back at least to the 3rd/9th century, and it may, like the planispheric astrolabe itself, have originated in Hellenistic

399. For description of the instrument's construction and use, see Michel (III), 212, 115–22.
400. Now in the Museum of the History of Science, Oxford. See Sotheby (II), lot 13; Maddison (II), *passim*.
401. In a private collection, see Canobbio.

times. There is, however, no clear evidence of any kind from this period. Suggestions[402] that there may have been an early inverted form of spherical astrolabe – analogous to the dial of the anaphoric clock – in which a celestial globe marked with stars and the ecliptic was surrounded by a grid of hour lines and almucantars which rotated around it, are totally hypothetical but not entirely impossible given that an analogous form is attested to in medieval Latin manuscripts.[403] Possibly the spherical astrolabe originated from the celestial globe and the armillary sphere;[404] it is equally possible that the instrument was developed as a deliberate 'de-planispherization' of the astrolabe, perhaps inspired by the thought that an instrument which resembled its subject more closely would be more useful for demonstration and teaching.[405]

The earliest clear evidence for the existence of the spherical astrolabe is a reference by al-Khwārizmī (d. *c.* A.D. 850)[406] that implies that the instrument existed before this time. Thereafter, descriptions of it were written by Qusṭā b. Lūqā (d. *c.* A.H. 300 [A.D. 912/3]), an-Nayrīzī (d. *c.* A.H. 310 [A.D. 922/3]), al-Bīrūnī, and Abū 'Alī al-Ḥasan b.'Alī b. 'Umar al-Marrākushī.[407] In 1276, like the two forms of universal astrolabe, the spherical astrolabe was described in the *Libros del Saber*,[408] but it is perhaps significant that the treatise there included had to be specially composed by Isaac b. Sid since no extant description of its construction could be found. The instrument does not seem to have become very popular. Few other treatises on it are known from Islam,[409] and references to it in Christian Europe are rare.[410] More difficult, and therefore more expensive to construct than a planispheric astrolabe, the spherical astrolabe offered no obvious advantage of portability or use and had a more-limited capacity than the planispheric model. If few examples of the spherical astrolabe have survived, this is probably because few were ever made.

402. Tannery, 53–5; Price (VI), 61.
403. See Poulle (XIV), 26–9.
404. Maddison (II), 102, n. 6.
405. Maddison & Turner, 131.
406. Maddison (II), 102.
407. Al-Marrākushī's description of the construction of the instrument is translated in Sédillot, 142–8. A detailed discussion of all the Islamic accounts can be found in Seeman.
408. Rico y Sinobas, ii, 113–222.
409. One spherical astrolabe is recorded as having been made by Abū Isḥāq aṣ-Ṣābī for the ruler Qābūs-i Washmgīr (Maddison (II), 103, n. 14). A treatise on the instrument by ar-Rūdānī (1037–1094 [A.D. 1627/8–1683]) has recently been edited and translated by Pellat, to which some additional comments have been added by Janin (II).
410. For some references, see Maddison (II), 103, n. 13.

figure 159. Eastern Islamic
spherical astrolabe signed 'Work of Musa
Year A.H. 885 [A.D. 1480/41].?
(Photograph: Museum of the History
of Science, Oxford)

Astrolabe Related Instruments

Astrolabe Related Instruments

During the Middle Ages and Renaissance, a number of instruments were developed which, while sharing characteristics of the astrolabe, were less versatile than their parent instrument, being intended to perform a limited number of operations either more conveniently or more accurately than was possible with the astrolabe itself. Such instruments attained a considerable popularity and may be considered as part of the history of the astrolabe. They include devices used in surveying (such as the Holland Circle, the circumferentor, and the graphometer) and a number of astrological and time-converting devices. Here, however, we are concerned only with instruments related to the astrolabe chiefly used for time measurement and which share the fundamental characteristic of the astrolabe, stereographic projection.

HORIZONTAL INSTRUMENTS

The horizontal instrument may be considered a special form of single-latitude astrolabe in which the plane of the projection is moved from the equator to that of the observer's horizon. The centre of the projection is the zenith of this horizon and normally only a part of the sphere, bounded by the tropics of Cancer and Capricorn, is drawn. The resulting diagram may be seen in fig. 160. The horizon projects as a full circle ESWN and provides the limb of the instrument. Within this horizon the projection is drawn. The tropic of Capricorn is the short horizontal arc drawn nearest to the horizon AB; the tropic of Cancer is the semicircular arc CD closest to the zenith point Z. The area between the two tropics is divided by the equator EFW, which cuts the meridian line of the instrument SN at F. The ecliptic is represented by the two arcs EGW and EHW, cutting the meridian line at G and H. The hours are marked on the instrument by vertical arcs radiating from the point PW, the pole of the world; for clarity these are not drawn in below the tropic of Cancer. The diagram is completed by horizontal arcs crossing the hour lines. These are lines of declination.

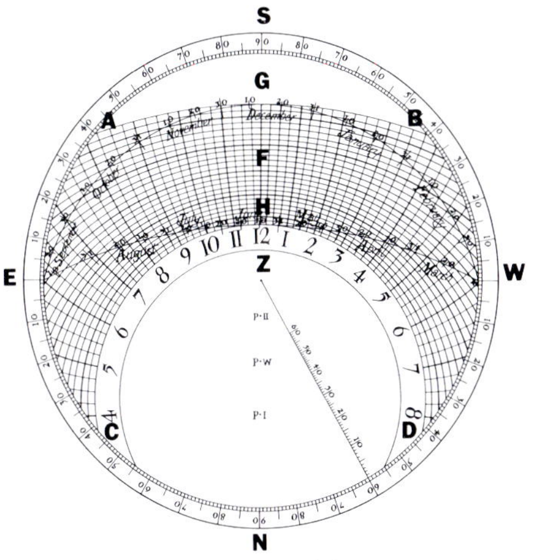

figure 160. The horizontal instrument.

In this projection the almucantars or lines of equal altitude emerge as circles concentric with the horizon. It is clear that if they were drawn in, it would be almost impossible to read the other scales beneath them. They are therefore marked on a rule which is pivoted at the centre Z. PI and PII are the two poles of the ecliptic

figure 161. William Oughtred (1574/?5-1660).

arcs, and there is a degree scale marked around the limb.[1]

The origins of this device, like those of the astrolabe itself, are obscure. Clearly, once the various possible applications of the stereographic projection of the sphere came to be investigated systematically, as to some extent they were during the development of universal projections in Islam, it was likely that the horizontal projection would be derived and investigated. The earliest description of the device known at present is by Ḥāmid b. al-Khidr al-Khujandī in the 4th/10th century, after which there is a gap until the description in the 13th-century instrument treatise by al-Marrākushī.[2] Nothing is known of the subsequent history of the device in Islam nor of its transmission to Christian Europe, where, however, it was certainly known by 1523, when Georg Hartmann signed a quadrant carrying a form of the projection.[3] Four years later Hartmann included a description and illustration of the projection in a manuscript treatise on instruments, and towards the end of the century it was comprehensively described under the name of 'Triens' by Philip Apian and published in 1586.[4] Apian ascribed the invention of the instrument to his father, Peter (1495–1552), who may have derived it – as he did his meteoroscope, an instrument which also uses stereographic projection – from his reading of the works of Johan Werner.[5] Since Werner lived at Nuremberg, his manuscripts may also have provided Hartmann (who, like Apian, is known to have possessed them) with the source for the projection. Be that as it may, Apian's account was certainly drawn up in the first half of the 16th century and thus predates an example of the 'Triens' made in 1569 by Christoph Schissler the Elder at Augsburg.[6] Three other examples of the instrument are known from the 16th century: one by J.A. Linden at Heilbronn in 1596,[7] one probably by Joshua Habermel,[8] and an English example signed 'An Azumoth dyall for the Latitude of 52 deg. inuented by S^r^ Ro: Duddeley Ano 1598.'[9]

Of the sources for Dudley's use of the horizontal projection, which, judging by its form, seems to be independent of the German examples, we know nothing, nor of what contact he had, if any, with the English mathematician William Oughtred (1574/5–1660), who, according to his own account, designed a pocket dial for himself

1. The fullest accounts of the horizontal instrument and its uses remain those published in the 17th century by Forster, Delamain, and Oughtred. The modern literature is limited. Turner (III), and Janin (I) offer the most detailed accounts and there are briefer notices in Michel (III), 129–30; Macrez; Gunther (I), 140–2, 388; Price (I), 10–1. It is evident from the diagram that the common horizontal garden sundial is an example of the horizontal projection reduced to the minimum needed for time determination. Further discussion of it will appear in Volume I, part 2 of this catalogue, which is devoted to sundials.
2. For al-Khujandi, see Cairo Ms DM 970 fols. 5^v–6^r, for knowledge of which I am indebted to Mme M. Archinard. For al-Marrākushi, see Sédillot, 151–3.
3. Germanisches National Museum, W1 26, Zinner (II), 362.
4. Philip Apian, *De Utilitate Trientis, instrumenti astronomici novi Libellus,* Tübingen, 1586.
5. North (I), 61–2.
6. Germanisches National Museum, Nuremberg, Zinner (II), 511.
7. *Ibid.*, 432.
8. Eckhardt, no. 187, pl. 37.
9. British Museum, London; Dept. of Medieval and Later Antiquities. Ward, 346, no. 120.

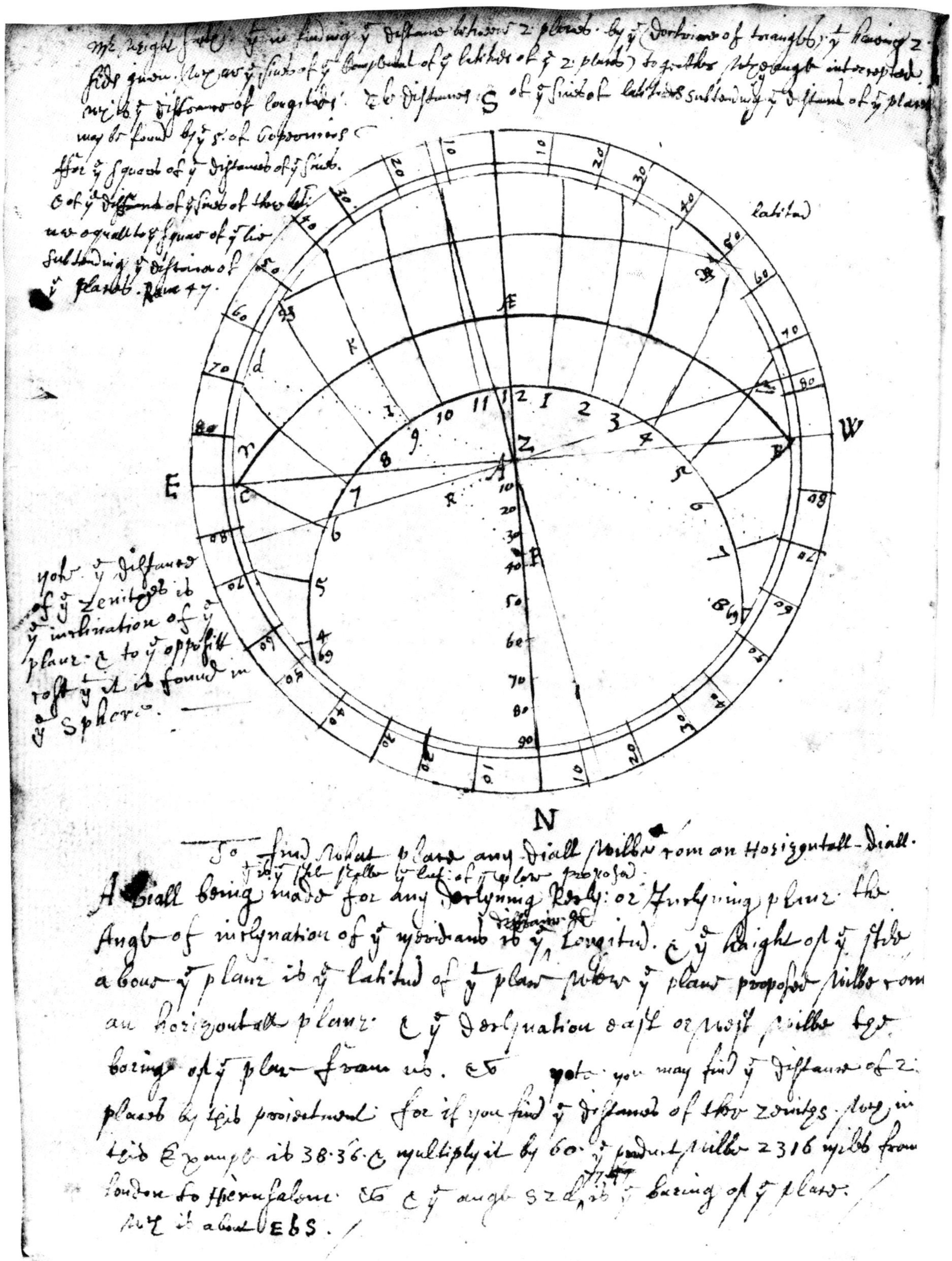

figure 162. The horizontal projection drawn (*c.* 1653) in the notebook of the provincial mathematical practitioner Thomas Brush 'Gardener & practissioner in ye mathematicks.' London, British Library Sloane Ms 3881 fol. 127v. (Photograph: British Library)

using the projection at about this time.[10] Some years later, in 1627, Oughtred described the dial to the London instrument-maker Elias Allen,[11] who began to make an example following Oughtred's instructions. This was promptly plagiarized by fellow instrument-maker Richard Delamain[12] and gave birth to a vituperative quarrel in the following months. Possibly the publicity occasioned by the dispute over the invention of the instrument helped to stimulate interest in it; a number of examples signed, but not dated, by Elias Allen have survived. Thereafter, the instrument was made sporadically throughout the 17th century in England, usually by men who can be shown to belong to a master-apprenticeship tradition deriving from Elias Allen. Indeed, the instrument almost seems to be the characteristic mark of this descent.[13]

The general form of the horizontal instrument is shown in figs. 160 & 162. As described by Oughtred it could be engraved on the back of an astrolabe or some other instrument, or it could be added to an ordinary horizontal garden sundial.[14] In this form, which was the most popular, it was engraved on one side of a square, octagonal, or circular plate. An hour scale for an ordinary horizontal dial was engraved on the edge and a gnomon was supplied. The gnomon has two shadow casting edges, one of which is the oblique edge normally found on horizontal dials, the other a chamfered vertical edge that works on the projected diagram in the centre (fig. 163). In this form the instrument was first published separately in 1636.[15]

What were its capabilities? Basically, apart from enabling its owner to know the time, the instrument allowed most problems concerning the position of the sun throughout the year to be determined graphically. In this lay part of its appeal, for such problems could thus be solved without calculation and for all but a very few men in the 17th century even the simplest problems of multiplication and division were considered difficult. With the instrument, however, all was easy. If, for example, one wished to know the time

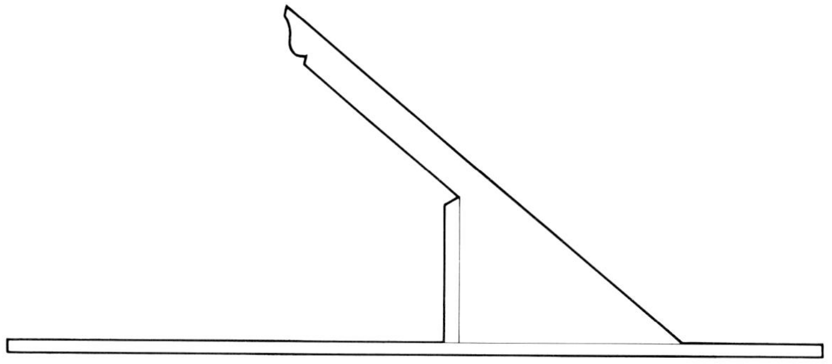

figure 163. Gnomon for a double horizontal dial.

10. Turner (III), which see for all that follows concerning the horizontal instrument in England.
11. For whom, see Volume I, part 2 of this catalogue.
12. In the form of a quadrant. For Delamain's career, see Turner (IV).
13. Turner (III), who includes a checklist of surviving instruments.
14. For horizontal dials in general, see Volume I, part 2 of this catalogue.
15. Oughtred.

of the sun's rising or setting for a particular day of the year (information that in a world without artificial illumination might well be of some moment), it was only necessary to find the declination line of the sun corresponding to the date required and to note the two points where it cut the horizon. The time of sunrise could then be read off from the hour scale on the limb on the east side of the instrument, and of sunset on the west side at these points. Finding the sun's declination in the first place was equally straightforward. All that had to be done was to find the date in the ecliptic and count how many days distant it was from the equator. Thus on 11 August there are six lines between the date and the equator in fig. 164. Since these lines are drawn for every 2°, the sun's declination is 12° north. On larger instruments the parallels are usually drawn in for each degree. The instrument could also be used for the laying out of fixed vertical and horizontal dials, and when being set up itself had the great virtue of being self-orienting.[16]

Like the astrolabe itself, by the early decades of the 18th century the horizontal instrument was beginning to fall out of use, although it was never entirely forgotten. In Germany its demise was more rapid – indeed, no examples later than the 16th century have been noted from any region of Europe except England and France.[17] Forgotten for nearly two centuries, a form of the instrument with the hour lines drawn as curves of the equation of time to give a mean time reading was invented *c.* 1898 by G. Howyan of Suczawa (Austria). It was later revived in 20th-century America as a simple device for indicating the position of the sun.[18]

16. Use of the dial for laying out other dials was briefly described by Oughtred, whose indications were greatly augmented by R.L.
17. Ozanam.
18. For Howyan, see *Bull. de la Société Astron. de France*, xvii, 1903, 162 & 208. For the American device, see cat. no. 27 below. The stimulus for this development was the usefulness to a photographer of knowing the position of the sun at a given time. The initial re-invention of the instrument was carried out by graphical methods by Commodore Harold Dodds, U.S.N. during World War II in association with Thomas Spencer. (Private communication from Thomas Spencer, 1 April 1981.)

Astrolabe Related Instruments Catalogue Horizontal Instruments

26. English Double Horizontal Dial

c. *1650*
Brass
Diameter 13 5/8 in. (346 mm)
Not signed, although a signature may have been erased between the VIII and IIII hour numerals. Inscribed at a later date 'The Gift of Eliza Hayton from Bledlow House Bucks.'
Inventory 1614

The octagonal plate has eight attachment holes and is engraved round the edge with a circular hour scale for the conventional horizontal dial (IIII–XII–VIII) reading to minutes. Within the hour scale is a circular scale of degrees (0°–90°–0°–90°–0°, reading from the top) enclosing the horizontal projection. Beneath the arc of the tropic of Capricorn is the exhortation, probably added at the same time as the dedicatory inscription, 'From the Rising of the Sun THE LORD'S NAME IS TO BE PRAISED unto the going down of the same.'[19] Engraved from the centre of the plate towards the right (west) side is the rule graduated with azimuths. The gnomon (not shown) is a modern and incorrect reconstruction.

This example of the double horizontal dial is exactly of the form described by William Oughtred in 1636. The calligraphy of the numerals and month names is very close to that of Elias Allen, but shows sufficient differences to suggest that it is either a very late product of Allen's workshop, or that it was made by one of his immediate successors.

Bledlow House, Buckinghamshire, referred to in the inscription and where this dial was originally placed, did not belong to the Hayton family in the 17th century but was acquired in the 18th century by Henry Cross, whose son, Thomas, held the manor in 1745, his heiress being Elizabeth (d. 1788, aged 86), who later married William Hayton of Ivinghoe (d. 1764).

19. Psalm 113:3. Eden & Lloyd, 260, no. 321.

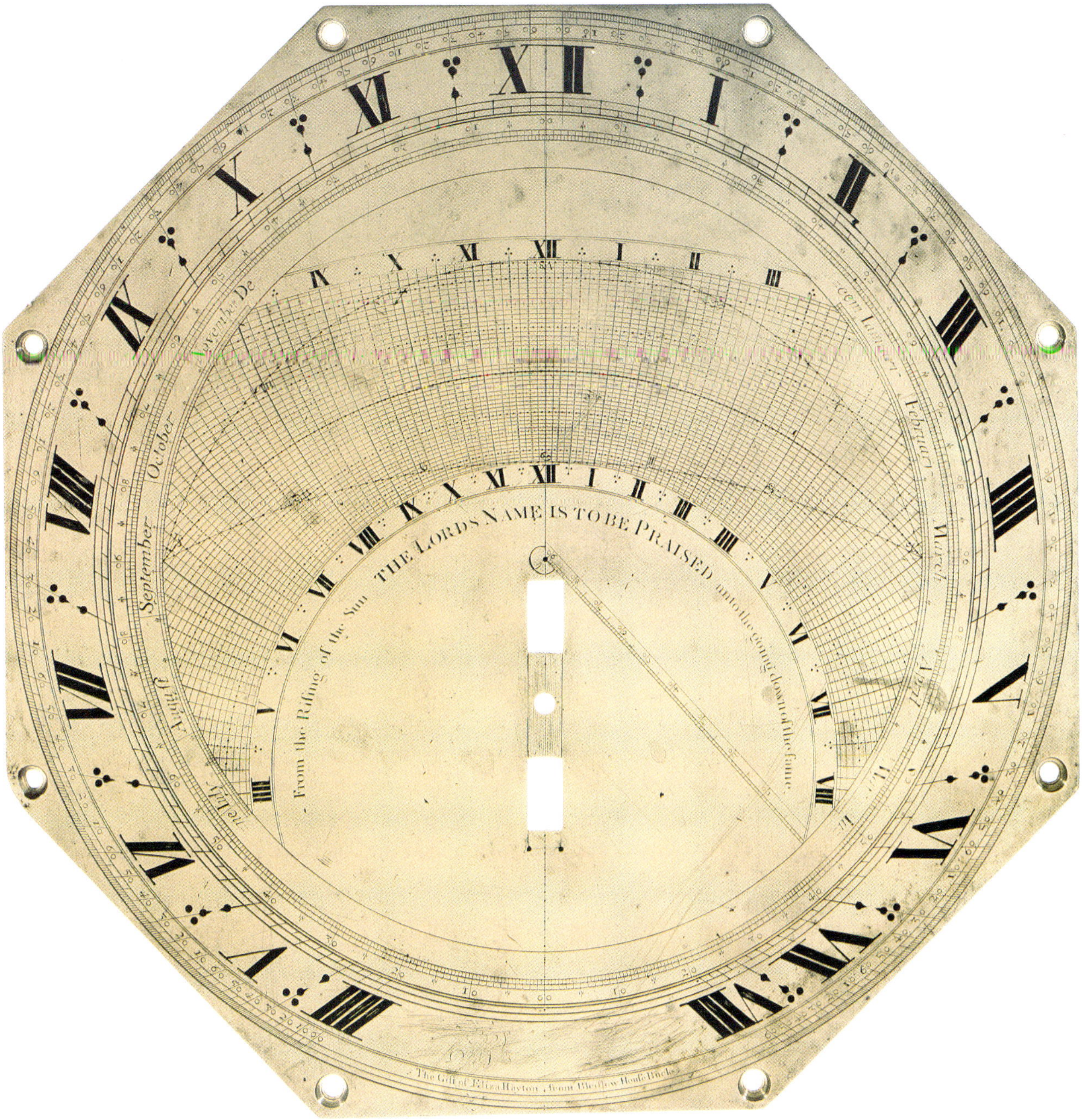

figure 164. Catalogue no. 26: English double horizontal dial.

27. American Horizontal Instrument

1973
Plexiglas and vinyl, with brass gnomon printed in blue and red
4½ in. square (114 mm)
Signed: SUNDICATOR/Copyright 1970, 1973/THOMAS SPENCER

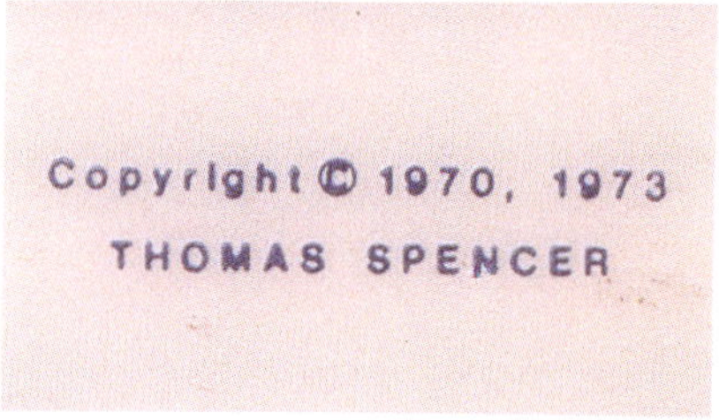

Inventory 1527

A simplified form of the horizontal projection is printed on a square tablet. The projection is contained within a circle divided into 360°, and the cardinal points are marked. Above the hour scale is a calendar table and at the centre of the base a small brass socket for the gnomon. Overlaying the base scale and rotatable is a transparent plastic disc printed with a degree scale for the almucantars. Inset in the top-left corner is a small magnetic compass with paper card showing eight directions and marked 'GERMANY.'

The principle of this instrument is the same as that of the double horizontal dial described in catalogue no. 26. It is used in a similar way. The purposes for which its 20th-century manufacturer intended it were: 'to help the *photographer* decide when the sun will be just right for his prize-winning photograph. It will assist the *home-owner* and *architect* with orientation for heat and light control. The Sundicator shows the *gardener* where to plant for the best combination of sun and shade. It is an emergency navigational aid to the *yachtsman*. With it *scouts* will learn to tell time by the sun and find their way without a compass.'[20] This example is the deluxe model made for latitude 40° north.

20. Spencer, 3.

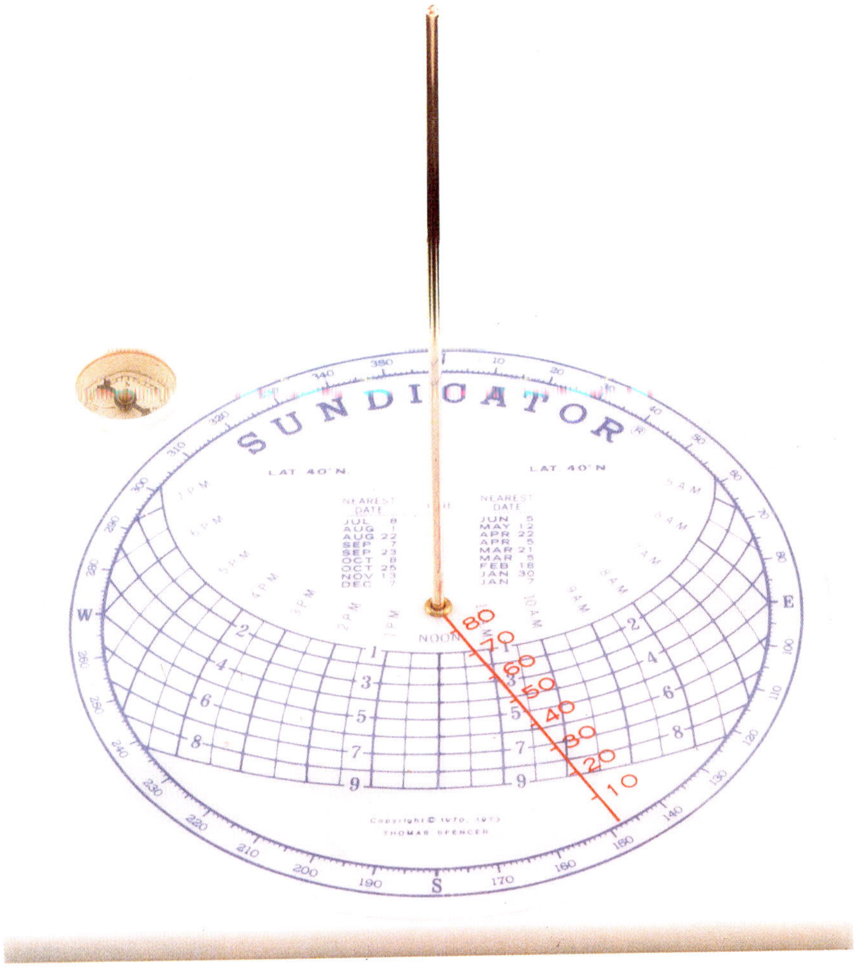

figure 165. Catalogue no. 27: American horizontal instrument – Sundicator.

ASTROLABE QUADRANTS

Although the degree scales engraved on astrolabes were often divided in a single sequence of 0°–360°, for the measurement of height or depth this was unnecessary, and a single quadrant scale of 90° would suffice. From early times quadrants had been used for simple altitude measurement, and these were later adapted to trigonometrical and gnomonic uses.[21] Stereographic projection offered still further possibilities and led to the development of a quadrantal form of the astrolabe.

If the *rete* of an astrolabe is (arbitrarily) supposed fixed so that the first point of Capricorn lies on the east-west line of the instrument, and the astrolabe is then twice folded, once about this line, and once about the vertical meridian (fig. 166), the path of the main lines traced on the instrument will be that shown in fig. 167. If this diagram, a quarter of a stereographic projection for a particular latitude, is then engraved on a flat piece of wood or metal, fitted with sights on one of the straight edges, and, at the apex, with a plumb-line carrying a sliding head or pearl, the result is the instrument known even in the Middle Ages as the *quadrans novus*.[22] Its capacities would be similar to those of an astrolabe, although perhaps it is less commodious. To find the time, for example, the user first measured the altitude of the sun above the horizon by means of the two sights set along one of the straight edges, allowing the plumb-line to hang freely. The point where the line cut the degree scale on

figure 166: Folding the stereographic projection to produce a quadrant.

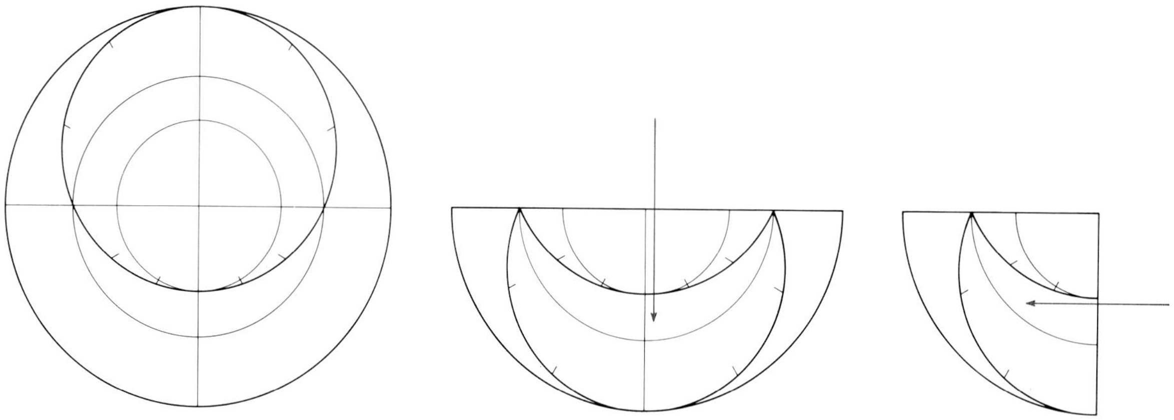

21. See Maddison (I), 13–4; this catalogue, Volume I, part 2; Anthiaume & Sottas, 71–5.
22. The name served to distinguish the instrument from earlier forms of quadrant. The principles of the construction and use of the *quadrans novus* have been extensively described by Anthiaume & Sottas, and more briefly by Michel (III), 123–6. The lucid description by Morley (II) provides the Arabic terminology for the scales, while Poulle (VI) supplies the best survey of the instrument in the medieval West.

the limb indicated the required altitude. Having determined the sun's position in the ecliptic for the date of observation – from a set of tables or an almanac if no scale for this purpose was provided on the instrument – the user placed the plumb-line so that the bead or pearl sliding on it could be set against that point of the ecliptic equivalent to the sun's position. He then rotated the plumb-line from the apex, keeping the pearl always in the same position on the thread, until it lay upon the almucantar equivalent to the sun's observed altitude. This operation is, of course, equivalent to rotating the *rete* of an astrolabe.[23] The position of the plumb-line on the limb then showed the time either directly in hours if an hour scale was marked on the limb, or in degrees which were to be converted into hours if no hour-scale was engraved.

The *quadrans novus*, as even this simple example shows, was more complicated to use than the astrolabe and also had the disadvantage of not showing directly the circular celestial movements. Since, however, it had no pierced or movable parts it was considerably easier to make than the astrolabe and was therefore cheaper. Being lighter it was also more easily carried. That it was more widely used than the astrolabe we do not know. Very few medieval examples have survived from either Islam or the Latin West.[24] The incidence of manuscripts, however, describing its use is considerably higher than the number of surviving examples. Commonly used or not, the *quadrans novus* was evidently taught.

The earliest description of the *quadrans novus* known at present is that by the Judaeo-Provencal mathematician, astronomer, and zoologist Ya'aqob ben Mahir ibn Tibbon, known in Latin as

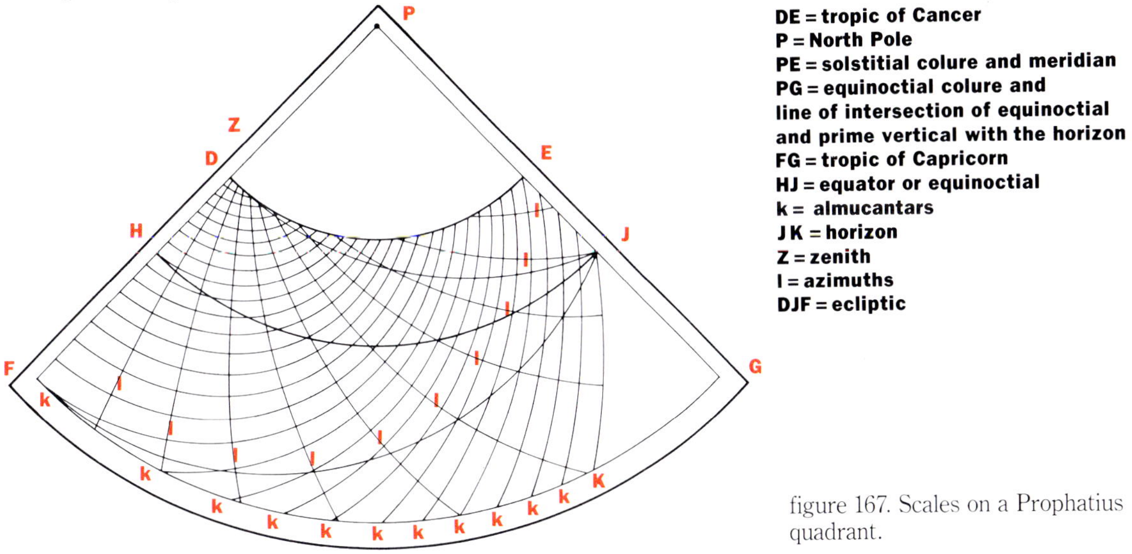

figure 167. Scales on a Prophatius quadrant.

23. *Cf.* the method of finding time with an astrolabe described above. It may be noted here that in the first version of his treatise Prophatius described a method for finding the time which used the sine quadrant usually engraved on the back of the *quadrans novus* and a short trigonometrical calculation (Michel (III), 128). The obvious practical disadvantages of this in everyday life sufficiently account for its disappearance.

24. Four European examples are known. These are preserved at Merton College, Oxford; Musée des Antiquités de la Seine Maritime, Rouen; Musée de Saint-Jean, Angers; and the Smithsonian Institution, Washington, D.C. For Islamic examples, see Maddison & Turner, 152–4.

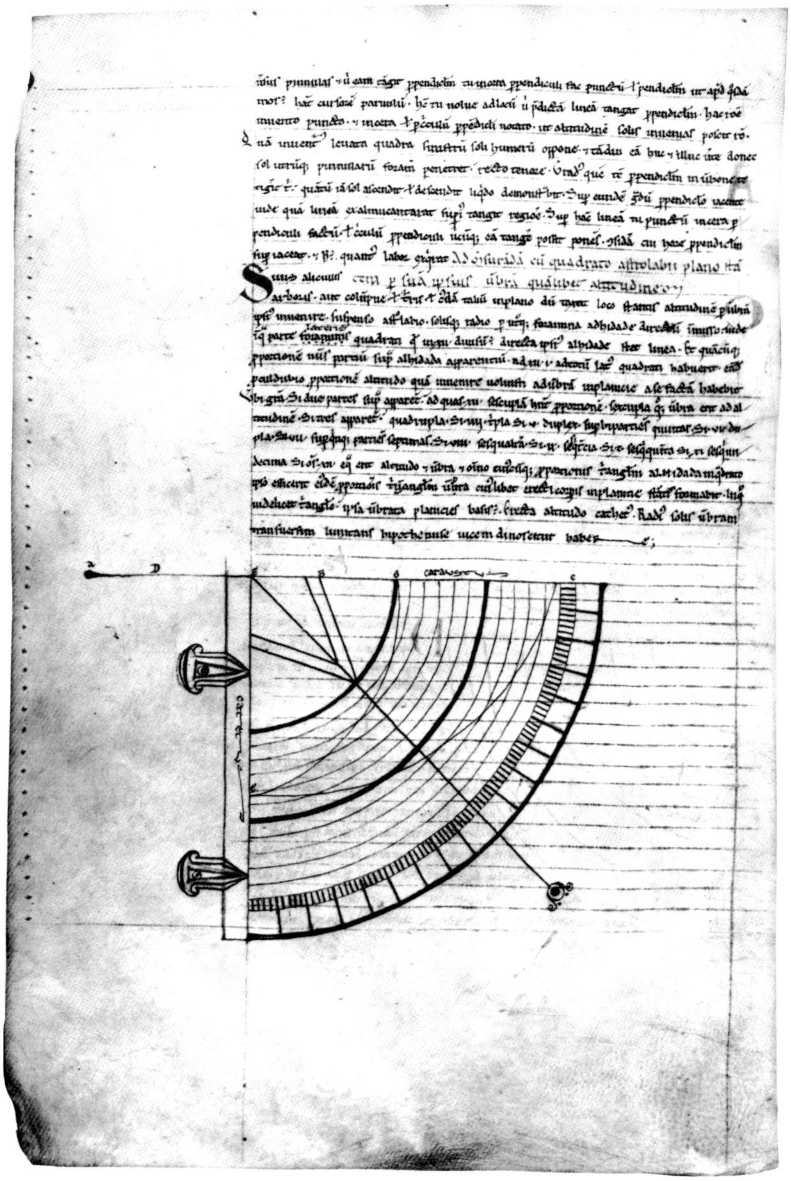

figure 168. A *quadrans novus*, or Prophatius quadrant, illustrated in an addendum describing various instruments to a late 13th-century copy of the treatise on the astrolabe of Hermannus Contractus. London, British Library Ms Royal 15.B.IX fol. 60v. (Photograph: British Library)

Prophatius Judaeus and in Romance as Profeit Tibbon (*c.* A.D. 1236–*c.* 1304).[25] Written in Hebrew in 1288, Prophatius' treatise was translated into Latin two years later by Ermengaud Blasius under its author's supervision.[26] Both versions were prepared at Montpellier. The treatise, however, was quickly disseminated. In 1293, Peter of St. Omer (Petrus de Sancto Audomaro) who may probably be identified with Peter Nightingale,[27] wrote his own treatise, *Tractatus novi quadranti*, at Paris, introducing a number of improvements to the instrument. Some of these were incorporated in a revised

25. For his life, see Millás (II).
26. Poulle (VI). We may note that a short text on the quadrant with an illustration is found among additions, describing instruments, to a late 13th-century copy of the treatise on the astrolabe of Hermanus Contractus (British Library Ms Royal 15 B ix f 60v.) The relationship of this text with that of Prophatius has, however, yet to be determined. I am indebted to D.L. d'Avray and Emmanuel Poulle for their advice on the date of this manuscript.
27. For the complex problem of Peter's identity, see Pedersen (I), 5.

version of Prophatius' treatise written in 1301. The interest that the instrument immediately aroused is shown by the number of treatises on it. At least fifteen manuscripts of Peter Nightingale's *Tractatus*... are known.[28] An anonymous account appeared in 1316; another, by Andalo di Negro, perhaps in 1324; and a third, by Eligerus of Gondesleren, in 1349. By this time, however, the instrument was also certainly known in Islam; two surviving examples by Muḥammad b. Aḥmad b. 'Abd ar-Raḥīm al-Mizzī (d. A.D. 1349) are dated A.H. 727 (A.D. 1326/7).[29] Al-Mizzī himself wrote a treatise on the instrument[30] as well as several on astrolabes. The closeness of the dates of the *quadrans novus* manuscripts in Islam and in Christian Europe raises the possibility of an earlier common source in Islam.[31]

The *quadrans novus* thus many times described in the early 14th century continued to be taught and studied throughout the 15th century. John of Gmunden devoted three treatises to it,[32] and *résumées* of the instrument included in treatises on quadrants in general were not uncommon.[33] The *quadrans novus* itself, however, shows no changes or development after the revised version of Prophatius' treatise of 1301. The embellishments found in the publications of such 16th-century scientific popularizers as Peter Apian and Oronce Fine added little that was new.[34] Indeed, after the 14th century in Europe the *quadrans novus* has no history – only a bibliography. Rivalled increasingly during the later 16th and 17th centuries by other forms of horary quadrants, it gradually fell out of use, disappearing quietly at about the same time and in the same way as its parent instrument, the astrolabe.

But if the *quadrans novus* had a relatively short life in the West, it endured far longer in Islam. From at least the 17th century onward in the Ottoman empire an attractive form of quadrant was produced in lacquered wood, which included a Prophatius quadrant on one of its two faces.[35] This form of instrument seems to have been made in considerable quantities and continued in use until the early 20th

28. Pedersen (I), 4.
29. They are now in the Museum of Islamic Art, Cairo (no. 3092) and the British Museum, London (95.11–16.1). Mayer (I), 61. Morley (II). Three other *quadrans novus* by al-Mizzī are also known, Brieux & Maddison.
30. *Ar-Rawḍāt al-muzhirāt fi-l-'amal bi-rub'al-muqanṭarāt* (The Blossoming Gardens of the Use of the Astrolabe-Quadrant). Maddison & Turner, 152.
31. Pedersen (I), 10, suggested the possibility that Peter Nightingale may have been working from an independent Latin translation of Prophatius made before 1293. He may equally well have been working from a translation made from an independent Islamic source. Clarification of the origins of the quadrant will no doubt be obtained when Ms Istanbul Haci Mahmud Ef. 5713 is examined. Dr. D.A. King (private communication) had indicated that fols. 10v–25v contain a treatise on the quadrant by an Egyptian astronomer, Abu'l-Haṣan 'Alī b. Muḥammad known as Ibn al-Ḥammamī. The date of the copy is illegible, but Dr. King estimates a mid-12th century date for the treatise.
32. This was widely used, especially in John's own University of Vienna. Mundy, 200. Perhaps one of the two quadrants willed by John to the university was a *quadrans novus*.
33. See, for example, the description of Bibl. Nat. Ms 7294 of 1433 in Anthiaume & Sottas, 78–9.
34. *Ibid.*, 80 ff.
35. See cat. no. 28 for a typical example.

century;[36] it was perhaps the typical astronomical time-telling instrument of the later Ottoman world. Numerous examples have survived, and a production of Prophatius quadrants (although in metal rather than wood) developed in Turkish-influenced regions of the Maghrib. Throughout all this time, however, the instrument showed neither change nor refinement.

Among the many quadrants specifically developed during the 16th and 17th centuries for time measurement which contributed to the decline of the *quadrans novus*, a few continued to use stereographic projection. Quadrant forms of the horizontal instrument produced in 16th-century Germany have already been discussed.[37] Thereafter, most examples of such instruments that have survived are English – a fact perhaps indicative of the still backward state of English mathematical practice. In 1618, Edmund Gunter (1581–1626) showed a newly finished quadrant of his own devising that employed stereographic projection to Henry Briggs and William Oughtred at Gresham College, and it was compared with the latter's horizontal instrument. Five years later Gunter described the instrument in his notable treatise, *The Description and Use of the Sector, Cross-Staff and other Instruments*, London, 1623. The quadrant generally named after him, and of which his invention has not been questioned, consists of a stereographic projection of part of the equator, tropics, ecliptic, and horizon onto the plane of the equator. Thus in fig. 169, point A being the centre, the arc ED represents either of the tropics of Cancer or Capricorn, and FG the equator. The ecliptic projects as the arc FD and on the edge at FE a declination scale of 23½° is marked, showing the distance of the sun from the equinoctial FG toward either of the tropics. On the limb BC, a scale of degrees is marked, and in the space between it and the tropic ED, a calendar scale JK is projected. The arc FH represents the horizon, which is divided into forty unequal parts. The hour lines are drawn as arcs on the left of the instrument, and numbered 1–12. Those lines curving to the left show the time in winter, those curving to the right, the time in summer. The arcs drawn on the right of the instrument and numbered 30–120 are azimuths. On the meridian AC, a pair of sights LM are mounted. The instrument is completed by hanging a plumb-line with a sliding bead from the centre A. If desired, some of the stars located between the equator and the tropic can also be projected on the instrument.

Like the *quadrans novus* and the horizontal instrument, indeed like any astrolabe quadrant, Gunter's quadrant enabled its user to solve mechanically most problems that related to the sun's position. Thus the sun's right ascension could be found simply by laying the plumb-line across the point of the sun's position in the ecliptic and

36. Maddison & Turner, 155–7, who list representative instruments from the 17th to the 19th centuries; Rohr & Janin, 115 for a Turkish example dated A.H. 1330 (A.D. 1911/12).
37. Above, p.202 . For non-stereographic forms of horary quadrant, see Volume I, part 2 of this catalogue.

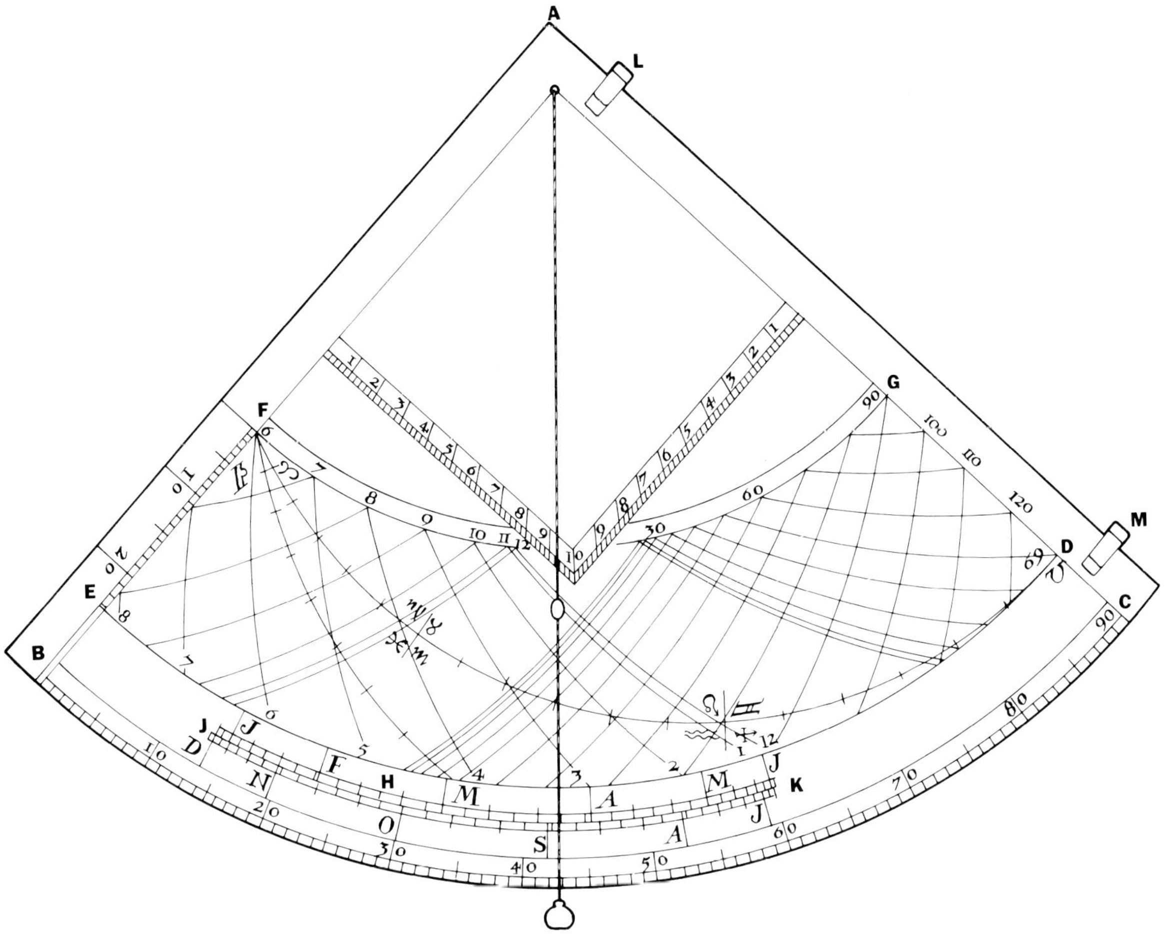

figure 169. Scales on a Gunter quadrant.

reading off the right ascension from the point where the plumb-line crossed the scale of degrees on the limb. Time-finding was also relatively simple. First the sun's altitude was found by observation using the sights, plumb-line, and degree scale on the instrument. The bead on the plumb-line was then set for the sun's declination. The plumb-line was then laid to the degree of the observed altitude, and the position of the bead among the hour lines showed the time.[38] This mechanical simplicity of the Gunter quadrant commended it to the lay public. The fact that it was relatively simple and quick to make commended it to the instrument-makers. These qualities combined with other advantages. The instrument did not need to be very large to work to an acceptable accuracy of perhaps 5 to 10 minutes error, and a radius of as little as 4 or 5 inches is common among surviving quadrants. Little metal was thus employed in its making, and it was therefore cheap and light. These factors ensured its popularity, which was further guaranteed by the attention drawn to it by the many descriptions and enlargements published in the decades following Gunter's early death in 1626.

38. Further examples of the construction and use of Gunter's quadrant can be found in Stone, 193–7 and in the works listed below.

figure 170 (opposite). An astronomer, supposed to be King Alfonso the Wise of Castile, shown calculating at his desk using a quadrant. Although the artist has made some effort to represent the scales and shadow square, these are still inaccurate. Vienna, Staatsbibliotek Codex Latinus 2352. (Photograph: Staatsbibliotek)

Gunter's treatise of 1623 was immediately reprinted in 1624, and a second edition appeared in 1636.[39] In 1624 also, Elias Allen published a pamphlet by Gunter, *The Use of the Quadrant*, to accompany examples of the instrument which he made.[40] Not only was the instrument thus immediately in commercial production, but it also seems to have offered a challenge to other mathematicians. In 1630, John Hulett (?1607–63), who in that year proceeded to take his B.A. degree from Hart Hall, Oxford, made a Gunter quadrant for himself which has survived.[41] Perhaps at the same time he wrote a short lucid tract on it, later published.[42] Its title sums up the nature of the instrument's appeal, which was essentially astronomical: *The Description and Use of a Quadrant by which all the most usefull and necessary Propositions of both the Globes are easily and exactly performed; as the right Ascension, Declinations, Altitude, Amplitude, Rising, Setting, Azimuth and Hour of the Day*. . . . The same year, 1630, another Gunter quadrant that has survived was completed by John Chatfield.[43] In the following years the possibilities of the instrument were explored by Samuel Foster, whose results were included in the third edition (1653) of Gunter's works by its editor, Henry Bond.[44] By this time, however, the instrument was well established and subsequent works on it merely reflect its popularity, both as a practical instrument for everyday use, and as a simple instrument that *amateurs* of mathematics could make for themselves. Several examples of such instruments have survived.[45]

In 1658, the most comprehensive English treatment of quadrants based on stereographic projection was published.[46] John Collins', *The Sector on a Quadrant*. . ., not only gave a comprehensive account of the Gunter quadrant, but also described other forms. One of these was a quadrant devised by Thomas Harvey, a friend of both John Collins and of the engraver and instrument-maker Henry Sutton (d. 1665). Sutton, whose mathematical competence seems to have been considerable, quickly saw the possibilities of the quadrant and began to make examples. As Collins explained, 'the said M. *Sutton* conceiving that it would be an advancement to their Trade in general, besides satisfactory to the desires of the studious in Mathematiques, to have the uses of a good Quadrant, prevailed with me, in regard M. Harvey was not at leisure. . . to write two or

39. STC, 12521–3.
40. Taylor (II), 345, no. 139.
41. Now in a private collection, London. It is illustrated in Wynter & Turner, 24.
42. Editions appeared in 1665 and 1672, but there may have been earlier ones because Hulett died in 1663.
43. It is now in the National Maritime Museum, Greenwich (Q25/66–21), see Stimson, 26–4. Little else is known of Chatfield except that in 1650 he wrote for Anthony Thompson a work entitled *The Trigonal Sector*, Taylor (II), 236, no. 236.
44. *Ibid*, 359, no. 210.
45. See, for example, cat. no. 29 below and the instrument by HK illustrated in Wynter & Turner, 25.
46. Taylor (II), 364, no. 238. Immediately reprinted in 1659 with a description of another quadrant added, it was further revised and enlarged as late as 1750.

three sheets. . . .'[47] Collins' 'two or three sheets' became a rather longer treatise accompanied by finely engraved examples of the instrument by Sutton. These were available separately in three different sizes – 4 inches, 5 inches, or 10 inches. They could be had in paper or already mounted on wooden boards.[48] They were also available, though more expensively, in brass.[49]

The quadrant devised by Thomas Harvey, which came to be known as 'Sutton's quadrant'[50] or occasionally as Collins', is in effect a simplified form of the *quadrans novus*. Since nothing is known of Thomas Harvey, however, we cannot know if he was aware of the earlier tradition or not. As may be seen in fig. 183, the quadrant consists of a network of altitude parallels (i.e. the almucantars of an astrolabe plate) and azimuths contained within the arcs of the tropics of Capricorn and Cancer. Crossing the grid thus formed are the arcs of the ecliptic, marked with the signs of the zodiac, and from the point on the left edge of the instrument whence these arcs arise, the horizon is also drawn. Scales for hours, minutes, and degrees are marked on the limb, and in the apex is a calendar scale. Use of the instrument, thanks to the positioning of these scales, is considerably simpler than that of its medieval predecessor. To find the hour, for example, the sun's altitude was measured, the plumb-line was laid across the appropriate date on the calendar and the bead adjusted to the proper declination. The plumb-line was then rotated until it lay over the appropriate parallel of altitude. The position of the bead among the azimuths then showed the place of the sun and the position of the plumb-line against the hour scale on the limb showed the time.

Gunter's and Sutton's quadrants both continued to be made until at least the middle of the 18th century, and they survived in textbooks and dictionaries even longer. Occasionally they were combined with other diagrams to form a composite quadrant instrument.[51] Such efforts, however, like the changes which were made to the scales on the reverse of the instrument, are perhaps evidence as much for the decline of interest in the instrument as for the makers' ingenuity. Although it survived its parent instrument, the astrolabe-quadrant did not survive the 18th century in any form, and the stagnation of interest in it is clearly shown by the fact that descriptions of the instrument in many late 18th-century encyclopedias were simply copied verbatim from the work of Edmund Stone.[52]

47. Fol. A2r.
48. For an example, see below no. 33. *Cf.* the fine set now in the Science Museum, London.
49. E.g. Science Museum, London. Museum of the History of Science, Oxford. Orrery Collection, 41. See Gunther (I), 771.
50. Brief descriptions of it under this name can be found in several 18th-century encyclopedias from Harris onward. Stone, 197, also describes it briefly under this name.
51. See, for example, the paper-on-wood quadrant by John Prujean now in the Museum of the History of Science, Oxford, which contains Gunter, Sutton, and horizontal quadrants.
52. E.g. Middleton's *The New Complete Dictionary of Arts & Sciences*. *S.V.* 'Quadrant': Alexander Jamieson, *A Dictionary of Mechanical Sciences, Arts, Manufactures and Miscellaneous Knowledge*, London, 1836, ii, 863.

Astrolabe Related Instruments Catalogue Astrolabe Quadrants

28. Turkish Quadrant

Late 18th or early 19th century
Wood, lacquered yellow with red edges; lines, script, and numerals written in black and red
Length of side 6⅝ in. (128 mm)
Not signed or dated
Inventory 985

On the face is a Prophatius astrolabe-quadrant surrounded by a scale of 90° on the limb. A point marked on the body of the instrument at approximately 43½° is named as 'the two festivals.'[53]

The two scales set above the projection are named as 'extra half' (upper) and 'first afternoon prayer' (lower). The inscriptions in the two cartouches read: 'After the sun's decline, the extra half [niṣf-i faḍla] is given to the first side of the arc of altitude. *Tamkīn*[54] is given to the other side' (left). 'In the winter signs, the extra half is given, before the sun's decline, to the other side of the arc of altitude. *Tamkīn* is given to the first side (right).'

On the back is a sexagesimal sinical quadrant marked with the arcs of sines, 'Versed sines,' and the arc of the obliquity of the ecliptic. The horizontal arc traversing the instrument from the end of the right-hand degree scale to the 42°50′ division is the line of the first afternoon prayer. If the plumb-line is held against the quadrant set to the degree on the limb equivalent to the sun's maximum altitude, the point at which it cuts the prayer arc indicates the height of the sun at the moment of prayer. Below this arc is a chord. On the limb is an hour scale and within this a scale of 90°. The inscriptions in the three cartouches read: 'In the summer signs, the polar distance is given to its first side, six months; in the winter signs, to its other side, six months' (left); 'After the sun's decline, the extra half and *Tamkīn* are given to the other side of the arc of altitude' (middle); 'In the summer signs, before the sun's decline, the extra half and *Tamkīn* are given to the first side of the arc of altitude' (right). At the apex is an inset brass eye through which passed the double plumb-line now missing. The quadrant is accompanied by a cloth bag that is probably original.

53. I am indebted to Dr. G.L. Lewis, Oxford, for the translations of the Ottoman Turkish inscriptions on this instrument and for the explanation of the 'Two Festivals': 'This was used to help fix the time of the prayer on the Festival of Sacrifice and the Festival which ends the Fast of Ramadan.'

54. Literally 'positioning,' 'stabilizing,' but used as a technical term relating to the Fast of Ramadan.

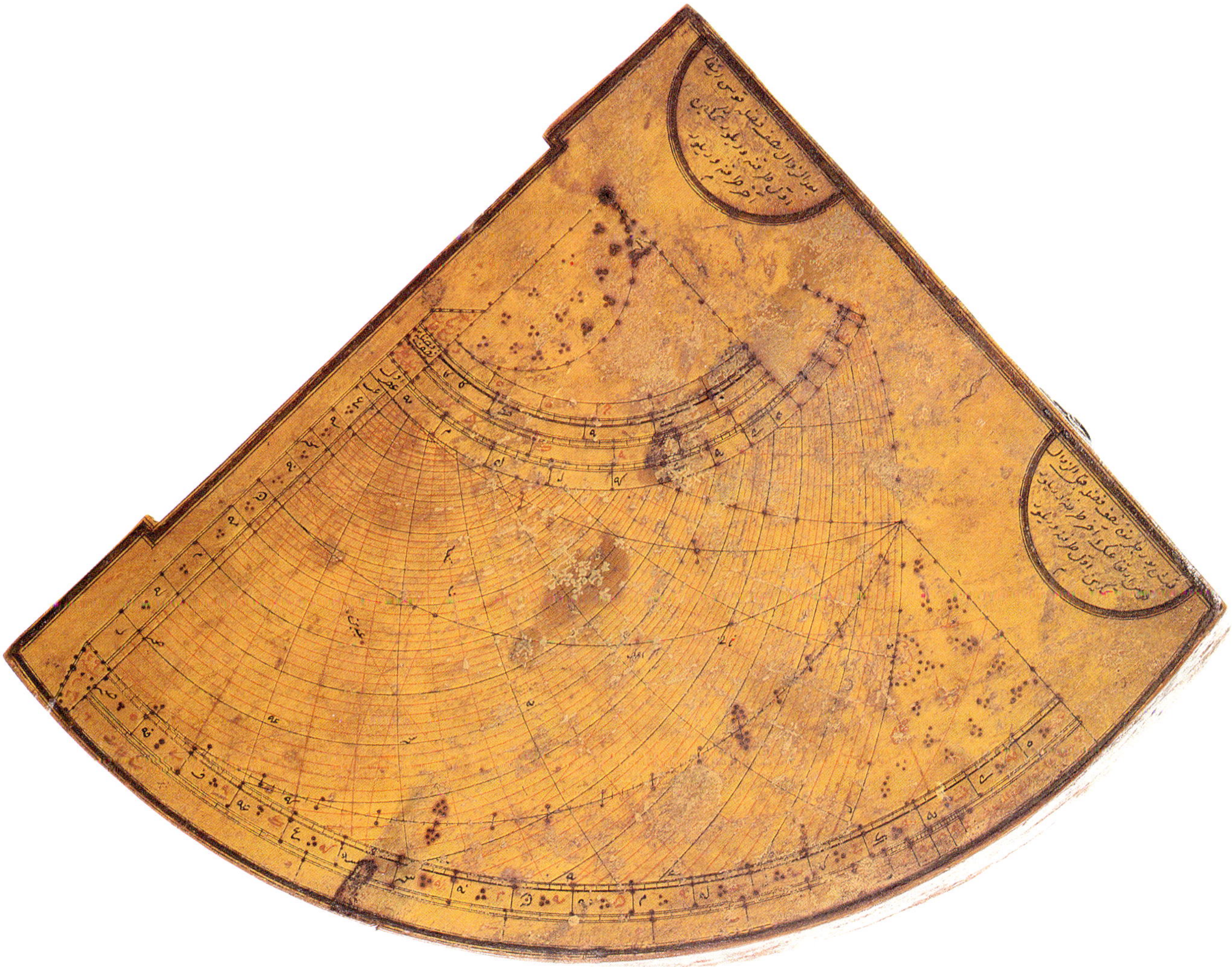

figure 171. Catalogue no. 28:
Turkish quadrant.

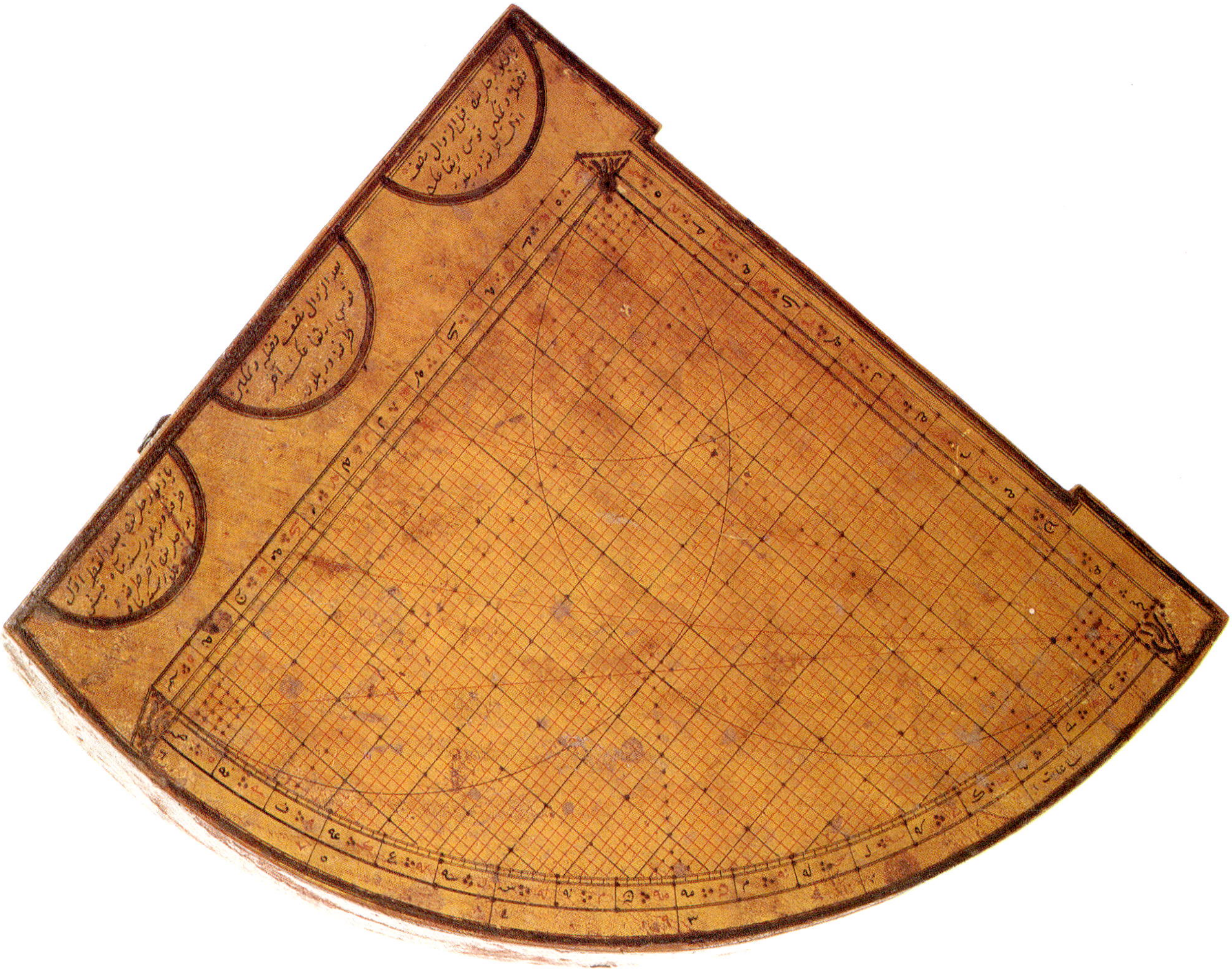

figure 172. Catalogue no. 28. Reverse.

Because of the manner of construction (fig. 167), a part of the ecliptic is hidden on the quadrant. The instructions are intended to enable the user to make the correction needed during the period of the year when direct readings cannot be made from the ecliptic scale. In the absence of any definition of the terms used, however, the exact meaning is uncertain.[55]

55. G.L. Lewis draws attention to a parallel passage in a 19th-century Turkish work 'On the Almucantars' *(Terjume-i Gedusi lil-Muqantarat*, 7), in which instructions are given for how to find the extra half: 'Put the plumb-line on the degree of the sun and mark the indicator on the degree of the sun, then move it until the indicator falls on the horizon. Then observe: if it is in the southern signs, from the degrees of the arc of altitude, however many degrees there are between the plumb-line and the line of sunrise and sunset, the extra half for that day will be so much.' But in this work also the terms are not defined.

29. English Quadrant

1652
Brass
Length of side (this is not a radius) 4 in.
(103 mm)
Not signed
Inventory 545

On the face is a Gunter quadrant with calendar scale and an arc of 90° on the limb. In the apex is a shadow square and within this a sun emblem above a scroll inscribed 'Elevatio 52°40'.' This latitude would be appropriate for several towns in the English Midlands (of which the most likely are Leicester, Norwich, and Great Yarmouth – although Shrewsbury, Lichfield, and Peterborough may also be considered possibilities). On one edge is a pair of vanes with pin-hole sights. On the back are five concentric rings containing (reading from the outermost toward the centre): 1) a scale of years from 1652–86, 2) Dominical Letters and the Epact, 3) Sunday dates, 4) months, and 5) zodiac signs which together form a zodiacal calendar (0° Aries = 10 March). At the centre is a lunar volvelle with a sixteen-point wind-rose at its centre. In the corners are emblems of the moon, a rose, and a thistle.

This crudely made instrument, of which the authenticity might be considered doubtful, is possibly the work of a 17th-century *amateur* working from a printed book.

figure 173. Catalogue no. 29. Reverse.

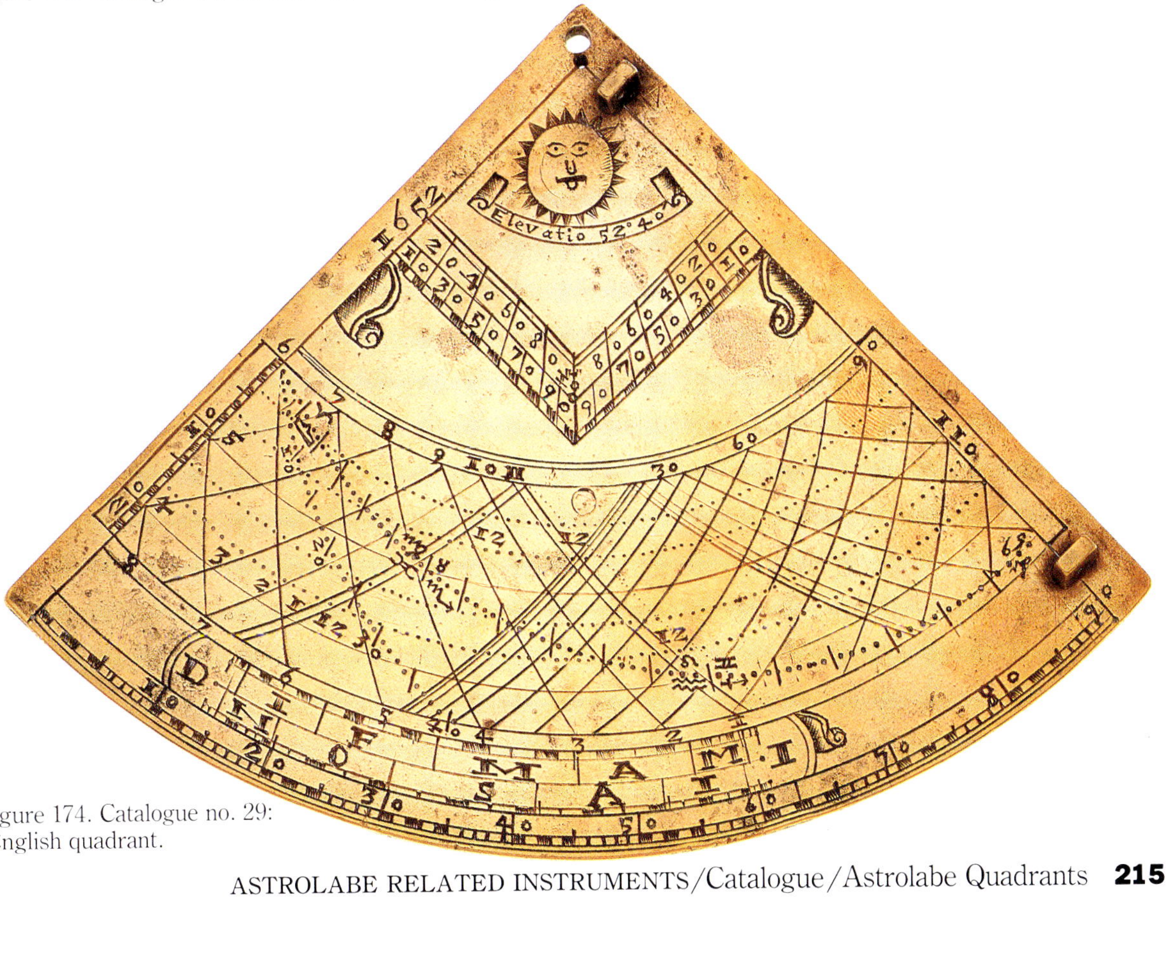

figure 174. Catalogue no. 29: English quadrant.

30. English Quadrant

Mid/late 17th century
Brass
Length of side (this is not a radius) 4½ in.
(114 mm)
Not signed
Inventory 377

On the face is a Gunter instrument with calendar scale and a table of seven named stars, the positions of which are marked on the projection by corresponding numbers. These are:

Ocul[us] Tauri
Dex[ter] hu[merus] Or[ionis]
Procyon
Regul[us]
Arctur[us]
Aquila
Ala[tus] peg[asus]

In the apex is an unequal-hour diagram *(quadrans vetus)*.[56] Set along one edge is a pair of sights used in conjunction with a plumb-line and bead (now missing). On the back is a 24-hour scale (1–12 x 2, 12 being indicated by a +), surrounded in the lower half by two quadrants of degrees. In the centre of the circle, otherwise blank, are several engraved doodles, including a star, the letters (? initials) I or J and M, and the words 'And,' 'In,' and 'my.' A small hole pierced at the centre suggests that originally there was, or was intended to be, a central volvelle, probably a nocturnal, for use in conjunction with the hour and degree scale.

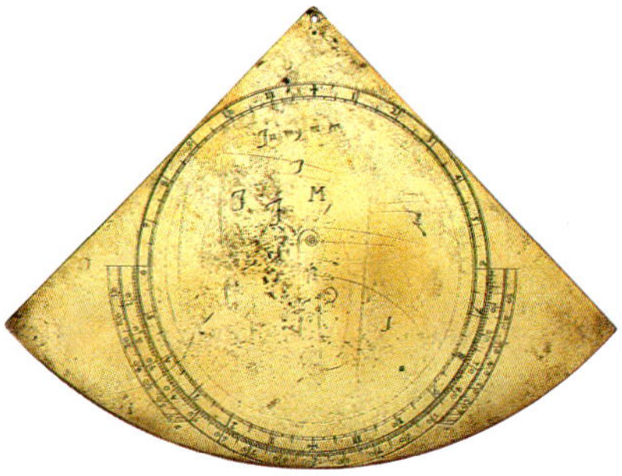

figure 175. Catalogue no. 30. Reverse.

56. See Volume I, part 2 of this catalogue for discussion of the *Quadrans Vetus*.

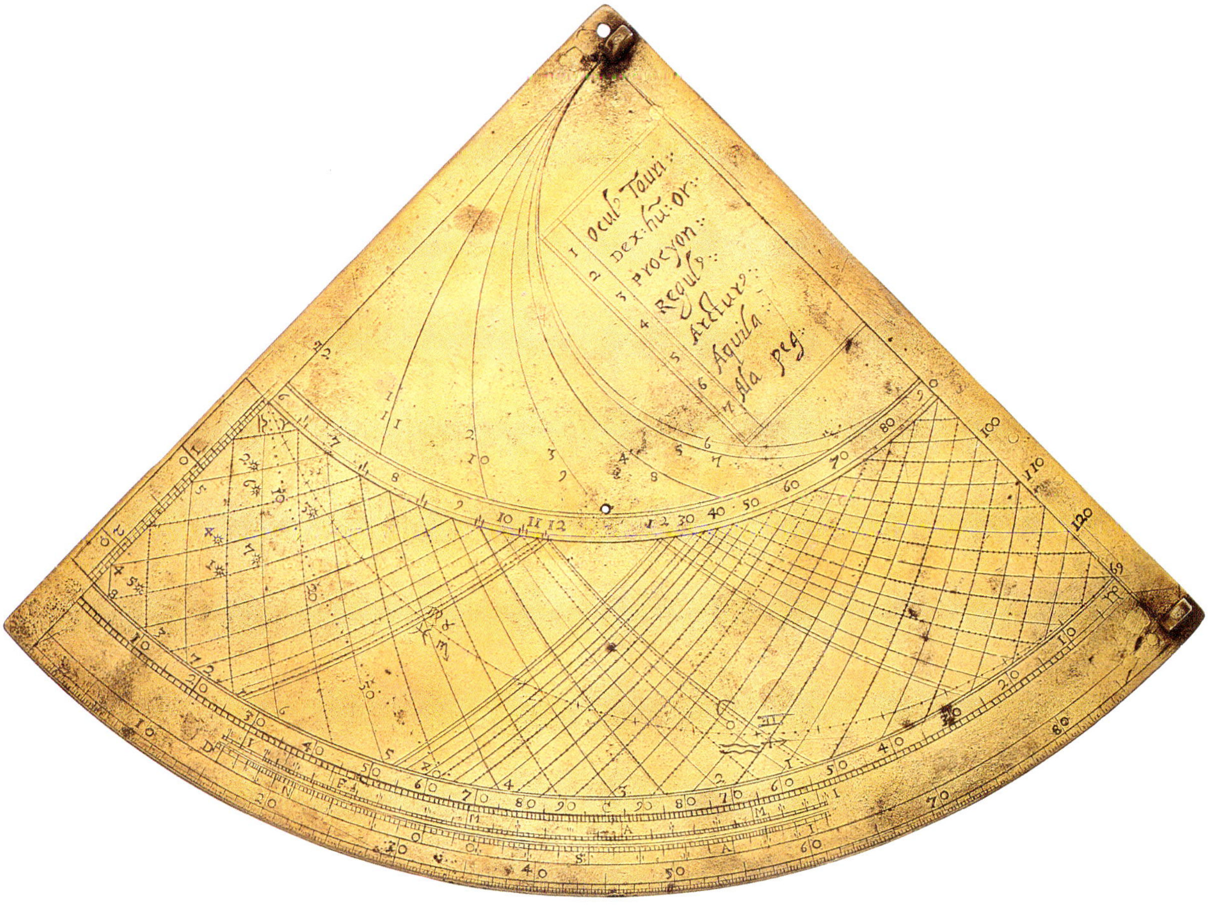

figure 176. Catalogue no. 30:
English quadrant.

31. English Quadrant

c. *1675*
Brass
Length of side (this is not a radius) 5⅞ in.
(149 mm)
Signed 'Hilkiah Bedford in Fleet Street'

Inventory 3528

The face is engraved with the following scales. On the limb is a scale of degrees reading to 30′. Concentric with this is a non-linear scale of co-tangents equivalent to a shadow square marked 1–100. Within the co-tangent scale is a calendar scale, and within this a Gunter quadrant. Five stars are marked on the instrument. These are:

Al[atus] Peg[asus]	23°54′
Arc[turus]	13°58′
Cor[Leonis]	9°48′
Oc[ulus] Tau[ri]	4°15′
Cor[Vulturi]	19°33′

The apex of the instrument is blank although it was probably intended that a shadow square should be engraved there, perhaps having other information within it.

On the back of the instrument is a circular scale of hours I–XII times two, reading to 15 minutes. Within is a volvelle engraved round the edge with a scale of months, and within this scale is a stereographic projection of seven of the circumpolar constellations seen from the front. These are:

Great Bear	14 stars marked
Bootes (part)	1 star marked
Capra (part)	1 star marked
Cassiopeia	7 stars marked
Cepheus	5 stars marked
Draco	14 stars marked
Little Bear	7 stars marked

Although both the elements that make up this instrument, the Gunter quadrant on the face and the constellation volvelle on the back, were described by Gunter in his *Description and Use of the Sector,* he had not there envisaged them as parts of a single instrument. Indeed there was no reason to combine them, for the task of the constellation volvelle – time-finding at night – could already be performed by the quadrant using the stars marked on the projection. The combination of the two instruments was therefore a duplication and was probably an initiative of the instrument-makers themselves, realizing that the volvelle was easier to use than the projected stars. Of these, as Gunter explained, the most convenient was *Alatus Pegasus,* which at the vernal equinox was only 6 minutes of time away from the meridian. It was, however, a star of the 2nd magnitude and was not therefore always easy to find. So Gunter

figure 177. Catalogue no. 31: English quadrant by Hilkiah Bedford.

figure 178. Catalogue no. 31. Reverse.

had added four further stars, one for each quarter of the year. One could thus use whichever was the most convenient. The times marked after each star position are explained by Gunter as 'the time when he cometh to the South at midnight.' This presumably means the time for a given date at which the star crosses the local meridian represented on the quadrant by the line for 12 o'clock, i.e. midnight. In his treatise, Gunter provides a table of his chosen five stars, showing their time of crossing the meridians for a given date, their right ascension, and their declination from the equator.[57] This table, drawn up for *c.* 1623, is as follows:[58]

Star	Date	Time of meridian passage		Right Ascention		Declination from the Equator	
		Hours	Minutes	Degrees	Minutes	Degrees	Minutes
Pegasus Wing	March 8	23	56	1	06	13	17
Arcturus	October 14	14	00	30	07	21	8
Lion's heart	August 7	9	50	32	28	13	42
Bull's eye	May 16	4	18	64	18	15	46
Vulture's heart	January 1	19	35	66	26	8	3

The stars are used to find the time at night in the following way. First the bead is placed over the star being used and its altitude is measured. The thread is then placed over the measured altitude on the limb when the position of the bead among the hour lines will indicate how many hours the star is away from the meridian. The right ascension of the star is now subtracted from the sun's right ascension (both having been converted into hours and minutes), and the result is added to the star's observed distance from the meridian. The result equals the number of hours elapsed since the sun left the meridian, and so the hour of the night.

Finding the time at night by this method was relatively lengthy and involved a good deal of arithmetic. It was for this reason that the volvelle was added. Being purely graphical, it was much easier to use. Indeed, Gunter had devised this 'nocturnal' specifically for seamen.[59] His description and explanation of it were as follows:

> 'It consists, as you see, of two parts, the one is a Plane divided equally according to the 24 hours of the day, and each hour into quarters or minutes, as the Plane will bear: the line from the Center to XII, stands for the Meridian, and XII stands for the hour of 12 at midnight. The other part is a rundle for such stars as are near the North-Pole, together with the twelvemonths, and the days of each month fitted to the right Ascension of the stars. Those that have occasion to see the South-Pole, may do the like for the Southern Constellations, and put them in a Rundle in the back of this Plane, and so it may serve for all the World.

57. Gunter, 100.
58. In the 5th edition, edited by William Leybourn, 1673, the text describing the quadrant (Bk III, 122) has corrected values for the meridian passage. These are 2 minutes less than those given in the table. They are therefore (with the exception of *Oculus Tauri*) the same as those given on the present instrument and help to substantiate the proposed date. The table, however, is uncorrected.
59. Gunter, 64.

> The Use of the Nocturnal
> The Use of this Nocturnal is easie and ready. For look up to the Pole, and see what Stars are near the Meridian: then place the Rundle to the like Situation, so the day of the month will shew the hour of the Night.'[60]

Addition of the nocturnal also made possible a simplified way of using the stars projected on the face. The bead was set to the star on the projection, the altitude of the star measured, and the position of the bead among the hour lines noted. The nocturnal, on the other side, was rotated until the star was aligned on the hour scale at the required hour. The time at night was then found on the hour scale from the date of observation on the rim of the volvelle of the nocturnal.[61]

Gunter included the 'Type' of his 'nocturnal,' as a leaf extraneous to the collation, tipped into his book. This could be cut out and mounted with thread on an hour scale printed on the page to create a usable instrument. The example on the present instrument is closely modelled on that in Gunter's work, although it is rather less well engraved and fewer stars are marked in the constellations.

Hilkiah Bedford (*c.* 1634–89) came from a Lincolnshire family. His father, Thomas Bedford, a gentleman of Sibsey, near Boston, was a Quaker, and Hilkiah may also have been. On 3 August 1646 he was apprenticed to the noted instrument-maker John Thompson of Hosier Lane, taking his freedom in the Stationers' Company on 27 June 1654.[62] Until his premises were destroyed during the fire of London in 1666, Bedford continued to live in Hosier Lane, moving from there to Fleet Street.[63] Although relatively few examples of his work have survived, those that do are well-executed, and the fact that John Collins intended to employ him to make a pantograph for James Gregory, and for other tasks, would suggest that he enjoyed a reputation for good work.[64] In February 1667/8 he was admitted to the Clockmakers' Company, and at least once took part in a company search for defective instruments.[65] He died on 6 May 1689, and was buried in the church of St Dunstan-in-the-West, Fleet Street.[66]

60. Gunter, 64–5.
61. Holland, 6. Immediately after this, Holland describes the use of the circumpolar constellations on the nocturnal, his method being the same as Gunter's. One advantage of having two sets of stars to perform the same operation on the same instrument was that even if the circumpolar stars were obscured, those projected might be visible.
62. McKenzie, 164.
63. Mayor, lix–lxi.
64. Turnbull, 239.
65. Atkins & Overall, 236.
66. Mayor, lx. For a fuller account of Bedford's life, Turner (VI).

32. English Quadrant

c. 1644
Ivory
Diameter 3 ⅝ in. (92mm)
Not signed (possibly by John Browne) but inscribed 'Londini fecit' (He made [it] at London)

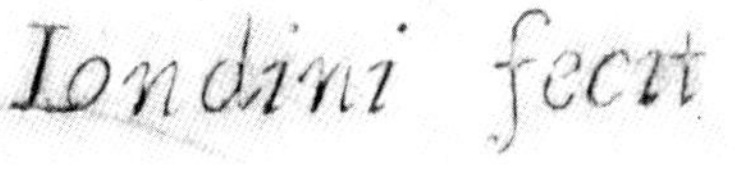

Inventory 1004

A circular disc is engraved on one side with a heart-shaped cartouche which contains, at the left, a Gunter quadrant simplified by the omission of the azimuth scales and the ecliptic, marked 'Latitude 51–32' (London), and, at the right, a perpetual almanac that includes a month calendar,[67] Dominical Letters, Epacts, and the dates of leap years from 1644 to 1668. Above the almanac is the inscription 'Londini fecit' (He made [it] at London). Below the almanac is the injunction 'Lege et intelige' [*sic*] (Read and understand). Outside the cartouche, along the edge of the instrument below the Gunter-quadrant, is a degree scale (0°–60°) and within this is a shadow arc (the equivalent of a shadow square on an astrolabe), divided 1 to 9. Both arcs are radial from a small hole on the opposite side, which is also the centre of the quadrant, and presumably was originally fitted with a plumb-line. Below this hole, along the edge of the instrument, is the slogan 'Perpetuus solis distinguet tempora motus' (the constant motion of the sun divides time). On the reverse is a series of concentric scales surrounding a geometrical design at the centre. The scales comprise a calendar for each half year; the time of sunrise; the sun's amplitude (i.e. the sun's azimuth at the moment of its rise); the sun's declination; the sun's position in the zodiac ('true place'; 0° Aries = 10 March); length of the longest day; and the sun's right ascension. An index that originally pivoted from the centre is now missing, as is the suspension piece.
Provenance: Sotheby & Co., London, 27 March 1972, lot 103.

The ascription of this instrument to John Browne (*fl. c.* 1644–*c.* 1697)[68] is based on its almost identical design with an instrument of boxwood signed 'J B Londini fecit' (fig. 180), now in the National Maritime Museum, Greenwich, and two, slightly larger, instruments of ivory in the British Museum, London, also signed 'J •⁘• B . . . Londoni fecit.'[69] One of the best-known makers of his day, Browne was the son of Thomas Browne (*fl.* 1627–53), joiner and instrument-maker, who made sets of Napier's Bones, and designed an

67. For discussion of this form of perpetual calendar, see Volume I, part 4 of this catalogue. The month lines of this example following the classification there set out are Turner type B1b.
68. These dates, which differ from those given by Taylor (II), 231, are based on the leap-year dates marked on the present instrument and others like it (assuming that the earliest date, 1644, represents the nearest leap-year date to that of the making of the instrument), and an entry in the Clockmakers' Company records 29 September 1697 when Browne was excused attendance at the Court because of his advanced years. Brown (I), 19.
69. Ward, 94–5.

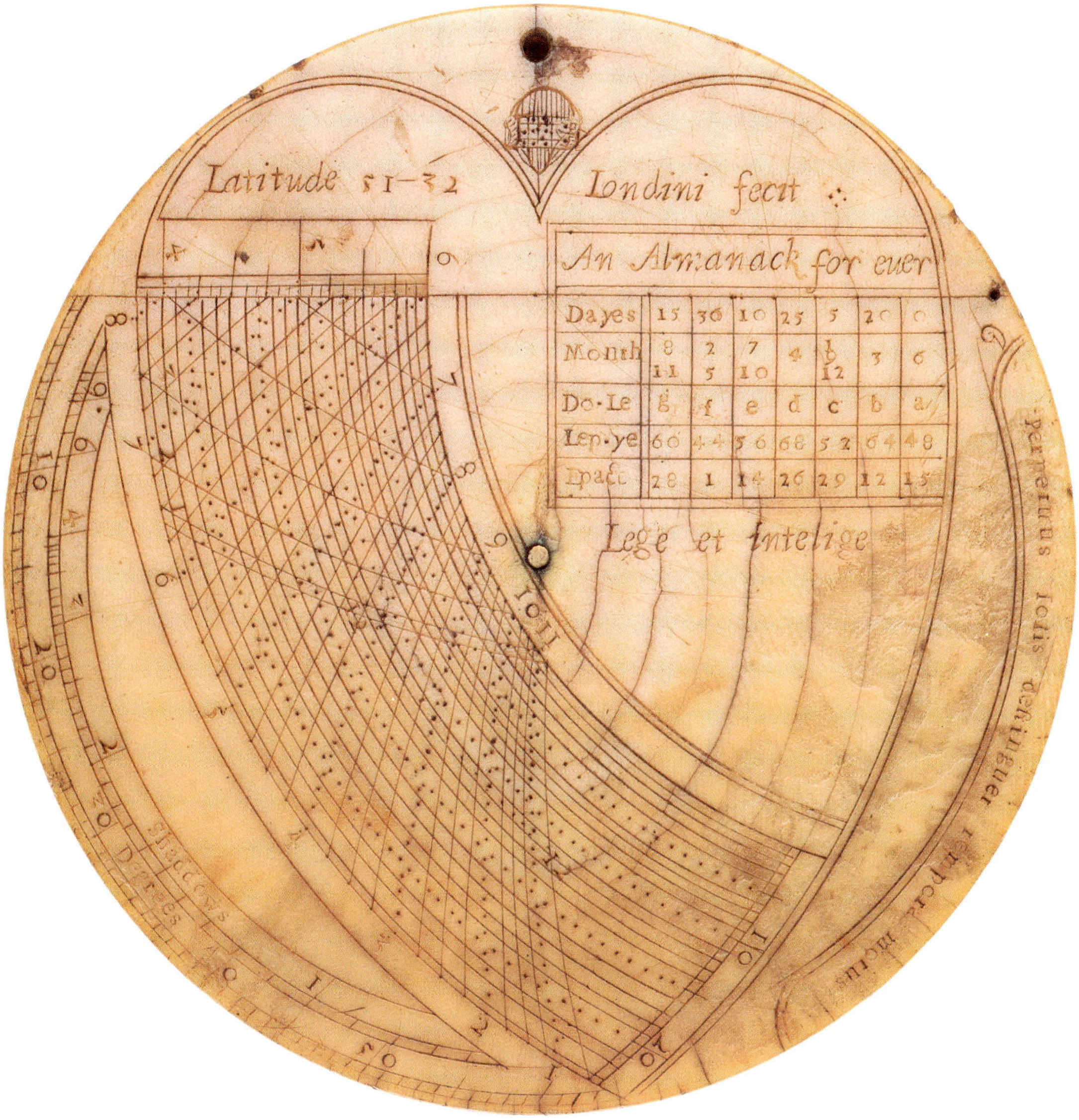

figure 179. Catalogue no. 32:
English quadrant.

early form of circular slide rule.[70] Examples of this slide rule, the 'serpentine instrument,' are occasionally to be found on the reverse of Browne-type Gunter quadrants, replacing the calendar and solar scales. Like his father, John Browne seems to have specialized in instruments made of wood, and particularly in scales and rules. Between 1661 and 1688 he published four books on these.[71] Two of these works were published by Browne jointly with Henry Sutton, the copper-plate engraver (catalogue no. 33), and it seems likely that the two men worked closely together, the one in wood, the other in metal. In 1661, Browne was working in Aldgate, at Duke's Place, later moving to the Minories. In February 1667/8 he was one of nineteen instrument-makers admitted as brothers of the Clockmakers' Company, of which he became Master in 1681.[72] 'On account of his great age'[73] he was excused from further attendance at the Clockmakers' Court in 1697, and may be assumed to have died soon after, although there was a reissue of his book on the carpenter's rule as late as 1704.

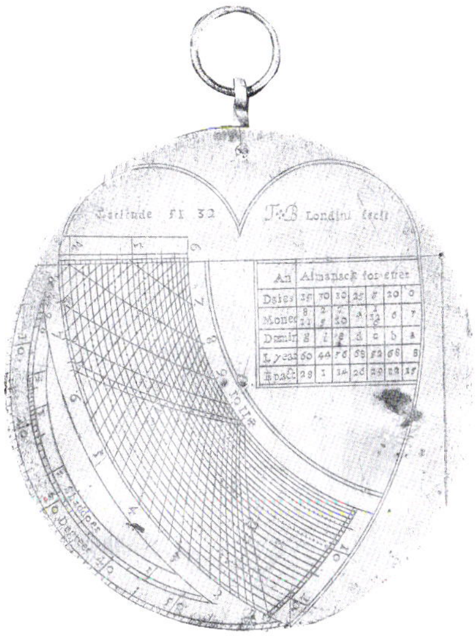

figure 181. Wooden quadrant signed 'J.B.' Greenwich, National Maritime Museum.

figure 180. Catalogue no. 32. Reverse.

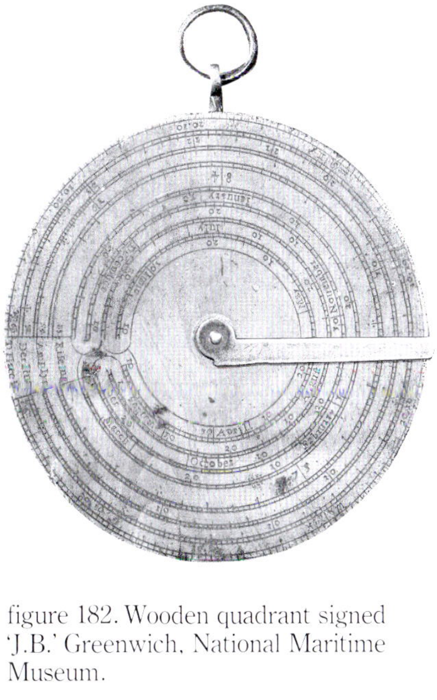

figure 182. Wooden quadrant signed 'J.B.' Greenwich, National Maritime Museum.

70. Taylor (II), 210. A more detailed treatment of this subject is promised by D.J. Bryden.
71. Taylor (II), nos. 262, 270, 430, and 466.
72. Brown (I), 16.
73. *Ibid.*, 19.

33. English Quadrant

1658
Varnished paper, printed from copper plates, on wood, with brass sights
Length of side (this is not a radius) 5¾ in. (146 mm)
Signed: 1658 Henr Sutton Londini fecit

Inventory 1445

On the face is an astrolabe quadrant for latitude 51°32′ (London) with the positions of twenty-one stars marked by lower-case alphabetical letters. These letters identify the stars on a table on the back of the instrument. In the apex of the quadrant, surrounding the signature, is a calendar scale (0° Aries = 10 March, the signs of the zodiac being indicated by small circles above the appropriate dates). Along one edge of the quadrant is a scale of hours (1 to 6) and on the other edge of latitude (1 to 90). Along the circumference is a double hour scale (VI–XII/ XII–I); a degree scale (90°–0°) marked 'Quadr' and following this a scale (1 to 50) marked 'Shadow' being the equivalent of a shadow square on an astrolabe. Two square vanes with pin-hole sights were set on one edge; one is now missing as are the plumb-line and the bob. On the back of the instrument, in the apex, is a perpetual calendar[74] with a table of leap years from 1656 to 1708. Beneath this is a table of the twenty-one stars marked on the face of the instrument with their co-ordinates and hour-arc. Beneath the table, in parallel arcs, are eleven trigonometric scales marked respectively 'Part sine, V[ersed] sine duode, V[ersed] sine quadr., Tangent, seTangent, Tangent, Secant V[ersed] sine, Hour, V[ersed] sine and sine.' Below these, on the edge of the circumference is a double hour scale (XII–VI) complementary to that on the face. Running down one edge is an azimuth scale, and on the other, an hour scale.

One of the leading technical illustrators of his day, Henry Sutton (*fl.* 1637–65) specialized in the graduation of scales and in engraving copper plates from which portable instruments, like the present example, could be printed. Associated with John Browne,[75] two of whose books he jointly published, Sutton was closely connected with Sir Samuel Morland (1625–95), whose calculating machines he made with the clockmaker Samuel Knibb (1625– ?1670),[76] and with John Collins.[77] From his premises at Threadneedle Street (behind the Royal Exchange in London), Sutton published several mathematical books, and filled special orders, such as printing the graph

74. See n. 67 above.
75. See cat. no. 32.
76. For Morland, see Dickinson, *passim*; for Knibb, see Beeson 124–5.
77. See above, pp. 208 & 210.

figure 183. Catalogue no. 33: English quadrant by Henry Sutton.

figure 184. Catalogue no. 33. Reverse.

paper used for the famous Down Survey of Ireland by William Petty in 1654.[78] A little earlier, he had invented a magnetic azimuth dial which he used as a trade card.[79] Inventive, highly competent, and much esteemed, Sutton was a valuable member of the English scientific community to which his sudden death during the plague of 1665 was a serious blow.[80]

78. Larcom, 48.
79. For an example, see Volume I, part 2 of this catalogue.
80. 'Wee all here [Oxford] are much troubled with the loss of poor Thomson & Sutton,' Sir Robert Moray to Henry Oldenburg, 10 October 1665, in Hall & Hall, 561.

ASTROLOGICAL ASTROLABES

That the earliest probable reference to a true planispheric astrolabe, and the only one which occurs in the works of Ptolemy as we have them, is to be found in his *Tetrabiblos,*[81] rather than in the *Planispherium* as might have been expected, is an indication of the importance throughout its history of the instrument's relation with astrology. For it was in astrology that tedious calculation which the astrolabe could shorten had most frequently to be made, and it was in astrology that the capacity of the astrolabe to represent the configuration of the heavens for dates in the past was most useful. That astrologers supplied an important proportion of the clientele for astrolabes both in Islam and in Europe cannot be doubted. Whether they actually used the instrument for calculation, or merely regarded it in order to impress clients is, however, an open question.[82] Illustrations of the use of astrolabes for astrological operations from different periods appear in the history which prefaces this volume. Also in this volume, the scales added to astrolabes for purely astrological purposes may be seen on several examples. What may be emphasized here is that from time to time the importance of the astrolabe for astrology led to the development of instruments intended purely for such purpose. An early example is provided by the instrument devised by Henry Bate of Malines and described in a short treatise written at the request of William of Moerbeke in 1274.[83] An instrument devised and published in the early 17th century is the subject of the following catalogue entry.

81. Robbins, 228–9.
82. That this was common in early Islam is suggested by the satirical account of a visit from a barber contained in the 'Tale of the Lame Young Man & the Barber of Baghdad,' a part of the 'Tale of the Hunchback' in the *Thousand and One Nights.* The barber not only consults his astrolabe to decide whether the moment is propitious for shaving or not, but also mentions that when asked to let blood he similarly first used his astrolabe. See Dawood, 33–5.
83. For Bate, see Sarton, ii, 994–5. The treatise is reprinted in Gunther (V), 368–76.

Astrolabe Related Instruments Catalogue Astrological Astrolabes

34. Remains of a French Universal Mathematical Instrument ('Pantocosme')

1612 and ? later
Paper scales printed from an engraved copper plate, mounted on a mahogany board, and varnished; brass alidade, index, and fittings Diameter 9 5/8 in. (244 mm)
Not signed (devised by Noel Leon Morgard and engraved by François Galand)
Inventory 556

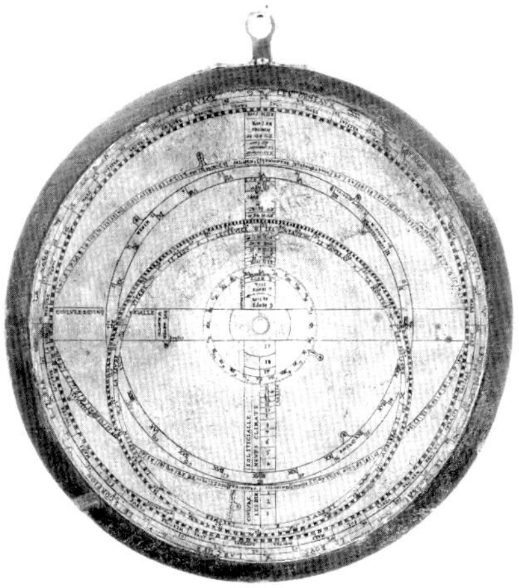

figure 185. Catalogue no. 34. *Rete* (uncut).

The instrument is incomplete. The *rete*, the unwanted portions of which have not been cut away, and two of the 'Indexes' from a 'Pantocosme' of Noel Leon Morgard have here been mounted on a base. The *mater*, back, two plates, and accessories (all of which are missing) are shown in fig. 187 c-g. The *rete* is that of a standard astrolabe with the addition of the equinoctial and solstitial colures, a polar circle, and two arcs representing the rise of the ecliptic above the horizon. The positions of twelve named stars are marked. Being primarily interested in astrological computations, Morgard imputes functions and even names to some parts of the *rete* which differ from those of the traditional instrument. According to him, the outermost set of circles represents the horizon, and is marked with the names, signs, degrees of the zodiac and the aspects of the planets for each of the houses of the heavens. The solstitial colure is marked on one side of the polar circle with the 19 'climates' following the system of Francesco Maurolyco, Henry Glareau, and Martin Borcha.[84] On the other side is a scale showing the length of the day and night according to latitude. The circle of the tropics[85] is divided with a 24-hour scale intended to serve as a scale for finding time throughout the world. The two arcs rising from 0° Aries and 0° Libra and forming a pointed oval represent the ecliptic and are divided into 23½° four times to show the declination of the sun. An eccentric circle marked with the names, signs, and degrees of the zodiac is intended for setting up the houses of the heavens following the methods described by Regiomontanus.[86] The alidade and rule, which have been executed in brass, were intended to be mounted on the back of the instrument (fig. 187h) for use with the scales there engraved. They have no relation with the *rete*. Although the script employed is very closely modelled on that of Galand, the 'Index Solaire' is incomplete

84. 'Clima est spatium terrae, interduos parallelos comprehensum, in quo porrectissime ab initio climatis, vsque ad finem eiusdem, dimidia hora variatio est diei.' (A climate is a portion of the world included between two parallels in which from the beginning of the climate to the end of the same the greatest variation of the day is a half-hour.) Henry Glareau, *de Geographia*, Cologne, 1581, 200. For lists of them, see *Ibid.*, 201, and the *de Sphaera* of Francesco Maurolyco in *Opuscula Mathematica*. . ., Venice, 1585, 18. Ultimately the *climata* derive from Ptolemy's *Geography* by way of Sacrobosco's *de Sphaera*.
85. The polar circle and the circle for the tropics represent the arctic or the antarctic, Cancer or Capricorn, as the north or south plate is placed beneath the *rete*.
86. Morgard, sig. e^{1}v. Methods of establishing the houses of the heavens are described by Regiomontanus in his *Tabulae Directionum et Profectionum*, problems XIV–XVI, XXI, XXIV. (Several editions of this work were published between 1475 and 1606; see Houzeau & Lancaster, 552–3. A French translation with commentary by D. Henrion appeared in 1625.)

figure 186. Catalogue no. 34:
An astrological instrument.

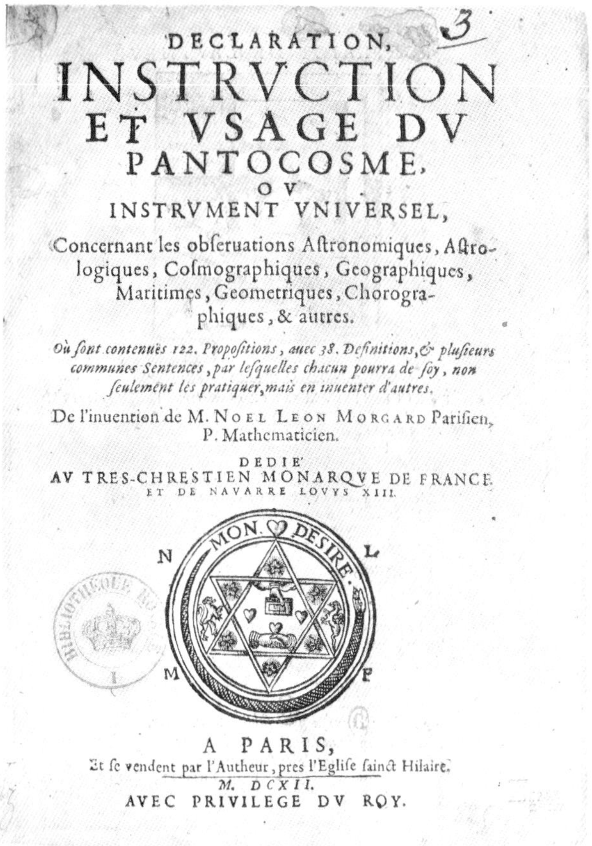

DECLARATION,
INSTRVCTION
ET VSAGE DV
PANTOCOSME,
OV
INSTRVMENT VNIVERSEL,
Concernant les obseruations Astronomiques, Astrologiques, Cosmographiques, Geographiques, Maritimes, Geometriques, Chorographiques, & autres.

Où sont contenuës 122. Propositions, auec 38. Definitions, & plusieurs communes Sentences, par lesquelles chacun pourra de soy, non seulement les pratiquer, mais en inuenter d'autres.

De l'inuention de M. NOEL LEON MORGARD Parisien, P. Mathematicien.

DEDIE'
AV TRES-CHRESTIEN MONARQVE DE FRANCE
ET DE NAVARRE LOVYS XIII.

A PARIS,
Et se vendent par l'Autheur, pres l'Eglise sainct Hilaire.
M. DCXII.
AVEC PRIVILEGE DV ROY.

figure 187a. Title-page.

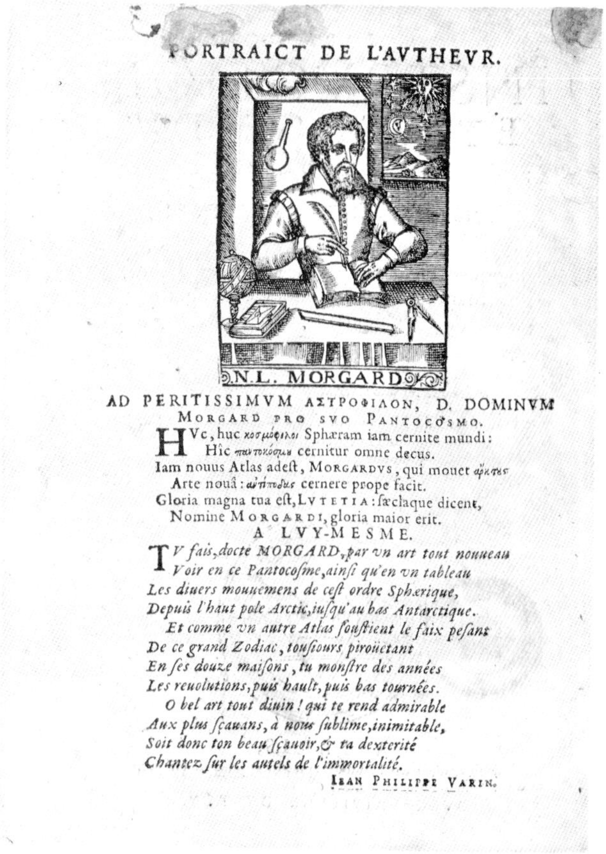

PORTRAICT DE L'AVTHEVR.

N.L. MORGARD

AD PERITISSIMVM ΑΣΤΡΟΦΙΛΟΝ, D. DOMINVM
MORGARD PRO SVO PANTOCOSMO.

HVc, huc κοσμόφιλοι Sphæram iam cernite mundi:
Hîc παντοκόσμῳ cernitur omne decus.
Iam nouus Atlas adest, MORGARDVS, qui mouet ἄρκτους
Arte nouâ: ἀντίποδας cernere prope facit.
Gloria magna tua est, LVTETIA: sæclaque dicent,
Nomine MORGARDI, gloria maior erit.

A LVY-MESME.

TV fais, docte MORGARD, par vn art tout nouueau
Voir en ce Pantocosme, ainsi qu'en vn tableau
Les diuers mouuemens de cest ordre Sphærique,
Depuis l'haut pole Arctic, iusqu'au bas Antarctique.
Et comme vn autre Atlas soustient le faix pesant
De ce grand Zodiac, tousiours pirouëtant
En ses douze maisons, tu monstre des années
Les reuolutions, puis hault, puis bas tournées.
O bel art tout diuin! qui te rend admirable
Aux plus sçauans, à nous sublime, inimitable,
Soit donc ton beau sçauoir, & ta dexterité
Chantez sur les autels de l'immortalité.

IEAN PHILIPPE VARIN.

figure 187b. Portrait of Noel Leon Morgard.

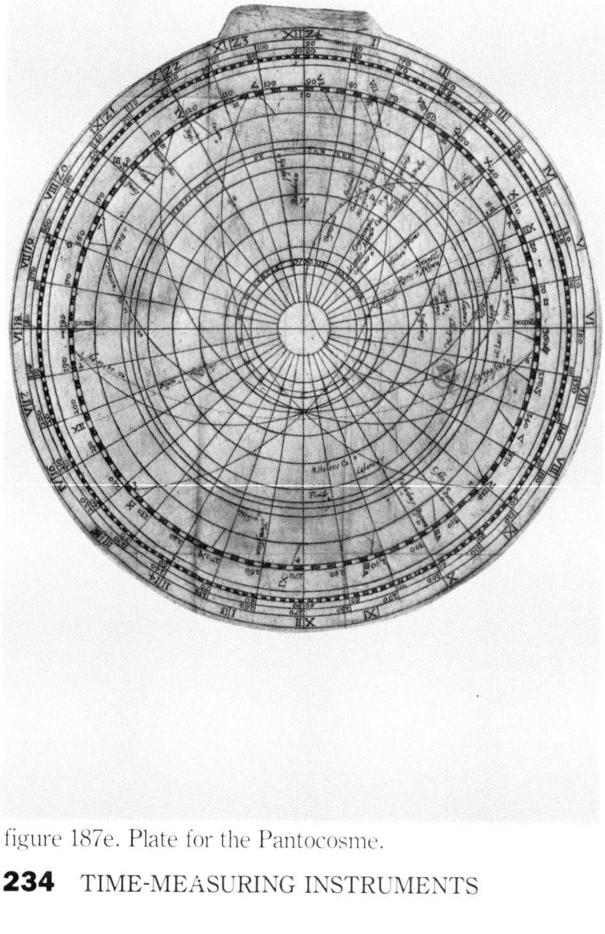

figure 187e. Plate for the Pantocosme.

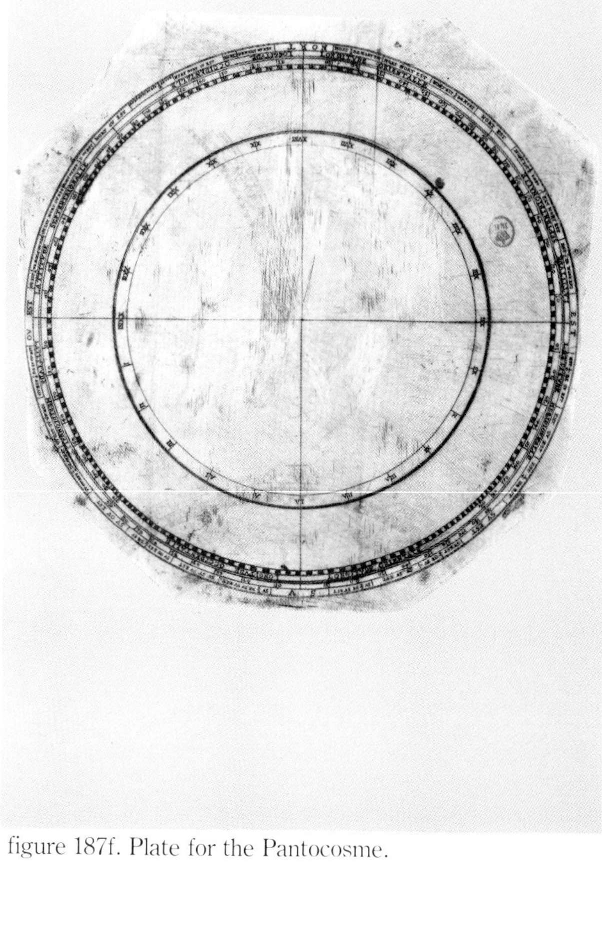

figure 187f. Plate for the Pantocosme.

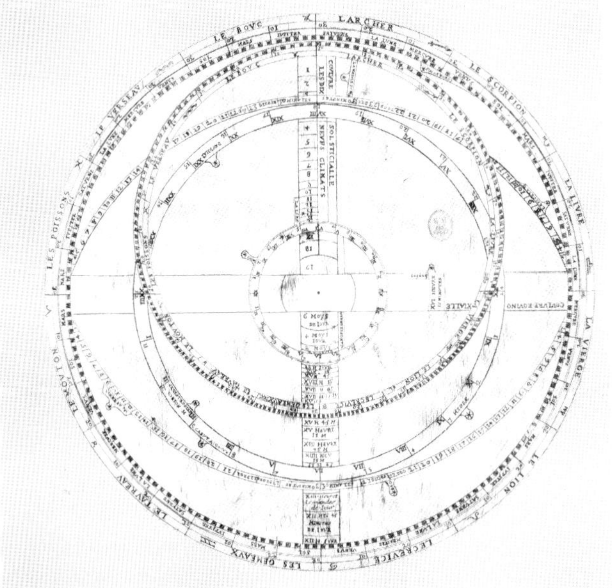

figure 187c. *Rete* of the Pantocosme.

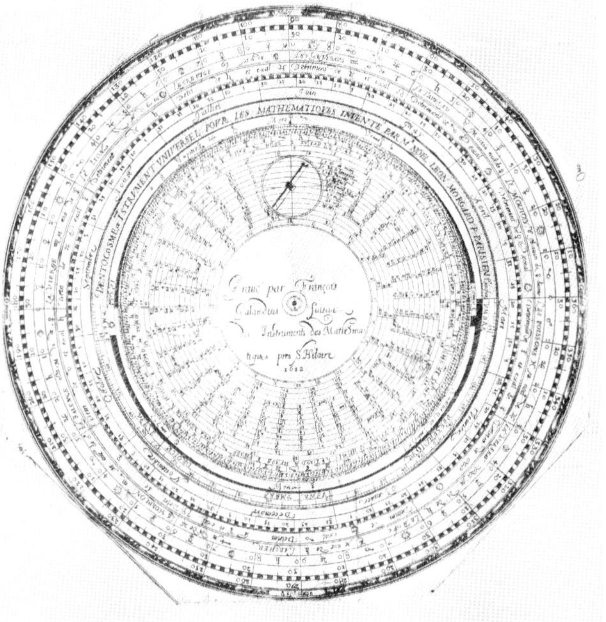

figure 187d. Back of the Pantocosme.

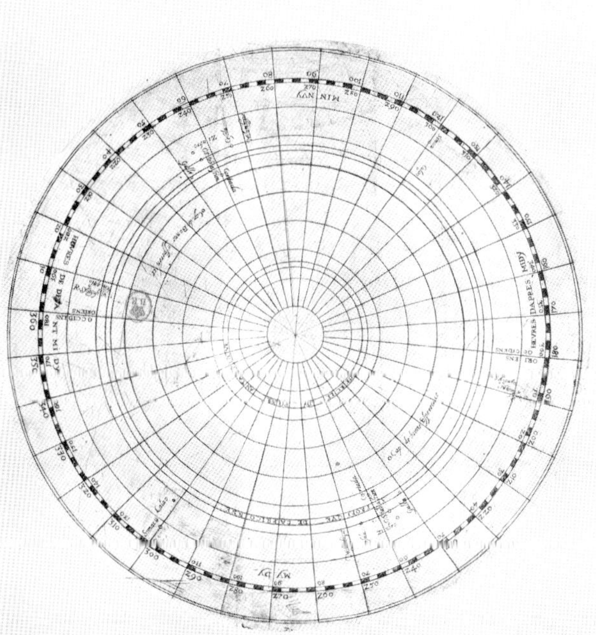

figure 187g. *Mater* of the Pantocosme.

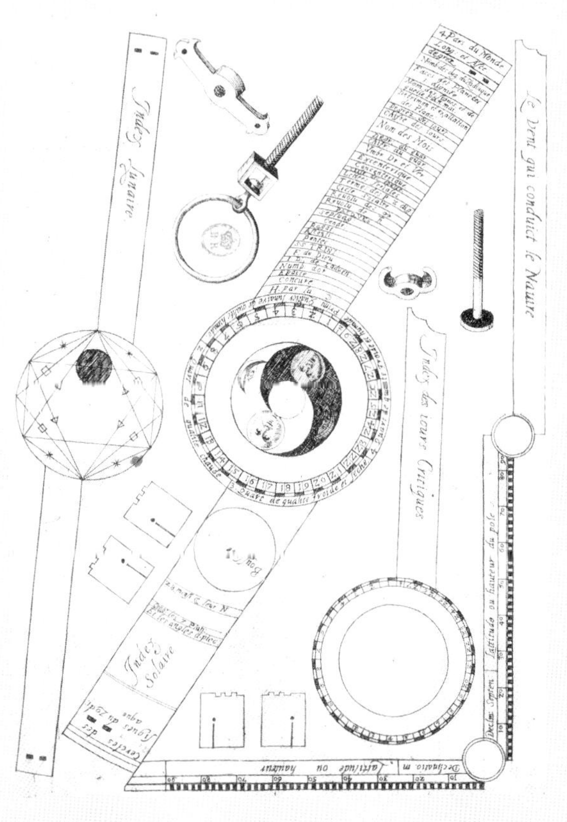

figure 187h. Rules for the Pantocosme.

(*cf.* fig. 186). The small circle on the arm of the 'Index Solaire' marks the position where a compass was to be mounted. At the centre of the two indexes is an *aspectarium* and a lunar-phase volvelle.

The 'Pantocosme' of Noel Leon Morgard was an astrolabe adapted to carry out a wider range of tasks than the traditional instrument, and to be more useful in astrology. As can be seen from the illustrations, the normal series of latitude plates carrying the horizon-zenith co-ordinate system was replaced by two plates, one for the northern hemisphere, one for the southern hemisphere, carrying stereographic projections of the terrestrial sphere. A great variety of astrological and calendarical scales were added to the back of the instrument, space being found for them by converting the shadow square to a circular form. A wind-rose was also added on a separate plate. By these modifications the user of the instrument was enabled to solve problems such as those relating to the time in any part of the world in relation to Paris, the position of planets in relation to a given place, or the time when a given planet or zodiacal sign would cross the meridian of such a place. As Morgard himself described it, 'This *Pantocosme* is made with such skill, so certain and easy, that one sees there the heavens revolve over the earth and the celestial influences at each town, province, or region of whatever country it may be: similarly by knowledge of the hour, place or other, [one can] know what time it is in all places and regions of the world and know what planets or stars rise acronically, heliacally and cosmically: [one can] also see eclipses of either the Sun or the Moon, and other things that an experienced astrologer can notice in his observations.'[87] In the long pamphlet which describes his device, Morgard gives numerous examples of how it was to be used and provides a detailed description in which no reference is made to the six finely engraved plates. These plates were intended to be removed from the book, cut out and glued to wood or pasteboard to form a usable instrument, as has been partially, though inaccurately, done with the present example.[88] The fact that very few copies of the book and, as yet, no other example of the instrument are known suggests that, like most such complex 'universal' instruments, the 'Pantocosme' did not achieve great popularity.

Morgard (fig. 187b), the deviser of this instrument, is an obscure figure. A teacher of 'mathematical sciences' at Paris (of which city he was perhaps a native),[89] he was an active astrologer, and all his

87. 'Ce Pantocosme est composé d'un tel artifice, si asseuré & aysé, qu'on y voit le Ciel faire ses reuolutions sur la terre, & les influences celestes agir sur chasque ville, province ou region de quelque pays que ce soit; Mesmes par la conoissance de l'heure, de l'habitation, ou autre, scavoir quelle heure il est en tous lieux & climats du monde, & conoistre quels planettes ou estoilles se leuent acroniquement, heliaquement & cosmiquement: voir aussi les Eclipses tant de Soleil que de Lune, & autres choses que peut remarquer vn experimenté Astrologue en ses observations.' Morgard, sig. ^{iii}r.

88. Of the two copies of the work in the Bibliothèque Nationale, Paris, one (pressmark Vz 825) retains its plates; the other (pressmark V 6369) does not have them.

89. He describes himself as 'Parisen' on the title-page and at the end of the dedication of the *Pantocosme* (sig. e^{iii}v).

publications which date between 1600 and 1619 are concerned with this subject. In 1614 he published a prognostication for the year[90] which provoked two replies.[91] Either this, or another astrological speculation, seems to have led to his being condemned to spend some years as a galley prisoner.[92] Of the rest of his life, however, as of the dates of his birth and death, nothing seems to be known.[93]

90. *Prédiction de Morgard pour le présente année 1614: avec les centuries pour le même année*, Paris, 1614. *Les centuries* were also issued separately in at least 2 editions.
91. *L'Anti-Morgard ou ses prédictions de la présente année Mill six cens quatorze*, Paris, 1614. *L'Anti-Mauregard ou le fantosme du bien public*, Paris, 1614.
92. *Le Manifeste de Noel Leon Morgard*, Paris, 1619, 4–5.
93. I am indebted to Emmanuel Poulle for first recognizing this instrument as a 'Pantocosme.'

Glossary

Listed here are the few technical expressions that it was impossible to avoid using in the present work. For a brief account of the astronomical basis of time measurement, see Charles H. Cotter, *A History of Nautical Astronomy*, London, 1968, 32–7.

Alidade	a sighting device with two vanes with pinhole sights (fig. 10)
Almucantars	lines of equal altitude above the horizon (fig. 4)
Azimuth	angular distance of a celestial body east or west of the meridian (*qv*)
Brachiolus	an articulated arm attached to the cursor (*qv*) of a universal astrolabe to facilitate the marking of a position on the curved meridians and parallels (fig.137)
Cancer, tropic of	northern circle of declination which touches the ecliptic (*qv*) at its greatest distance from the equator at the beginning of the sign of Cancer
Capricorn, tropic of	southern circle of declination which touches the ecliptic (*qv*) at its greatest distance from the equator at the beginning of the sign of Capricorn
Climata (climates)	latitudinal divisions of the earth into seven regions derived from Greek geography, regions usually counted from the equator to the pole
Cosines, quadrant of	See 'Sinical quadrant'
Counter-changed	term pertaining to an alidade or the east-west bar of a *rete*: having the thickness of the bar placed alternately above and below the central line (fig.54)
Crepuscular line	see 'Hour of twilight'
Cursor	a short straight edge which slides along a rule or alidade and forms a right angle with it (fig.157)
Declination	angular distance of a celestial body from the equator
Dominical letter	the letter which falls on Sunday when letters A through G are applied to the first seven days of each year; for successive years follow each other in retrograde series and repeat in a cycle of 28 years
Ecliptic	the great circle which the sun appears to describe in the heavens during one year, caused by the earth's motion around the sun
Epact	a number, repeating every 19 years, which represents the excess length of the solar year over the lunar year, and so indicates the age of the moon on 1st January
Equal hours	see 'Hours, equal'
Equator	the circle in the heavens which would be traced out if the earth's equator were extended to the celestial sphere; also called equinoctial

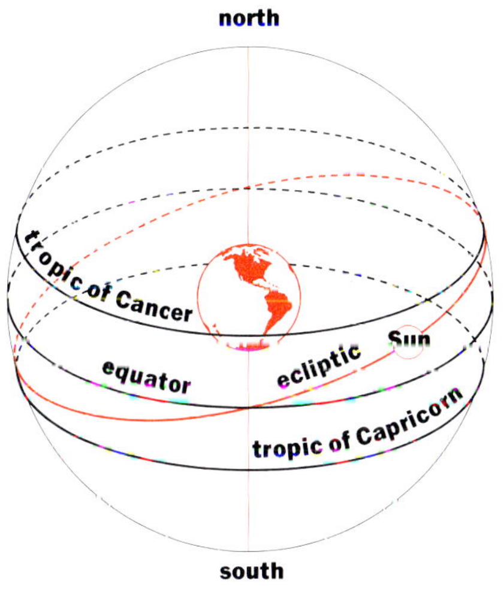

fig. 188. Fixed celestial co-ordinates

Equinox	the two points of intersection of the ecliptic and the equator, so called because on the sun's arrival at either of them night and day are equal in length throughout the world
Ghaṭi	Indian unit of time measurement: 60 *ghaṭika* = 24 hours
Horizon plate	an astrolabe plate on which are projected the horizon lines for a series of different latitudes, all other lines being omitted; useful for problems concerning the rising and setting of sun or stars and the length of day or night
Hour angle, arc	angular distance of a heavenly body east or west of the meridian
Hour of twilight	period of illumination immediately before sunrise or after sunset, caused by reflection and scattering by the earth's atmosphere; cannot be precisely measured but no perceptible twilight can be seen after the sun has sunk about 18° below the horizon; period is conventionally marked on the astrolabe plates by a line (sometimes dotted) drawn horizontally across the plate beneath the horizon; indicated period of Muslim morning and evening prayer
Hours, equal	hours obtained by dividing the entire period of day and night into 24 equal parts, which thus each equal 60 minutes
Hours, unequal	hours obtained by dividing the periods of daylight and darkness each into 12 equal parts, periods thus not equal to each other except at the equinoxes *(qv)*; change in length throughout the year
Hours, Italian	equal hours counted up to 24 starting from 30 minutes after sunset
Hours, Babylonian	equal hours counted up to 24 starting from 30 minutes after sunrise
Houses of the Heavens	astrological division of the celestial sphere into 12 parts, formed by dividing into 3 equal parts the 4 quarters of the sky made up by the intersection of horizon and meridian of any given place; numbered from the horizon in the opposite direction from that of the earth's motion
'ilaqa	a cord attached to the suspension ring of an astrolabe
Inḥirāf	azimuth of the *Qibla (qv)*
Jihat	the direction of the azimuth of the *Qibla (qv)* in terms of the four cardinal points
Khaṭṭ al-'aṣr	the period of the Muslim afternoon prayer
Khaṭṭ az-zawāl	line marking the moment of true noon
Khaṭṭ aẓ-ẓuhr	line marking a time shortly after midday, thus the period of the Muslim midday prayer
Kursī	triangular projection on an astrolabe, which carries the suspension apparatus
Limb	the raised circumference of the astrolabe, which is engraved with a degree or hour scale
Mansions of the moon	the 28 constellations through which the moon passes during its course across the sky
Mater	the body of the astrolabe, which contains the plates

Meridian	great circle which passes through an observer's zenith *(qv)* and the celestial poles
Meridian altitude of the sun, graph of	may be drawn for any requisite latitude (or a series of latitudes); in conjunction with the arcs of the signs of the zodiac *(qv)* provides a reading of the midday height of the sun for any day of the year; on Indo-Persian astrolabes arcs are drawn in a characteristic sigmoid form (fig. 189)
Mudīr	a small knob, sometimes of silver, set on the *rete*, and which serves as a handle to turn it
Muwaqqit	the mosque official charged with the calculation of the times of prayer
Pole, celestial	the two ends of the axis about which the celestial sphere appears to turn
Qibla	the direction of Mecca from any point on the earth's surface; the direction a Muslim must face while praying
Qibla, azimuth	the angle between the great circle of the meridian of any point on the earth's surface and another great circle passing through the zenith *(qv)* of the point and the zenith of Mecca
Qibla, graph of azimuths of	a diagram found on some Islamic astrolabes (mainly Persian) from the Safavid period onward, usually in the top-right quadrant on the back of the instrument (arcs of the azimuth for selected places are drawn across the arcs of the signs of the zodiac *[qv]* and give the azimuth by indicating for any day of the year the height of the sun as it crosses the circle passing through the zenith *[qv]* of the place of observation and that of Mecca)
Rete	skeletal plate marking the ecliptic and a number of stars, which may be rotated over the plates of an astrolabe
Saphea Azarchelis	Latin name of the form of universal stereographic astrolabe designed by Ibn az-Zarqellu
Shadow square	square divided along two sides into 7 or 12 radial divisions enabling the direct solution (with the aid of the alidade) by proportions of some problems which would otherwise require use of elementary trigonometry
Sines, quadrant of	see 'sinical quadrant'
Sinical quadrant	quadrant with a grid of horizontal and vertical lines which give respectively the sines and cosines of the angles marked in degrees along the edge; frequently engraved in the upper-left quadrant on the back of astrolabes
Solstices	the two points where the tropics meet the ecliptic
Stereographic projection	a perspective and conformable projection of the three-dimensional celestial sphere onto a plane surface (fig. 1)
Tablet of horizons	see 'Horizon plate'
Triplicities	the division of the signs of the zodiac by groups of three among the four elements

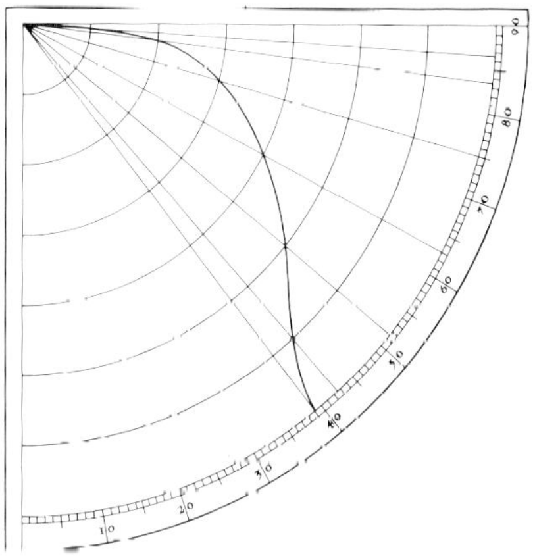

fig. 189. Sigmoid curve.

Unequal hours	see 'Hours, unequal'
Tropics	see 'Cancer' and 'Capricorn'
Zenith	point on the celestial sphere vertically over a place on the earth
Zodiac	zone of the heavens extending 9° on each side of the ecliptic within which the annual revolutions of the sun and the planets are contained
Zodiacal signs	the 12 constellations which divide the zodiac: Aries Libra Taurus Scorpio Gemini Sagittarius Cancer Capricorn Leo Aquarius Virgo Pisces
Zodiacal signs, arcs of	stereographic projections of circles of solar declination drawn on the back of the astrolabe, usually in the upper-right quadrant; on Indo-Persian astrolabes drawn equidistant not in stereographic projection

Bibliography of works cited

This list contains only those works referred to in abbreviated form in the notes. It is not intended to be a complete bibliography of the astrolabe.

D'ALVERNY — MARIE THERESE D'ALVERNY, 'Avendauth?,' in *Homenaje a J. M. Millás-Vallicrosa*, 2 vols., Barcelona, 1954–6, i, 17–43.

ANTHIAUME & SOTTAS — A. ANTHIAUME & J. SOTTAS, *L'astrolabe-quadrant du Musée des antiquités de Rouen. Recherches sur les connaissances mathématiques, astronomiques et nautiques au moyen âge*, Paris, 1910.

ARBERRY — A.J. ARBERRY, *The Koran Interpreted*, London, 1964.

ARCHINARD — MARGARIDA ARCHINARD, *Astrolabe*, Geneva, 1983.

ARMSTRONG — C.A.J. ARMSTRONG, 'An Italian Astrologer at the Court of Henry VII,' in E.F. Jacob (ed.), *Italian Renaissance Studies*, London, 1960, 433–54.

ATKINS & OVERALL — SAMUEL ELLIOTT ATKINS & WILLIAM HENRY OVERALL, *Some account of the Worshipful Company of Clockmakers of the City of London*, London, 1881.

BEAUJOUAN — GUY BEAUJOUAN, 'L'enseignement du quadrivium' in *La Scuola nell' Occidente Latino dell' alto Mediaevo* (Settimane di Studio del Centro Italiano di Studi sull' altro Mediaevo XII), Spoleto, 1972.

BEESON — C.F.C. BEESON, *Clockmaking in Oxfordshire, 1400–1850*, 2nd edit., Oxford, 1967.

BEHRENDS — FREDERICK BEHRENDS, *The Letters and Poems of Fulbert of Chartres*, Oxford, 1976.

BEHRENDS & McVAUGH — FREDERICK BEHRENDS & MICHAEL McVAUGH, 'Fulbert of Chartres' Notes on Arabic Astronomy,' *Manuscripta*, xv, 172–7.

BELL — AUBREY F.G. BELL, *Benito Arias Montano* (Hispanic Notes & Monographs, v), Oxford, 1922.

BENSAUDE — JOAQUIM BENSAUDE, *Histoire de la science nautique portugaise à lépoque des grandes découvertes. Collection de documents publiés par ordre du Ministère de l'instruction publique de la République portugaise. I. Regimento do estrolabio e do quadrante. Tractado da spera do Mundo. Reproduction facsimilé du seul exemplaire connu appartenant à la Bibliothèque Royale de Munich*, Munich, 1914.

BION — [NICOLAS] BION, *L'usage des astrolabes, tant universels que particuliers. . .*, Paris, 1702

BLAGRAVE — JOHN BLAGRAVE, *The mathematical iewel, shewing the making, and most excellent use of a singuler instrument. . .* London, 1585.

BLUNDEVILLE — THOMAS BLUNDEVILLE, *M. Blundeville, his Exercises, containing sixe treatises, which treatises are verie necessarie to be read and learned of all young gentlemen that have not bene exercised in such disciplines, and yet are desirous to have knowledge as well in cosographie, astronomie, and geographie, as also in the art of navigation. . .*, London, 1594.

B.M.C. *Catalogue of Books Printed in the XVth Century Now in the British Museum*, London, 1908–62.

BONELLI MARIA LUISA RIGHINI BONELLI, *Il Museo di storia della scienza a Firenze*, Milan, 1968.

BONELLI & SETTLE M.L. RIGHINI BONELLI & T. SETTLE, *The Antique Instruments of the Museum of the History of Science in Florence*, Florence, n.d.

BRIEUX (I) ALAIN BRIEUX, *Histoire des Sciences: Livres–Instruments* (catalogue), Paris, 1973.

BRIEUX (II) ALAIN BRIEUX, 'Communication' [concerning Muḥammad Laḥbābī], *L'arabisant*, i, January 1974, 16–8.

BRIEUX (III) ALAIN BRIEUX, 'Sur un astrolabe nord-sud.' Communication to the XIVth International Congress of the History of Science, 1975.

BRIEUX (IV) ALAIN BRIEUX, *Histoire des sciences: livres, instruments, autographes* (catalogue), Paris, 1977.

BRIEUX & MADDISON ALAIN BRIEUX & FRANCIS MADDISON, *Répertoire des facteurs d'astrolabes et de leurs oeuvres: première partie: Islam plus Arménie, Géorgie, et Inde Hindou*, Paris, 1985.

BROWN (I) JOYCE BROWN, 'Guild Organisation and the Instrument-making Trade, 1550–1830: the Grocers' and Clockmakers' Companies, *Annals of Science*, xxxvi, 1979, 1–34.

BROWN (II) JOYCE BROWN, *Mathematical Instrument-Makers in the Grocers' Company 1688–1800 with Notes on Some Earlier Makers*, London, 1979.

BUCHNER (I) EDMUND BUCHNER, 'Antike Reiseuhren,' *Chiron*, i, 1971, 457–82.

BUCHNER (II) EDMUND BUCHNER, 'Römische Medaillons als Sonnenuhren,' *Chiron*, vi, 1976, 329–48.

CANOBBIO ERNESTO CANOBBIO, 'An Important Fragment of a West Islamic Spherical Astrolabe,' *Annali dell'Istituto e Museo di Storia della Scienza di Firenze*, i, 1976, 37–41.

CARDONER ANTONIO CARDONER, 'La Medicina astrólogica durante el siglo XIV en la Corona de Aragon,' *Actes du IXe Congres International d'Histoire des Sciences, Barcelona – Madrid, 1–7 September 1959*, 2 vols., Barcelona & Paris, 1960, 341–6.

CARRA DE VAUX BARON CARRA DE VAUX, 'l'Astrolabe linéaire ou Baton d'et-Tousi,' *Journal Asiatique*, 9th ser., v, 1895, 464–516.

CHARDIN JEAN CHARDIN, *'Voyages . . . en Perse et autres lieux de l'Orient,'* new edition, 3 vols., Amsterdam, 1735.

CHAUCER GEOFFREY CHAUCER, *A Treatise on the Astrolabe* (ed. W.W. Skeat), London, 1872.

CHRISTIE CHRISTIE, MANSON & WOODS LTD., *Important Scientific Instruments . . . , 9 April 1975*, London, 1975.

CLAIR COLIN CLAIR, *Christopher Plantin*, London, 1960.

COCHRANE LOUISE COCHRANE, 'Adelard of Bath and the Astrolabe,' *Somerset Archaeology & Natural History*, cxxiv, 1980, 141–50.

COOPLAND G.W. COOPLAND, *Nicole Oresme and the Astrologers: A Study of his Livre de Divinacions*, Liverpool, 1952.

COPERNIC *Nicolas Copernic ou la Révolution Astronomique*, Paris, 1973.

COSMAN MADELEINE PELNER COSMAN, 'Machaut's Medical Musical World,' in Madeleine Pelner Cosman & Bruce Chandler (eds.), *Machaut's World: Science and Art in the Fourteenth Century* (Annals of the New York Academy of Sciences, 314) New York, 1978, 1–36.

COWPER H. SWAINSON COWPER, *Through Turkish Arabia: A journey from the Mediterranean to Bombay by the Euphrates and Tigris valleys and the Persian Gulf*, London, 1894.

CUMMING A.S. CUMMING, 'Gordon of Straloch's Astrolabe,' *Scottish Geographical Magazine*, xxxxii, 1926, 79–82.

DALTON O.M. DALTON, 'The Byzantine Astrolabe at Brescia,' *Proceedings of the British Academy*, xii, 1926, 133–46.

DAWOOD N.J. DAWOOD (tr.), *Tales from the Thousand and One Nights*, Harmondsworth, 1973.

DELACHENAL R. DELACHENAL, *Histoire de Charles V*, 5 vols., Paris, 1909–31.

DELAMAIN RICHARD DELAMAIN, *The making, description, and use of a small portable Instrument . . . Called a Horizontall Quadrant . . .*, London, 1632.

DELISLE LEOPOLD DELISLE, *Recherches sur la librairie de Charles V*, 2 vols., Paris, 1907.

DESTOMBES (I) MARCEL DESTOMBES, 'Un astrolabe carolingian et l'origine de nos chiffres arabes,' *Archives Internationales d'Histoire des Sciences*, lviii–lix, 1962, 3–45.

DESTOMBES (II) MARCEL DESTOMBES, 'La diffusion des instruments scientifiques du haut moyen âge au XVe siècle,' *Journal of World History*, x, 1966, 31–51.

DICKINSON H.W. DICKINSON, *Sir Samuel Morland, Diplomat and Inventor, 1625–1695*, Cambridge, 1970.

DICKS D.R. DICKS, *Hipparchus' Geographical Fragments*, London, 1960.

DODGE BAYARD DODGE (ed. & tr.), *The Fihrist of al-Nadim; a Tenth Century Survey of Mulsim Culture*, 2 vols., New York, 1970.

DRACHMAN A.G. DRACHMAN, 'The Plane Astrolabe and the Anaphoric Clock,' *Centaurus*, iii, 1953–4, 183–9.

D.S.B. CHARLES COULSTON GILLISPIE (ed.-in-chief), *Dictionary of Scientific Biography*, 16 vols., New York, 1970–80.

DURAND DANA BENNETT DURAND, *The Vienna-Klosterneuburg Map Corpus of the Fifteenth Century; a Study in the Transition from Medieval to Modern Science*, Leiden, 1952

ECKHARDT WOLFGANG ECKHARDT, 'Studien zu Erasmus Habermel,' *Jahrbuch der Hamburger Kunstsammlungen*, xxi & xxii, 1976–7, 55–92 & 13–74.

EDEN & LLOYD H.K.F. EDEN & ELEANOR LLOYD, *The Book of Sun-dials; originally compiled by the late Mrs. Alfred Gatty*, London, 1900.

EELES SUSAN EELES, 'Sundials and Nocturnals,' in *An Inventory of the Navigation and Astronomy Collections in the National Maritime Museum, Greenwich*, 3 vols., London [1971–2], sect. 29.

ESTREICHER KAROL ESTREICHER, *Copernican relics in Jagellonian University*, trans. by Marianna Abrahamowicz, (Uniwersytetu Jagiellońskiego, Zeszyty naukowe. CCCVIII Prace Historyczne, zeszyt 41: Copernicana Cracoviensia II), Cracow, 1973.

EVANS R.J.W. EVANS, *Rudolph II and his World; a Study in Intellectual History 1576–1612*, Oxford, 1973.

EXP. DES INST. *Exposition des instruments et outils d'autrefois* (Musée des Arts Décoratifs), [Paris, 1936].

FASTES Les Fastes du Gothique. *Le Siècle de Charles V*, Paris, 1981.

FEBVRE & MARTIN LUCIEN FEBVRE & HENRI-JEAN MARTIN, *L'apparition du livre*, Paris, 1958.

FERNANDEZ VILLAR MIGUEL ANGEL FERNANDEZ VILLAR, *Sobre el astrolábio firmado por G. Frisius y G. Arsenius* (Museo Nacional de Historia: Castillo de Chapultepec), n.p., 1976.

FINE ORONCE FINE, *Protomathesis. . . [de arithmetica practica libri IIII; de geometria libri duo; de cosmographia sive mundi sphaera libri V; de Solaribus horologiis et quadrantibus libri IIII]*, Paris, 1532.

FITZGERALD AUGUSTINE FITZGERALD, *The Letters of Synesius of Cyrene. Translated into English with Introduction and Notes*, Oxford & London, 1926.

FORSTER WILLIAM FORSTER (tr.), *The Circles of Proportion and the Horizontal Instrument. Both invented and the Uses of both written in Latin by Mr. W.O. . .*, London, 1632.

FRANK JOSEF FRANK, *Zur Geschichte des Astrolabs*, Erlangen, 1920.

GABB G.H. GABB, 'The Astrological Astrolabe of Queen Elizabeth,' *Archaeologia*, lxxxvi, 1936.

GALLOIS LUCIEN J. GALLOIS, *Les géographes allemands de la renaissance* (Bibliothèque de la Faculté des Lettres de Lyon, xiii), Paris, 1890.

GANDZ SOLOMON GANDZ, 'The Astrolabe in Jewish Literature,' *Hebrew Union College Annual*, iv, 1927, 469–86.

GARCIA SALVADOR GARCIA FRANCO, *Catálogo crítico de astrolabios existentes en España*, Madrid, 1945.

GIBBS SHARON L. GIBBS, *Greek and Roman Sundials*, New Haven & London, 1976.

GIBBS & SALIBA SHARON GIBBS WITH GEORGE SALIBA, *Planispheric Astrolabes from the National Museum of American History*, Washington, D.C., 1984.

GINGERICH [OWEN GINGERICH] *Collector's Choice. A Selection of Books and Manuscripts given by Harrison D. Horblit to the Harvard College Library*, Cambridge (Mass.), 1983.

GOLDSTEIN BERNARD R. GOLDSTEIN, 'Levi Ben Gerson: On Instrumental Errors and the Transversal Scale,' *Journal of the History of Astronomy*, viii, 1977, 102–12.

GRASSI GIOVANNA GRASSI, *Union Catalogue of Printed Books of the XV and XVI Centuries in Astronomical European Observatories*, Rome, 1977.

GRENVILLE G.S.P. FREEMAN-GRENVILLE, *The Muslim and Christian Calendars*, London, New York, & Toronto, 1963.

GUILLEVIC JEANNE C. GUILLEVIC, *Horlogerie et instruments de mesure de temps passé (Musée Paul Dupuy), Toulouse, 1979.*

GUNTER EDMUND GUNTER, *The Description and Use of the Sector,* London, 1623.

GUNTHER (I) R.T. GUNTHER, *Astronomy (Early Science in Oxford, ii),* Oxford, 1923.

GUNTHER (II) R.T. GUNTHER, 'The Great Astrolabe and other Scientific Instruments of Humphrey Cole,' *Archaeologia,* lxxvi, 1926–7, 273–317.

GUNTHER (III) R.T. GUNTHER, *Chaucer and Messahalla on the Astrolabe (Early Science in Oxford,* v*),* Oxford, 1929.

GUNTHER (IV) R.T. GUNTHER, 'The Uranical Astrolabe and other Inventions of John Blagrave of Reading,' *Archaeologia,* lxxix, 1929, 55–72.

GUNTHER (V) R.T. GUNTHER, *The Astrolabes of the World,* 2 vols., Oxford, 1932.

GUNTHER (VI) R.T. GUNTHER, 'The astrolabe of Queen Elizabeth,' *Archaeologia,* lxxxvi, 1937, 65–72.

HAIN LUDWIG FRIEDRICH THEODOR HAIN, *Repertorium bibliographicum . . . ad annum MD,* Stuttgart, 1826–38.

HALL & HALL A. RUPERT HALL & MARIE BOAS HALL, *The Correspondence of Henry Oldenburg, volume II, 1663–1665,* Madison, Milwaukee, & London, 1966.

HAMILTON (I) GERTRUDE HAMILTON, 'Old Scientific Instruments,' 1930. Unpublished typescript with illustrations.

HAMILTON (II) HENRY HAMILTON, *The English brass and copper Industries to 1800,* London, 1926.

HAMMOND ELEANOR PRESCOTT HAMMOND, *Chaucer; a Bibliographical Manual,* New York, 1908.

HARRISON KENNETH HARRISON, 'Vitruvius and Acoustic Jars in England during the Middle Ages,' *Transactions of the Ancient Monuments Society,* N.S. 15, 1967–8, 49–58.

HARTNER (I) W. HARTNER, 'The Principles and Use of the Astrolabe' in Arthur Upham Pope (ed.), *A Survey of Persian Art,* London & New York, 1939. Reprinted in Hartner (II), 287–311, and also privately by the Société Internationale de l'Astrolabe, as *Astrolabica I,* 1978.

HARTNER (II) W. HARTNER, *Oriens Occidens: Ausewählte schriften zur Wissenschafts-und Kulturgeschichte. Festschrift zum 60. Geburtstag* (Collectanea III), Hildesheim, 1968.

HASKINS CHARLES HOMER HASKINS, *Studies in the History of Medieval Science,* Cambridge, (Mass.) 1924.

HAY DENYS HAY, *Europe in the Fourteenth and Fifteenth Centuries,* London, 1966.

HEIBERG JOHAN LUDVIG HEIBERG (ed.), *Claudii Ptolemaei: Opera quae extant omnia,* Leipzig, 1907.

HIND A.M. HIND, *Engraving in England in the Sixteenth and Seventeenth Centuries,* 2 vols., Cambridge, 1952–5.

HIST. TECH. CHARLES SINGER, E.J. HOLMYARD, A.R. HALL, & TREVOR I. WILLIAMS (eds.), *A History of Technology,* 5 vols., New York & London, 1954–8.

HOLLAND RICHARD HOLLAND, *An Explanation of Mr. Gunter's Quadrant as it is enlarged with an Analemma*, Oxford, 1676.

HONIG PETER S. HONIG, 'The Construction of an Astrolabe,' *Horological Dialogues, Journal of the American section of the Antiquarian Horological Society*, i, 1979, 50–5.

HOUZEAU & LANCASTER J.C. HOUZEAU & A. LANCASTER, *Bibliographie générale de l'astronomie jusqu'en 1880*. . ., (new edit. ed. by D. W. Dewhirst), 2 vols. in 3, London, 1964.

HUTCHIESON A.R. HUTCHIESON, 'Bequest to the Royal Scottish Museum of the Astrolabe of Robert Gordon of Straloch,' *The Mariners Mirror*, xxxiv, 1948, 122–3.

IC/CCA SHARON L. GIBBS, JANICE A. HENDERSON, & DEREK J. PRICE, *A Computerized Checklist of Astrolabes*, New Haven, 1973.

INSTRUMENTS SCIENTIFIQUES *Instruments Scientifiques, Livres Anciens: 1 Collection Leonard* Linton. . . *2 Objets appartenant à divers amateurs. . . dont la vente aux enchères publiques aura lieu au Nouveau Drouot*. . . 9 et 10 Octobre 1980. . . *Messrs. Etienne Libert et Alain Castor Commissaires-Priseurs M. Alain Brieux Expert*, Paris, 1980.

JANIN (I) LOUIS JANIN, 'Un text d'Ar-Rudani sur l'astrolabe sphérique,' *Annali dell'Istituto e Museo di Storia della Scienza di Firenze*, iii, 2, 1978, 71–5.

JANIN (II) LOUIS JANIN, 'L'astrolabe et cadran solaire en projection stéreographique horizontale,' *Centaurus*, xxii, 1979, 298–314.

[JOMARD] [EDM. FR. JOMARD], *Description de l'Egypte ou recueil des observations et de recherches qui ont été faites en Egypte pendant l'Expedition de l'Armée Française*. . ., 21 vols., Paris, 1809–28.

JORDAN P. JORDAN, *La Collection Greppin: Instruments Scientifiques* (sale catalogue), Zurich, 1975.

JOSTEN C.H. JOSTEN, *Scientific Instruments, 13th–19th century; the collection of J. A. Billmeir C.B.E.*, Oxford, 1954.

JOURDAIN CHARLES JOURDAIN, 'Nicole Oresme et les astrologues de la cour de Charles V, *Revue des Questions Historiques*, xvii, 1875, 136–59.

KAYE (I) G.R. KAYE, *The Astronomical Observatories of Jai Singh*, (Archaeological Survey of India: New Imperial Series, xl), Calcutta, 1918.

KAYE (II) G.R. KAYE, *Astronomical instruments in the Delhi Museum*, (Memoirs of the Archaeological Survey of India, 12) Calcutta, 1921.

KELLENBENZ HERMANN KELLENBENZ, 'Augsberger Wirtschaft 1530 bis 1620,' in *Welt im Umbruch: Augsburg zwischen Renaissance und Barock*, 2 vols., Augsburg, 1980, i. *Zeughaus*, 50–71.

KENNEDY (I) E.S. KENNEDY, 'A Survey of Islamic Astronomical Tables,' *Transactions of the American Philosophical Society*, N.S., xlvi, 1956.

KENNEDY (II) E.S. KENNEDY (tr. & comm.), *The Exhaustive Treatise on Shadows by Abū al-Rayḥān Muḥammad b. Aḥmad al-Bīrūnī*, 2 vols., Aleppo, 1976.

KENNEDY & HADDAD E.S. KENNEDY & F.I. HADDAD, 'Geographical Tables of Medieval Islam,' in E.S. Kennedy et al., *Studies in the Islamic Exact Sciences*, Beirut, 1983, 636–51.

KING (I) DAVID A. KING, 'An Analog Computer for Solving Problems of Spherical Astronomy: the *Shakkāzīya* Quadrant of Jamāl ad-Dīn al-Māridīnī,' *Archives Internationales d'Histoire des Sciences*, xxiv, 1974, 219–42.

KING (II) DAVID A. KING, 'On the Astronomical Tables of the Islamic Middle Ages,' *Colloquia Copernicana III (Studia Copernicana XIII)*, Wroclaw, Warsaw, & Cracow, 1975, 37–56.

KING (III) DAVID A. KING, 'Notes on the Astrolabist Nasṭūlus/Basṭulus,' *Archives Internationales d'Histoire des Sciences*, xxviii, 1978, 117–20.

KING (IV) DAVID A. KING, 'On the Early History of the Universal Astrolabe in Islamic Astronomy and the Origin of the Term "Shakkāzīya" in Medieval Scientific Arabic,' *Journal for the History of Arabic Science*, iii, 1979, 244–57.

KING (V) DAVID A. KING, 'The Astronomy of the Mamluks,' *Isis*, lxxiv, 1983, 531–55.

KING (VI) DAVID A. KING, *The Astronomical Instruments of Ibn al-Sarrāj*, Athens (forthcoming).

KLARWILL VICTOR VON KLARWILL, *The Fugger news-letters*. . . (tr. Pauline de Chary), 2 vols., London, 1924.

KLEBS A.C. KLEBS, 'Incunabula scientifica et medica,' *Osiris*, iv, 1938, 1–359.

KNECHT PIERRE KNECHT, *I Libri astronomici di Alfonso X in una versione Fiorentina del trecento. Tesi di leaurea presentate alla Facolta di Lettere dell' Universita di Zurigo*, Zaragoza, 1965.

KREN CLAUDIA KREN, 'Hermann (Hermannus) the Lame,' *D.S.B.*, vi, New York, 1972, 301–3.

KUNITZCH (I) PAUL KUNITZCH, *Arabische Sternnamen in Europa*, Wiesbaden, 1959.

KUNITZCH (II) PAUL KUNITZCH, 'On the authenticity of the Treatise on the Composition and Use of the Astrolabe ascribed to Messahalla,' *Archives Internationales d'Histoire des Sciences*, xxxi, 1981, 42–62.

KUNITZCH (III) PAUL KUNITZCH, 'Observations on the Arabic Reception of the Astrolabe,' *Archives Internationales d'Histoire des Sciences*, xxxi, 1981, 244–52.

KUNITZCH (IV) PAUL KUNITZCH (IV), 'Remarks regarding the Terminology of the Astrolabe.' Unpublished paper prepared for presentation to the 7e Assemblée Générale de la Société Internationale de l'Astrolabe, Oxford, 1982.

R.L. R.L. *Dialling performed Instrumentally by our Hemisphere in Plane*. . . , London, 1652.

LABARTE JULES LABARTE, *Inventaire du mobilier de Charles V, roi de France* (Collection de documents inédits sur l'histoire de France, 3e sér., Archéologie), Paris, 1879.

LACOMBRADE CHRISTIAN LACOMBRADE (ed. & tr.), *Synésios de Cyrène. I Hymnes*, Paris, 1978.

LARCOM THOMAS AISKEW LARCOM (ed.), *The History of the Survey of Ireland, commonly called the Down Survey, by Doctor William Petty, A.D. 1655–6*, Dublin, 1851.

LATTIN (I) H.P. LATTIN, 'Lupitus Barchinonensis,' *Speculum. Journal of Medieval Studies,* vii, 1932, 58–64.

LATTIN (II) H.P. LATTIN, 'The Eleventh Century Ms Munich 14436: Its Contribution to the History of Coordinates, of Logic, of German Studies in France,' *Isis*, xxxviii, 1948, 205–25.

LATTIN (III) H.P. LATTIN, *The Letters of Gerbert with his Papal Privilege as Sylvester II, translated with an Introduction*, New York, 1959.

LECERF & LABANDE GEORGES LECERF & E-R. LABANDE (eds.), *Instruments de musique du XVe siècle; les traités d'Henri-Arnaut de Zwolle et de divers anonymes (ms B.N. latin 7295)*, Paris, 1932.

LEVY RAPHAEL LEVY, 'The Authorship of a Latin Treatise on the Astrolabe,' *Speculum*, xvii, 1942.

LEWIS (I) BERNARD LEWIS, 'An Epistle on Manual Crafts,' *Islamic Culture*, xxvii, 1943, 142–51.

LEWIS (II) P.S. LEWIS, *Later Medieval France: the Polity*, London, 1968.

LINDBERG DAVID C. LINDBERG, 'The Transmission of Greek and Arabic Learning to the West,' in David C. Lindberg (ed.), *Science in the Middle Ages*, Chicago & London, 1978, 52–90.

LUNARDI HEINRICH LUNARDI, *900 Jahr Nürnberg: 600 Jahre Nürnberger Uhren*. Vienna & Stuttgart, 1974.

McKENZIE D.S. McKENZIE, *Stationers' Company Apprentices 1641–1700* (Oxford Bibliographical Society, N.S. XVII), Oxford, 1974.

MACCAGNI CARLO MACCAGNI, 'The Florentine Clock and Instrument-Makers of the della Volpaia Family,' *Actes du XIIe Congrès International d'Histoire des Sciences, Paris, 1968: t. Xa Histoire des Instruments Scientifiques*, Paris, 1971, 65–73.

MACREZ C. MACREZ, 'Cadrans solaires d'azimut (projections orthographique et stéreographique),' *L'astronomie et Bulletin de la Société Astronomique de France*, Oct. 1976, 435–8.

MADDISON (I) [FRANCIS MADDISON], *A Supplement to a Catalogue of Scientific Instruments in the Collection of J.A. Billmeir Esq. C.B.E.*, Oxford & London, 1957.

MADDISON (II) FRANCIS MADDISON, 'A 15th Century Islamic Spherical Astrolabe,' *Physics*, iv, 1962, 101–9.

MADDISON (III) FRANCIS MADDISON, 'Hugo Helt and the Rojas Astrolabe Projection,' *Rivista de Faculdade de Ciencias*, xxxix, (Agrupamento de Estudos de Cartografia Antiga XII Secção de Coimbra), Coimbra, 1966.

MADDISON (IV) FRANCIS MADDISON, 'Medieval Scientific Instruments and the Development of Navigational Instruments in the XVth and XVIth Centuries,' *Revista de Universidade de Coimbra*, xxiv (Agrupamento de Estudos de Cartografia Antiga XXX: Secção de Coimbra), Coimbra, 1969.

MADDISON (V) FRANCIS MADDISON, 'Astrolabists mentioned by Ibn an-Nadīm in the *Fihrist*, arranged to show Apprenticeship and Family Relationships.' Chart presented with commentary to the 2e Assemblée Générale de la Société Internationale de l'Astrolabe, Paris, 1977.

MADDISON & BRIEUX FRANCIS MADDISON & ALAIN BRIEUX, 'Basṭūlus or Nasṭūlus? A Note on the Name of an Early Islamic Astrolabist,' *Archives Internationales d'Histoire des Sciences*, xxiv, 1974, 157–60.

MADDISON & TURNER FRANCIS MADDISON & ANTHONY TURNER, 'Science and Technology in Islam. Catalogue of an Exhibition held at the Science Museum, London, April–August 1976, in association with the Festival of Islam. Not published: privately circulated in a xerographic edition of 50 copies.

MARTIN THOMAS-HENRI MARTIN, 'Astronomia' in Charles Daremberg & Edm. Saglio, *Dictionnaire des Antiquités Grecques et Romains d'après les textes et les monuments*, 9 vol., Paris, 1877, i, 476–504.

MASCART JEAN MASCART, 'Clavius et l'astrolabe,' *Bulletin Astronomique*, 1905.

MAULA E. MAULA, 'The Spider in the sphere. Eudoxus Arachne,' *Philosophia*, v–vi, 1975–6.

MAURICE & MAYR KLAUS MAURICE & OTTO MAYR (eds.), *The Clockwork Universe. German Clocks and Automata 1550–1650*, New York, 1980.

MAYER (I) L.A. MAYER, *Islamic Astrolabists and Their Work*, Geneva, 1956.

MAYER (II) L.A. MAYER, 'Islamic Astrolabists: Some New Material,' in Richard Ettinghausen (ed.), *Aus der Welt der Islamischen Kunst. Festschrift für Ernst Kühnel zum 75. . . Geburtstag an 26. 10.1957*, Berlin, 1959, 293–6.

MAYOR J.E.B. MAYOR, *Admissions to the college of St. John the Evangelist in the University of Cambridge, Part II, July 1665–July 1715*, Cambridge, 1893.

METRAUX & CROUZET GUY S. METRAUX & FRANCOIS CROUZET (eds.), *The Evolution of Science*, New York, 1963.

MICHEL (I) HENRI MICHEL, 'Un astrolabe de Lambert Damery,' *Ciel & Terre*, iii, 1939.

MICHEL (II) HENRI MICHEL, 'Méthodes de trace et d'exécution des astrolabes Perses,' *Ciel & Terre*, xii, 1941.

MICHEL (III) HENRI MICHEL, *Traité de l'astrolabe*, Paris, 1947 (augmented re-impression with preface by Francis Maddison, Paris, 1976).

MICHEL (IV) HENRI MICHEL, 'Astrolabistes, Géographes et Graveurs Belges du XVIe Siècle,' *La Science au seizième siècle: Colloque Internationale de Royaumont 1–4 Juillet, 1957*, Paris, 1957, 15–27.

MICHEL (V) HENRI MICHEL 'Thomas Lambert ou Lambrechts dit Gemini,' *Biographie Nationale Belgique*, xxxi, Brussels, 1961, cols. 385–95.

MICHEL (VI) HENRI MICHEL, 'Beauté archaïque des astrolabes,' *Sefunim. Bulletin du Musée Nationale Maritime, Haifa*, iv, Haifa, 1972.

MIGNE J-P. MIGNE, *Patrologia Graece*, lxvi, Paris, 1864.

MILLAS (I) J.M. MILLAS VALLICROSA, 'Estudios sobre Azarquiel: el Tratado de la Azafea,' *Archeion*, xiv, 1932, 393–419.

MILLAS (II) J.M. MILLAS VALLICROSA, *Don Profeit Tibbon: Tractat de l'assafea d'Azarquiel*, Barcelona, 1933.

MILLAS (III) J.M. MILLAS VALLICROSA, *Estudios sobre Azarquiel*, Madrid & Granada, 1943–50.

MILLAS (IV) J.M. MILLAS VALLICROSA, 'Un Ejemplar de azáfea arabe de Azarquiel,' *Al-Andalus: Revista de Escuelas de Estudios Arabes de Madrid y Granada*, ix, 1944, 111–9.

MILLAS (V) J.M. MILLAS VALLICROSA, 'Los Primeros Tratados de Astrolábio en la España Arabe,' *Revista del Instituto Egipcio de Estudios Islamicos en Madrid*, iii, 1955, 35–84.

MILLAS (VI) J.M. MILLAS VALLICROSA, 'Translations of Oriental Scientific Texts (to the End of the Thirteenth Century), *Journal of World History*, ii, 1957. Reprinted in Metraux & Crouzet, 128–67.

MØLLER JORGEN MILERT MØLLER, 'Det plane Astrolabium,' *Nordisk. Astronomisk Tidsakrift. Ugvit af Astronomisk Selskab*, iv, 1964, 113–34.

MORLEY (I) WILLIAM H. MORLEY, *Description of a planispheric astrolabe constructed for Shah Sultan Husain Safawi, King of Persia and now preserved in the British Museum. Comprising an account of the astrolabe generally*. . . , London, Edinburgh, Paris, & Leipzig, 1856. Reprinted in Gunther (V).

MORLEY (II) WILLIAM H. MORLEY, 'Description of an Arabic Quadrant,' *Journal of the Royal Asiatic Society*, 1860, 1–11

MUNDY J. MUNDY, 'John of Gmunden,' *Isis*, xxxiv, 1942–3, 196–205.

NADVI (I) SAYYID SULAYMAN NADVI, 'Some Indian Astrolabe Makers,' *Islamic Culture*, ix, 4, 1935, 621–31.

NADVI (II) SAYYID SULAYMAN NADVI, 'Indian Astrolabe Makers,' *Islamic Culture*, xi, 4, 1937, 537–9.

NAU F. NAU, 'Le Traité sur l'astrolabe plan de Sévère Sabokt . . . ,' *Journal Asiatique*, 1899.

NEUGEBAUER (I) O. NEUGEBAUER, 'The Early History of the Astrolabe,' *Isis*, xl, 1949.

NEUGEBAUER (II) O. NEUGEBAUER, *A History of Ancient Mathematical Astronomy*, 3 vols., Berlin, Heidelberg, & New York, 1975.

NEUGEBAUER & HOESEN OTTO NEUGEBAUER & H. B. VAN HOESEN, *Greek Horoscopes* (Memoirs of the American Philosophical Society, 48), Philadelphia, 1959.

NICOLAS SIR HARRIS NICOLAS, *The Chronology of History*, London, 1833.

NMM Staff of the Department of Navigation & Astronomy, National Maritime Museum, Greenwich, *The Planispheric Astrolabe* [Greenwich], 1976.

NORTH (I) J.D. NORTH, 'Werner, Apian, Blagrave, and the Meteoroscope,' *British Journal for the History of Science*, iv, 1966–7, 57–65.

NORTH (II) J.D. NORTH, 'The Astrolabe,' *Scientific American*, ccxxx, Jan. 1974, 96–106.

NORTH (III) J.D. NORTH, 'Kalenderes enlumyned ben they. Some astronomical themes in Chaucer,' *The Review of English Studies*, N.S. xx, 1969, 129–54, 257–83, 418–44.

NORTH (IV) J.D. NORTH, *Richard of Wallingford: An edition of his writings with introductions, English translation, and commentary*, 3 vols, Oxford, 1976.

NORTH (V) J.D. NORTH, 'The Alfonsine Tables in England,' in Y. Maeyama & W.G. Saltzer, *Prismata: Naturwissenschaftsgeschichtliche Studien. Festschrift für Willy Hartner*, Wiesbaden, 1977, 269–302.

NORTH (VI) J.D. NORTH, 'Astrolabes and the Hour-line Ritual,' *Journal for the History of Arabic Science*, v, 1981, 113–4.

OMONT H. OMONT, 'Maître Arnault: astrologue de Charles VI et les Ducs de Bourgogne,' *Bibliothèque de l'Ecole de Chartes*, xlii, 1881, 127–8.

VAN ORTROY FERNAND VAN ORTROY, *Bio-bibliographie de Gemma Frisius, fondateur de l'école belge de géographie, de son fils Corneille et de ses neveux les Arsenius*, Brussels, 1920. Reprinted Amsterdam, 1966.

OSLEY A.S. OSLEY, *Mercator: A monograph on the lettering of maps, etc. in the 16th century Netherlands with a facsimile and translation of his treatise on the italic hand and a translation of Ghim's Vita Mercatoris*, London, 1969.

OUGHTRED W. O[UGHTRED], *The Description and Use of the Double Horizontall Dyall: whereby not onely the hower of the Day is shewne; but also the Meridian Line is found: And most Astronomical Questions, which may be done by the Globe are resolved*, London, 1636.

OZANAM JACQUES OZANAM, *Recreations mathematiques et physiques . . .*, Paris, 1694.

PEDERSEN (I) OLAF PEDERSEN, 'The Life and Work of Peter Nightingale,' *Vistas in Astronomy*, ix, Oxford, 1967, 3–10.

PEDERSEN (II) OLAF PEDERSEN, 'The Corpus Astronomicum and the Traditions of Mediaeval Latin Astronomy: A tentative interpretation' in *Colloquia Copernicana III (Studia Copernicana*, XIII), Wroclaw, Warsaw, & Cracow, 1975, 57–96.

PELLAT CHARLES PELLAT, 'L'astrolabe sphérique d'al-Rūdānī,' *Bulletin des Etudes Orientales*, xxvi, 1973 (Arabic text 7–81), and xxviii, 1975, 83–165.

PEREIRA DA SILVA LUCIANO PEREIRA DA SILVA, 'O astrolábio universale da Sociedade de Geografía de Lisboa,' *Jornal de Ciências Matemáticas, Fisicas et Naturais da Academia das Ciências de Lisboa*, 3ª serie, v, 1925. Reprinted in *Obras Completas* III, Lisbon, 1946, 331–52.

PHILIPS C.H. PHILIPS (ed.), *Handbook of Oriental History*, London, 1963.

PINGREE (I) DAVID PINGREE, 'The Fragments of the Works of al-Fazarī,' *Journal of Near Eastern Studies*, xxix, 1970, 103–23.

PINGREE (II) DAVID PINGREE, 'History of Mathematical Astronomy in India,' in *D.S.B.*, xv, 533–633.

PINGREE (III) DAVID PINGREE, 'Islamic Astronomy in Sanskrit,' *Journal for the History of Arabic Science*, ii, 1978, 315–30.

PINTELON P. PINTELON, *Chaucer's treatise on the astrolabe, Ms. 4862–4869 of the Royal Library in Brussels* (Rijskuniversiteit te Gent werken Witgegeven door de Faculteit von de Wijsbegeerte en Letteren 89e Aflevering, Antwerp's-Gravenhage, 1940.

PLENDERLEITH ROBERT W. PLENDERLEITH, 'Discovery of an old Astrolabe,' *The Scottish Geographical Magazine*, lxxvi, 1960, 25.

POGO ALEXANDER POGO, 'Gemma Frisius, his method of Determining Differences of Longitude by transporting Timepieces (1530), and his Treatise on Triangulation (1533),' *Isis*, xxii, 1934, 469–85.

POULLE (I) EMMANUEL POULLE, 'L'astrolabe médiévale d'après les manuscrits de la Bibliothèque Nationale,' *Bibliothèque de l'Ecole de Chartes*, cxii, 1954, 81–103.

POULLE (II) EMMANUEL POULLE, 'La fabrication des astrolabes au moyen âge,' *Techniques et Civilizations*, iv, 1955, 117–28.

POULLE (III) EMMANUEL POULLE, 'L'equatoire de Guillaume Gilliszoon de Wissekereke,' *Physis*, iii, 1961, 223–51.

POULLE (IV) EMMANUEL POULLE, *La bibliothèque scientifique d'un imprimeur humaniste au XVe siècle: catalogue des manuscrits d'Arnaud de Bruxelles à la Bibliothèque Nationale de Paris*, Geneva, 1963.

POULLE (V) EMMANUEL POULLE, *Un constructeur d'instruments astronomiques au XVe siècle: Jean Fusoris*, Paris, 1963.

POULLE (VI) EMMANUEL POULLE, 'Le quadrant nouveau médiéval,' *Journal des Savants*, 1964, 148–67, 182–214.

POULLE (VII) EMMANUEL POULLE, 'Le traité d'astrolabe de Raymond de Marseille,' *Studi Medievali*, 3e sér., v, 1964, 866–900.

POULLE (VIII) EMMANUEL POULLE, 'Les Instruments astronomiques du moyen âge,' *Le Ruban Rouge*, xxxii, 1967. Reissued as Museum of the History of Science, Oxford, selected off-print, no. 7, Oxford, 1969.

POULLE (IX) EMMANUEL POULLE, 'Remarques sur deux astrolabes du moyen âge,' *Physis*, ix, 1967, 161–4.

POULLE (X) EMMANUEL POULLE, 'Un instrument astronomique dans l'occident latin: la "saphea,"' in *A Giuseppe Ermini* (Centre Italiano di Studi sull 'Alto Medioevo), Spoleto, 1970, 491–510.

POULLE (XI) EMMANUEL POULLE, 'Les instruments astronomiques de l'occident latin au XIe et XIIe siècles,' *Cahiers de civilisation médiévale X–XIIe siècles*, xv, 1972, 27–40.

POULLE (XII) EMMANUEL POULLE, *Les instruments de la théorie des planètes selon Ptolémée: équatoires et horlogerie planetaire du XIIIe au XVIe siècle*, 2 vols., Geneva & Paris, 1980.

POULLE (XIII) EMMANUEL POULLE, 'Walcher de Malvern et son astrolabe (1092),' *Rivista da Universidade de Coimbra*, xxviii, 1980, 47–54, (Centro de Estudos de Cartografia Antiga: Secção de Coimbra. Serie Separatas, CXXXII), Coimbra, 1980.

POULLE (XIV) EMMANUEL POULLE, *Les sources astronomique: textes, tables, instruments*, (Typologie des sources du Moyen Age Occidental. Fasc. 39), Turnhout, 1981.

PRICE (I) DEREK J. PRICE, 'The Early Observatory Instruments of Trinity College Cambridge,' *Annals of Science*, viii, 1952, 1–12.

PRICE (II) DEREK J. PRICE, 'Clockwork Before the Clock,' *Horological Journal*, Dec. 1955, 810–4, & Jan. 1956, 31–4. Reprinted as supplement to *Antiquarian Horology*, 1955–6 and in *Antiquarian Horology*, i, 1953–Sept. 1956 (Antiquarian Horological Society Monograph 14), Wadhurst, 1977.

PRICE (III) DEREK J. PRICE, 'An International Checklist of Astrolabes' (2 parts), *Archives Internationales d'Histoires des Sciences*, xxxiv, 1955, 243–63 & 363–81.

PRICE (IV) DEREK J. PRICE, 'The First Scientific Instrument of the Renaissance,' *Physis*, i, 1959, 26–30.

PRICE (V) DEREK J. PRICE, 'Portable Sundials in Antiquity, including an Account of a New Example from Aphrodisias,' *Centaurus*, xiv, 1969, 242–66.

PRICE (VI) DEREK J. PRICE, 'Proto-Astrolabes, Proto-clocks and Proto-calculators: the Point of Origin of High Mechanical Technololgy,' in Denise Schmandt-Bessarat (ed.), *Early Technologies: Invited Lectures on the Middle East at the University of Texas at Austin*, iii, 1979, 61–3.

PROCTOR DAVID PROCTOR, 'An Unusual French Astrolabe, date *c.* 1702,' *Physis*, x, 1968, 235–40.

REEVES E.A. REEVES, 'New Instruments for Travellers and Surveyors,' *The Geographical Journal*, xxxii, 1908, 607–10.

RENAUD H.J.P. RENAUD, 'Quelques constructeurs d'astrolabes en occident musulman,' *Isis*, xxxiv, 1942–3, 20–3.

RETI LADISLAO RETI (ed.), *The Unknown Leonardo*, London, 1974.

RICHARD J. RICHARD, 'Henri Arnault de Zwolle à Dijon et son Influence,' *Publication du Centre Européen d'études Burgondo-Médianes*, vi, 1964, 71–80.

RICHE PIERRE RICHE, *Les écoles et l'enseignement dans l'occident chrétien de la fin du Ve siècle au milieu du XIe siècle*, Paris, 1979.

RICO Y SINOBAS MANUEL RICO Y SINOBAS (ed.), *Libros del Saber de Astronomía del Rey D. Alfonso X de Castilla*, 4 vols., Madrid, 1863–6.

ROBBINS F.E. ROBBINS (ed. & tr.), *Ptolemy: Tetrabiblos*, London & Cambridge (Mass.), 1948.

ROBINSON F.N. ROBINSON, *The Works of Geoffrey Chaucer*, 2nd edit., London, 1957.

ROGERS J.E. THOROLD ROGERS, *A History of Agriculture and Prices in England . . .*, 8 vols., Oxford, 1866–1902.

ROHR & JANIN RENE R.J. ROHR & LOUIS JANIN, 'Deux astrolabes-quadrants turcs,' *Centaurus*, xix, 1975, 108–24.

ROSINSKA GRAZYNA ROSINSKA, *Instrumenty astronomiczne na Uniwersytecie Krakowskim w XV wieku* (Studia Copernicana xi), Wroclaw, Warsaw, & Cracow, 1974.

RYAN W.F. RYAN, 'Some Observations on the History of the Astrolabe and of Two Russian Words: *astroljabija* and *matka*,' in *Studies in Slavic Linguistics and Poetics in Honor of Boris O. Unbegaun*, New York & London, 1968, 155–64.

SACHAU EDWARD SACHAU (tr. & ed.), *The Chronology of Ancient Nations: An English Version of the Athār-ul-Bākiya of Albīrūnī. . .*, London, 1857.

SAFADI YASIN HAMID SAFADI, *Islamic Calligraphy*, London, 1978.

SAMSO JULIO SAMSO, 'Maslama al-Majrītī and the Alphonsine Book on the Construction of Astrolabes,' *Journal for the History of Arabic Science*, iv, 1980, 3–8.

SARTON GEORGE SARTON, *Introduction to the History of Science, Volume II, From Rabbi Ben Ezra to Roger Bacon*, 2 parts, Baltimore, 1931.

SAUNDERS HAROLD N. SAUNDERS, *The Astrolabe*, Bude, 1971.

SAUVAIRE & REY PAILHADE H. SAUVAIRE & J. DE REY PAILHADE, 'Sur une mère d'astrolabe arabe du XIIIe siècle (609 de l'Hégire) portant un calendrier perpétuel avec correspondence musulmane et chrétienne,' *Journal Asiatique*, 1893, 1–121.

SAYILI AYDIN SAYILI, *The Observatory in Islam and its Place in the general History of the Observatory*, (Publications of the Turkish Historical Society, ser. vii, No. 38), Ankara, 1960.

SEDILLOT L. AM. SEDILLOT, 'Mémoire sur les instruments astronomiques des Arabes,' *Memoires présentés. . . à l'Académie Royale des Inscriptions et Belles-Lettres de l'Institut de France*, 1e ser. (sujets divers d'érudition), t. I, Paris, 1844.

SEEMAN HUGO SEEMAN, *Das kugelförmige Astrolab nach den Mitteilungen von Alfons X von Kastilien und der vorhandenen arabischen Quellen* (Abhandlungen zur Geschichte der Naturwissenschaften und der Medizin, VIII), Erlangen, 1925.

SEGONDS (I) ALAIN SEGONDS, 'Aperçu nouveau sur l'origine de l'astrolabe?' *Comptes Rendus au 5e Réunion de la Société Internationale de l'Astrolabe*, Brussels, 1980 (unpublished).

SEGONDS (II) ALAIN SEGONDS, (ed. & tr.) *Jean Philopon: Traité de l'astrolabe* (Société Internationale de l'Astrolabe, Astrolabica 2), Paris, 1981.

SERGEYEVA & KARPOVA N.D. SERGEYEVA & L.M. KARPOVA, 'Al-Farghānī's Proof of the Basic Theorem of Stereographic Projection,' *Voprosy Istorii Estestvoznania i Tekniki*, iii, 40, 1972. English translation by Sheila Embleton in Thomson, 210–7.

SEZGIN FUAD SEZGIN, *Geschichte der arabisches Schriftums Band VI, Astronomie bis ca. 430 H*, Leiden, 1978.

SHARMA VIRENDRA NATH SHARMA, 'The Great Astrolabe of Jaipur and its Sister Unit," *Archaeoastronomy*, vii, *Supplement to the Journal for the History of Astronomy*, xv, 1984, f126–8.

SHERMAN CLAIRE RICHTER SHERMAN, *The Portraits of Charles V of France (1339–1380)*, New York, 1969.

SIDDIQI A. SIDDIQI, 'Construction of Clocks and Islamic Civilization,' *Islamic Culture*, i, 1927, 245–51.

DE SMET (I) ANTOINE DE SMET, 'Note sur Gemma Frisius,' *Bulletin de la Société Royale Belge de Géographie*, xxvi, 1956, 81–97.

DE SMET (II) ANTOINE DE SMET, 'L'orfèvre et graveur Gaspar van der Heyden et la construction des globes à Louvain dans le premier tiers du XVIe siècle,' *Der Globusfreund*, xiii, 1964, 38–48.

DE SMET (III) ANTOINE DE SMET, 'Louvain et la construction des instruments scientifiques au XVIe siècle,' *Actes du XIIe Congrès Internationale d'Histoire des Sciences, Paris, 1968, Xa, Histoire des Instruments Scientifiques*, Paris, 1971, 33–9.

SMITH & GNUDI CYRIL STANLEY SMITH & MARTHA TEACH GNUDI, *The Pirotechnia of Vannoccio Biringuccio*, New York, 1942.

SOLENTE S. SOLENTE (ed.), *Le livre des fais et bonnes meurs du sage roy Charles V par Chistine de Pisan* (Société de l'histoire de France, 256), 2 vols., Paris, 1936.

SOMMERARD E. DU SOMMERARD, *Musée des Thermes et de l'Hotel de Cluny: Catalogue et déscription des objets d'art de l'Antiquité, du Moyen Âge et de la Renaissance*, Paris, 1881.

SOTHEBY (I) SOTHEBY & CO., *The Sir John Findlay Collection: Catalogue of an Important Collection of Scientific Instruments* 2 parts, London, 1961–2.

SOTHEBY (II) SOTHEBY & CO., *Catalogue of a Highly Important Spherical Astrolabe and Other Scientific Instruments . . .*, 26 Feb. 1962, London, 1962.

SOTHEBY (III) SOTHEBY PARKE BERNET & CO., *Catalogue of Scientific Instruments, Watches, and Fine Clocks . . . sold by Auction . . .*, 31 Mar. 1978, London, 1978.

SOTHEBY (IV) SOTHEBY PARKE BERNET, *Important Contemporary and Period Jewelry, and Watches, Clocks, and Instruments*, 8 & 9 Dec. 1982, New York, 1982.

SOUBIRAN JEAN SOUBIRAN (ed.), *Vitruve: de l'Architecture Livre IX. Texte établi, traduit et commenté*, Paris, 1969.

SOUTHERN R.W. SOUTHERN, *Medieval Humanism and Other Studies*, Oxford, 1970.

SPENCER THOMAS SPENCER, *The Sundicator*, Santa Barbara [c. 1973].

STC A.W. PALLARD & G.R. REDGRAVE, *A Short-Title Catalogue of Books printed in England, Scotland & Ireland . . . 1475–1640*, London, 1946.

STEINBERG S.H. STEINBERG, *Five Hundred Years of Printing*, 3rd edit., Harmondsworth, 1974.

STERN S.M. STERN, 'A Treatise on the Armillary Sphere by Dunas Ibn Tamīm' in *Homenaje a Millás-Vallicrosa*, 2 vols., Barcelona, 1954–56, ii, 373–82.

STILLWELL MARGARET BINGHAM STILLWELL, *The Awakening Interest in Science during the First Century of Printing 1450–1550*, New York, 1970.

STIMSON ALAN STIMSON, 'Quadrants (hand-held),' in *An Inventory of the Navigation and Astronomy Collections in the National Maritime Museum, Greenwich*, 3 vols., London [1971–2], sect. 26.

STOFFLER JOHAN STOFFLER, *Elucidatio fabricae ususque astrolabii*, Oppenheim, 1513.

STONE EDMUND STONE, *The construction and principal uses of mathematical instruments. Translated from the French of M. Bion*. . . , 2nd edit., London, 1758.

TANNERY PAUL TANNERY, *Recherches sur l'histoire de l'astronomie ancienne*, Paris, 1893.

TAQIZADEH S.H. TAQIZADEH, 'Various Eras and Calendars Used in the Countries of Islam,' *Bulletin of the School of Oriental Studies*, ix, 1937–9, 903–22 & x, 1940–2, 107–32.

TARDY TARDY [HENRI LANGELLE], *La Pendule Française*, 3 vols., 5th edit., Paris, 1981–82.

TAYLOR (I) E.G.R. TAYLOR, 'Early Charts and the Origin of the Compass Rose,' *Journal of the Institute of Navigation*, iv, 1951, 351–6.

TAYLOR (II) E.G.R. TAYLOR, *The Mathematical Practitioners of Tudor & Stuart England, 1485–1714*, Cambridge, 1954.

THOMAS KEITH THOMAS, *Religion and the Decline of Magic: Studies in Popular Beliefs in Sixteenth and Seventeenth Century England*, rev. edit., Harmondsworth, 1973.

THOMPSON J.W. THOMPSON, 'The Introduction of Arabic Science into Lorraine in the Tenth Century,' *Isis*, xii, 1929, 184–94.

THOMSON RON B. THOMSON, *Jordanus of Nemore and the Mathematics of Astrolabes: de Plana Spera, an Edition with Introduction, Translation, and Commentary*, Toronto, 1978.

THORNDIKE (I) LYNN THORNDIKE, *A History of Magic and Experimental Science*, 8 vols., New York, 1923–58.

THORNDIKE (II) LYNN THORNDIKE, 'Robertus Anglicus,' *Isis*, xxxiv, 1943, 467–9.

THORNDIKE (III) LYNN THORNDIKE, *Michael Scot*, London, 1950.

TOMBA (I) T[ULLIO] TOMBA, 'Due Astrolabi latini del XIV secolo conservati a Milano,' *Physis*, viii, 1966, 295–306.

TOMBA (II) TULLIO TOMBA, 'Nuove Osservazione sui due Astrolabi Latini del XIV secolo conservati a Milano,' *Physis*, x, 1968, 119–22.

TOMBA (III) T[ULLIO] TOMBA, 'Gli Astrolabi della Collezione Settala nella Pinacoteca Ambrosiana' *Atti della Fondazione Giorgi Ronchi*, xxxiii, 1978, 297–314 & 636–47.

TURNBULL H.W. TURNBULL (ed.), *James Gregory. Tercentenary memorial volume*. . . , London, 1939.

TURNER (I) A.J. TURNER, 'Mathematical Instruments and the Education of Gentlemen,' *Annals of Science*, xxx, 1973, 51–88.

TURNER (II) A.J. TURNER, *Paper & Brass: Scientific Instruments and the Art of Printing. A Catalogue of an Exhibition held [by Harriet Wynter] June 13th–22nd*. . . , London, 1974.

TURNER (III) A.J. TURNER, 'William Oughtred, Richard Delamain and the Horizontal Instrument in Seventeenth Century England,' *Annali dell'Istituto e Museo di Storia della Scienza di Firenze*, vi, 1981, 99–125.

TURNER (IV) A.J. TURNER, 'The Mathematical Practitioner Richard Delamain and his son Richard Delamain the Younger, Mathematician and Radical Preacher' (in preparation).

TURNER (V) A.J. TURNER, *Time Measuring Instruments, part 3: Water-clocks, Sand-glasses, Fire-clocks* (The Time Museum, Catalogue of the Collection – Vol. I), Rockford, 1984.

TURNER (VI) A.J. TURNER, 'A Note on the Life of Hilkiah Bedford,' *Bulletin of the Scientific Instrument Society*, iv, 1984.

TYACKE & HUDDY SARAH TYACKE & JOHN HUDDY, *Christopher Saxton and Tudor Map-making*, London, 1980.

URSCHLECHTER ANDREAS URSCHLECHTER, *500 Jahre Regiomontan: 500 Jahre Astronomie*, Nuremberg, 1977.

VIELLARD JEANNE VIELLARD, 'Instruments d'astronomie conservés à la bibliothèque du college de Sorbonne au XVe et XVIe siècles,' *Bibliothèque de l'Ecole des Chartes*, cxxxi, 1973, 587–93.

VOGT & SCHRAMM JOSEPH VOGT & MATTHIAS SCHRAMM, 'Synesius vor dem planisphaerium,' in *Das Altertum und jedes neue Gute. Für Wolfgang Schadewaldt, zum 15. März 1970*, Stuttgart, Berlin, Cologne, & Mainz, 1970.

VOIGT J. VOIGT (ed.), *Briefwechsel der berühmtesten Gelehrten des zeitalters der Reformation mit Herzog Albrecht von Preussen 1541–1544*, Königsberg, 1841.

VYVER A. VAN DER VYVER, 'Les premières traductions latines (X–XIe siècles) de traités arabes sur l'astrolabe,' *1e Congrès Internationale de Géographie & Histoire, Tome II, Mémoires*, Brussels, 1931.

WARD F.A.B. WARD, *A Catalogue of European Scientific Instruments in the Department of Medieval and Later Antiquities of the British Museum*, London, 1981.

WATERS DAVID W. WATERS, *The Art of Navigation in England in Elizabethan and Early Stuart Times*, London, 1958.

WEBSTER RODERICK S. WEBSTER, *The Astrolabe. Some Notes on its History, Construction and Use*, Lake Bluff, 1974.

WEBSTER & WEBSTER R[ODERICK] S. & M[ARJORIE] K. WEBSTER, *An Index of Western Scientific Instrument Makers to 1850*, Winnetka, 1968 (in progress).

WEDEL T.O. WEDEL, *The Medieval Attitude toward Astrology, particularly in England* (Yale Studies in English), New Haven, 1920.

WELBORN M.C. WELBORN, 'Lotharingia as a Center of Arabic and Scientific Influence in the 11th Century,' *Isis*, xvi, 1931.

WHITE LYNN WHITE, JR. 'Medical Astrologers and Late Medieval Technology,' *Viator*, vi, 1975, 295–308. Reprinted in Lynn White, Jr., *Medieval Religion and Technology: Collected Essays*, Berkeley, Los Angeles, & London, 1978, 297–315.

WICKERSHEIMER ERNEST WICKERSHEIMER (ed.), *Recueil des plus célèbres astrologues et quelques hommes doctes faict par Symon de Phares du temps Charles VIIIe, publié après le manuscrit unique de la Bibliothèque Nationale*, Paris, 1929.

WILLERS & HOLZAMER JOHANN WILLERS & KARIN HOLZAMER, *Treasures of Astronomy, Arabic and German Instruments of the German National Museum*, Nuremberg, 1983.

WRIGHT R. RAMSEY WRIGHT, (ed. & tr.) *The Book of Instruction in the Elements of the Art of Astrology by Abu'l-Rayḥān Muḥammad Ibn Aḥmad al-Bīrūnī*. . . , London, 1934.

WYNTER HARRIET WYNTER, *A Catalogue of Scientific Instruments*, II, pt. 3, 1974.

WYNTER & TURNER HARRIET WYNTER & ANTHONY TURNER, *Scientific Instruments*, London, 1975.

ZAMBELLI PAOLO ZAMBELLI, *Astrologia, magia e alchimia nel Rinascimento fiorentino ed europeo* (catalogue of an exhibition forming part of *Firenze e la Toscana dei Medici nell' Europa del Cinquecento*), Florence, 1980.

ZINNER (I) ERNST ZINNER, *Geschichte und bibliographie der astronomischen literatur in Deutschland zur zeit der renaissance*, Leipzig, 1941.

ZINNER (II) ERNST ZINNER, *Deutsche und Niederländische astronomische Instrumente des 11–18 Jahrhunderts*, Munich, 1967.

Concordance

Inventory numbers	Catalogue numbers
308	4
377	30
430	11
507	1
545	29
556	34
610	18
765	13
985	28
1004	32
1025	25
1056	12
1433	3
1436	7
1437	8
1438	20
1445	33
1527	27
1614	26
1752	14
1849	6
1935	23
2292	19
2393	10
2505	24
3123	17
3170	9
3171	21
3172	5
3173	15
3392	16
3407	2
3528	31
3529	22

Index

Numbers in *italics* are those of illustrations or their captions. In the alphabetization of Arabic names, which have been indexed under the *Ism('Alam)* except where some other element is used, e.g. the *nisba* al-Bīrūnī, the particles *al-*, *ar-*, etc., have been ignored, as have the apostrophe, de, du, le, la, l', etc., in European names.

'Abbas II, Shah of Persia 20 n.63
'Abd al-A'imma 90
'Abd al-Ghafūr 102
'Abd (?Allah) b. Yūsuf 177
Abelard 30
Abīyūn al-Baṭrīq 13 n.20
Abraham ibn 'Ezra 18
Abraham bar Hiyya 18
Abū 'Alī al-Ḥasan b. 'Alī b. 'Umar al-Marrākushī *see* al-Marrākushī
Abu'l-Ḥasan Muḥammad b. al-Ḥusayn, Shaykh 14
Adelard of Bath 32
Aḥmad b. Ibrahīm 22, 63
Aḥmad b. Khalaf 60, 62, 63
Aḥmad b. as-Sarrāj 156 n.327
'Alamgīr I 84
Albrecht, Duke of Prussia 43
Alfaqui, Abraham su 156
Alfonso X (the Wise) of Castile 32, *170*
'Alī, Ḥājjī 96–9, *76–9*
'Alī b. Ibrahīm al-Harrār 155 n.314
'Alī b. Ibrāhīm of Taza 27 n.84
'Alī b. 'Isā al-Asṭurlābī 20 n.64
'Alī b. Khalaf b. Aḥmar, 152–6, 160
'Alī al-Wadā'ī 168, 178
Allāh-dād Āsṭurlābī Humāyūnī Lāhūrī 26
Allen, Elias 49, 194, 198, 208
Alphonsine tables 156
Ammonius 13
anaphoric clocks 10 & n.11, 12, 13 n.19, 44
Andalo di Negro 52, 205
Antoninus Pius, medallion of 10, *13*, 10–11
Antwerp: instrument-making centre 45
Apian, Peter 158, 192;
  Cosmographia 132, *117*, 205
Apian, Philip 192
al-Aráuigo, Bernardo 156
Arialdus 19
armillary sphere, observing 11 n.13, 13
Arnaud of Brussels 32
Arnault, Henri 37, 40 & n.130, 83 n.233
Arsenius, Ferdinand 48, 139 n.295
Arsenius, Walter (Gualterus) 47, 48 & n.163, 49, 50, 64, 140, 158;
  astrolabe by 132–9, *115–20*;
  identified with 'Regnerus Arsenius' 139
Astrolabe (son of Héloise and Abelard) 30
astrolabe quadrants 202–10;
  history of 203–10;
  operation of 202–3, *170*;
  treatises on 203–5, 206, 208;
  and Islam 205–6;
  quadrans novus (Prophatius' quadrant) 202 & n.22, 203–6, *167*, *168*;
  Gunter's quadrant 206–8, *169*, 210, 215, 216, 218, *177*, 223, *179*;
  Sutton's 210;
  English 215–28, *173–84*;
  Turkish 212–4, *171–2*
astrolabes 1–187;
  as clock dials 44;
  functions of 1, 7 & n.6;
  –, astrological 7, 22, *19*, *20*, 30 & n.99, 32 & nn.102 & 103, 33 & n.111, 40, 44–5, 45 n.150, *see also* astrolabes, astrological;
  –, navigational 40, 57, 132–8;
  history of 10–57;
  instruments related to 191–228;
  made of paper 43–4, 44 n.146, 50, *36*, *39*, 143–9, *131–3*, 160 & n.348, 166;
  as 'mathematical jewel' 40 & n.129, *138*, 160;
  and medieval universities 30–1, 33, 38;
  metal for 41, 45;
  operation of 7, *12*;
  parts of *see under* astrolabes, planispheric, for single latitudes, *and Glossary*;
  production of in medieval Europe 36–7, 37–41, 42, 43, 44–50;

superseded 56–7;
transmission of knowledge of 13–23, 25, 26, 29, *see also* astrolabes, treatises on;
–, Jewish role in 18–9;
treatises on: Arabic 13, 15, 19, 21, 22, 23, 155–6;
–, Castilian 156, *see also* –, Spanish;
–, Dutch 55–6;
–, early printed 50, 52–3 (Table 1), 54;
–, English 34, 40 n.129, 54, 160;
–, French 34;
–, German 54–5;
–, Greek 12, 13;
–, Hebrew 18–9;
–, Indian 26–8;
–, Italian 55, 156;
–, Latin 16, 17, 17–8, 19–20, 50–4, 156–8;
–, Persian 156;
–, Spanish 15, 23, 55;
–, Syriac 13;
–, Turkish 156;
types of 1, *see also* astrolabes, linear, *etc.*;
(by origin) Byzantine 14–15 & n.30;
European 29–50, 56–7;
English 148–9, 131–3;
Flemish 132–9, *115–20*;
French (?) 124–7, *105–9*, 146–7, *127–30*;
German 128–31, *110–4*;
Italian 124–7, *105–9*, 140–5, *121–6*;
Spanish 23, *22*, *23*, 64–9, *44–8*, *see also* –, Maghribi;
Indian 25–9, 112–23, *93–104*;
Indo-Persian 25–6, 74–85, *54–63*;
Islamic *14*, 7, 13, 14, 15–16, 20–8, 29, *22*, *23*, 30, 60–3, *41*, *42*, *see also* –, Magribi *and* –, North African;
Maghribi 70–3, *49–53*, 108–11, *89–92*;
North African 23;
Persian 21, 22, 24–5, *20*, 86–107, *64–88*
astrolabes, astrological 229, 232–7, 185–7
astrolabes, linear 184–5, *158*
astrolabes, mariner's 1 n.2, 56 n.196; *see also* astrolabes, functions of, navigational
astrolabes, 'particular' *see* astrolabes, planispheric, for single latitudes
astrolabes, planispheric, for single latitudes 1–149, 151;
construction of 1–5, *1–3*, *4*, *5*;
terminology of 2–5 & nn.3 & 4, 15;
parts of, alidade 5, *9*, *10*, *11*, 21;
–, back *5*, *6*;
–, horse *(faras)* 5, *9*;
–, *kursi* (throne) 3 & n.4;
–, limb *(ḥajra)* 3;
– *mater (umm)* 2–3;
–, *rete ('ankabūt)* 5, *7*;
–, –, different designs of 34–7, *27*, 42;
–, rule 5, *9*;
–, scales 3, *5*, *8*;
–, shackle *('urwa)* 3;
–, suspension ring *(ḥalqa)* 3
astrolabes, planispheric, for general use, *see* astrolabes, universal
astrolabes, spherical 185–6, *159*
astrolabes, 'Stoffler's' 50, 54 n.183;
astrolabes, universal 151–83, *140*;
operation of 155;
treatises on 152, 154, 155–60;
English 182–3, *156–7*;
Flemish 180–1, 154–5;
Maghribi 174–9, *147–53*;
Syro-Egyptian 168–73, *145–6*
Aubrey, John 45 n.150, 50, 54 n.183, 160
Augsburg: instrument-making centre 41, 45

Basṭūlus 14 & n.28, 60, 62, 63
Bate, Henry 20, 52, 229
Batecumbe, William 156
al-Battānī 63
al-Baṭṭuṭī 23, 178
Beausard, Pierre 47 n. 157
Bedford, Hilkiah 218, 222
Berle, Jean de 37, 40
Bion, Nicholas 50, 56 & nn.194 & 195, 166
al-Bīrūnī 2 n.3, 21, 22 & n.69, 26, 27, *21*, 28, 83, 162, 172, 186
Blaeu, W.J. 50
Blagrave, John 40 n.129, 158, *138*, 160
Blasius, Ermengard 204
Blundeville, Thomas 1, 50, 132, 160
Boissard, J.J. *35*
Bond, Henry 208
Bonetus de Latis *40*, 52
Borcha, Martin 232

Bornmann, Zacharias 55
Bottens, F. 56
Bregny, Jehan de 34
Briggs, Henry 206
Browne, John 223, 225, 226
Browne, Thomas 223–5
Brush, Thomas *162*
de Bry, Theodore 35
Bylica, Martin 29, 42, 158, 164

Campanatius, Vincentius 44
Campanus de Novara 19
Cassini, J. 56 n.195
Chardin, Sir John 20 n.62, *17*, 25 & n.73
Charles V of France 30 n.99, 32 & n.110, 34, 157
Charles VIII of France 32
Chatfield, John 208
Chaucer, Geoffrey 7 n.5, 34 & n.120, *38*, 54
Clavius, Christopher *37*, 54, 163
clepsydrae 10, 14
Coignet, Michel 48, 50
Cole, Humphrey 48, 49, 132
Collections: Bahari 94;
 Brieux 94, 98;
 Devlay 98;
 Greppin 143;
 Hoffman 106;
 Linton 143;
 Nissen 106;
 Price 102;
 Riva 122;
Collins, John 208–10, 222, 226
Commandino, Federico 162–3
Cools, Jacob 139
Copernicus, N. 56
Copp, Johann 54–5
Cross family (of Bledlow House) 198
Cues, Nicholas of *see* Nicholas of Cusa
Cuningham, William 132

Damery, Lambert 48, 50, 83 n.233
Danfrie, Phillippe 49, 50
Danti, Egnatio 49, 50, 55, 165
dastūr 17
David, Johannes 18
Delamain, Richard 194
Demonstrationes pro astrolapsu 19
Descrolières, Adrien: astrolabe by 140–5, *121–6*
devenagāri script 27 n.84, 28
dioptra 13 & n.18
Ḍīyā' ad-Dīn Muḥammad b. Qā'im Muḥammad *see* Muḥammad b. Qā'im Muḥammad
Dorn, Hans 29, 42, 158, 164
Dryander, J. *40*
Dudley, Sir Robert 192
Dunas ibn Tamīm 14

Eligerus of Gondesleren 205
equatoria 41 & n.132

Faber Barduvicensis 52
Fabritius, Sebastian *136, 141*
Faḍl Allāh as-Sabawarī 20 n.63
al-Farghāni 14
Fatḥ 'Alī, Shāh 104
al-Fazārī 13
Fine, Oronce 158, 162, 205
Focard, Jacques *6, 8, 39*, 55
Foster, Samuel 208
Frederick II, Emperor 32
Fugger family 45
Fulbert of Chartres 17
Fusoris, Jean 34, 37, 38–40, 50, 83 n.233; astrolabe from workshop of *28*

Galand, François 232
Galileo Galilei 56
Gallucci, Giovani Paolo 55
Gaurico, Luca 44
Gemini, Thomas 48, 49, 132
Gemma, Cornelius 47 n.154
Gemma Frisius 45–7, *34*, 48, 49, 132, 138, 139, 140, 160 & n.347, 164;
 de Astrolabo 136–8, 158;
 astrolabes by *142, 143*
Gerbert (Pope Sylvester II) 16 & n.37, 29 nn.92 & 95; 32 n.109
Gilliszoon, Guillaume 40–1
Glareau, Henry 232
Godby, Thomas 54
Gordon, Robert, of Straloch 50
Graevius, Bartholemew 47
Gregoras, Nicephoras 52, 54
Gregory, James 222

Grüninger, J. 50
Guicciardini, Luigi 139
Gunter, Edmund 206, 218–22;
quadrant named after *see* under astrolabe quadrants

Ḥabash al-Ḥāsib 172
Habermel, Erasmus 45 n.150, 49, 132
Habermel, Joshua 192
Ḥāmid b. al-Khiḍr al-Khujandī *see* al-Khujandī
al-Harīrī: miniature from *Maqāmat 19*
Ḥarrān: astrolabe-making centre 13, 14
Harris, John 56 n. 196
Hartmann, Georg 42–3, 50, 148, 192;
astrolabe by 128–31, *110–14*
Harvey, Thomas 208, 210
Hayton family (of Bledlow House) 198
Héloise 30
Helt, Hugo 49, 164
Henry II of England 32
Henry VII of England 32
Hermannus Contractus 17, 204
Hermannus Dalmatus (Hermann the Dalmatian) 10 n.10, 18
Hero of Alexandria 162
Heyden, Gaspar van der (Gaspar a Myrica) 47, 138
Hipparchus of Nicaea 10 & n.7, 12
Hire, Philippe de la 166, *144*
horizontal instruments 191–5, *160, 162, 163,* 206;
uses of 194–5, 200;
known by name of 'Triens' 192;
American ('Sundicator') 200–1, *165*;
English 198–9, *164*
Howyan, G. 195
Hulett, John 208
Humāyūn 26

Ibn an-Nadīm 13 & n.20
Ibn aṣ-Ṣaffār 15 & n.33, 16, 18, 19, 23
Ibn ash-Shāṭir 177–8
Ibn aṣ-Ṣufī 16
Ibn Tibbon, Ya'aqob b. Mahir 19, 156, 158;
and Prophatius quadrant 203–5, *167, 168*
Ibrāhīm ad-Dimashqī 177
Isaac b. Sid 186

Jacquinot, Dominicque *7, 10,* 44, 55
Jāḥiẓ 14
Jaipur: astrolabes at 28–29
Jai Singh II, Mahārāja 28
Jamāl ad-Dīn b. Muḥammad 78
Johannes Hispalensis *see* John of Seville
Johannes de Machlinea 158
John of Brescia 156
John of Gmunden 157, 205
John of Lignères 156, 158
John Philoponus 13
John of Seville 18, 19, 50
Jordan, Peter *36*
Jordanus de Nemore 19

Kepler, J. 56
Khafīf 60, 62, 63
al-Khamā'irī 50, 162, 174 n.383, 177;
astrolabe by 64–9, *44–8*
al-Khujandī 20 n.62, *16*, 22, 63, 192
al-Khwārizmī 21, 186
Knibb, Samuel 226
Köbel, Jacob 54
Krabbe, Johann 50, 55
kufic script 21 & n.68, 22, 25, 27 n.84

Lahore: astrolabe-making centre 26, 28, 78, 82–3, 84, 122
lamina universal 153, 156 & n.327, 160
Lansberg, Philip 56
Laud, Archbishop William 42
Levi ben Gerson 29 n.91
Libros del Saber 153, 156, 162, 165, 186
Linden, J.A. 192
Llobet of Barcelona 16, 17
Louvain: instrument-making centre 45, 47, 48, 49, 132
Lucas de Heere 43, *31*
Lutf'alī b. 'Abd an-Nabī 94

Madana Sūri 26
Maelcote, Odon van 83 n.233, *142, 143*
Mahendra Sūri 26–7
al-Majrīṭī 10 n.10, 16 & n.34, 18
Malayendu Sūri 27
Maqṣūd Hirawī 26
Marcel, Jean-Joseph 73
Maricourt, Pierre de *see* Petrus Peregrinus

al-Marrākushī 83, 152, 156, 184, 186, 192
Māshā'allāh 18, 19
Maslama b. Aḥmad al-Majrīṭī *see* al-Majrīṭī
Maurolyco, Francesco 232
Mercator, Gerard 47, 138;
  humanistic script of 47, *32*, *33*, 48
Metius, Adrian 55–6
al-Mizzī 205
Montanus, Arias 49
Morgard, Noel Leon 232, 236–7, *187*
Morland, Sir Samuel 226
Mosé Sefardi (Pedro Alfonso) 18 & n.48
Moxon, Joseph 160
Muḥammad b. Aḥmad b. 'Abd ar-Raḥīm al-Mizzī
  see al-Mizzī
Muḥammad b. Aḥmad al-Baṭṭuṭī *see* al-Baṭṭuṭī
Muḥammad Akbar: astrolabe by 104–7, *85–8*
Muḥammad Bāqir Iṣfahanī 90
Muḥammad b. Fattūḥ al-Khamā'irī *see* al-Khamā'irī
Muḥammad b. Ibrāhīm 22, 63
Muḥammad b. 'Īsà 78, 82–3
Muḥammad Khalīl 90
Muḥammad Laḥbābī 23
Muḥammad Mahdī al-Khādim al-Yazdī: astrolabe
  by 86–90, *65–70*
Muḥammad b. Mufaḍḍil b. Aḥmad b. Kirān: astro-
  labe by 108–111, *89–92*
Muḥammad b. Muḥammad b. Hudayl 162, 177
Muḥammad b. Muḥammad aṭ-Ṭanūkhī 156 n.327
Muḥammad Muḥsin b. Muḥammad 'Alī ash-Sharīf
  al-Kirmānī: astrolabe by 92–5, *71–5*
Muḥammad Muqīm b. 'Īsà 78
Muḥammad b. Muqīm al-Yazdī 20 n.63, 90
Muḥammad b. Qā'im Muḥammad 78, *60*;
  astrolabe by 84–5, *62–3*
Muḥammad b. aṣ-Ṣaffar 15, *14*, 16, 23, 29, 177
Muḥammad Shāfī 20 n.63
Muḥammad Shāh 28
al-Muḥsin b. Muḥammad: astrolabe by 60–3, *41–3*
al-Mulk, Niẓām 92 n. 242
Mūsà: spherical astrolabe by 185, *159*
Myrica, Gaspar a *see* Heyden, Gaspar van der

nashkī script 21 n.68, 22, 25
nasta'līq script 25
an-Nayrīzī 186
Newton, Sir Isaac 56
Nicholas of Cusa 37, 41
Nightingale, Peter 204
nocturnal (constellation volvelle) *178*, 221–2
Nuremberg: instrument-making centre 41, 42, 45, 49

Olearius, Adam 55
Oliver, Thomas 54
Oresme, Nicholas 37 & n.125
Ortelius, A. 139
Ottheinrich, Duke 43
Oughtred, William *161*, 192–4, 198

Palmer, John 160
'Pantocosme' *see* astrolabes, astrological
Parron, William 32
Pepys, Samuel 54 n.183
Pedro Alfonso *see* Mosé Sefardi
Peter of St. Omer 204; *see also* Nightingale, Peter
Petrus Peregrinus (Pierre de Maricourt) 19–20, 156
Petty, William 228
Peurbach, Georg 41–2, 158
Phares, Symon de 32
Piquet, G. 48
Pisan, Christine de 32 n. 110
Pisan, Thomas de 32 & n.102
Plantin, Christopher 47, 138
Plato of Tivoli (Plato Tiburtinus) 15 n.33, 18
Poblacion, J.M. 54, 55
projections: horizontal 191–4, *162*;
  stereographic 1–2, *1*, *3*, 10, *13*, 11, 12, 151, 191;
  –, on quadrants *166*, *167*, 202–3, 206, 208;
  universal (de la Hire) *144*;
  universal orthographic *139*, 160–5, *141*
  universal stereographic *134*, 151–3
Prophatius Judaeus *see* Ibn Tibbon
Prophatius quadrant *see under* Ibn Tibbon
Prosdocimo de Beldomandi 52
Prujean, John 50, 148
Prusse, Pelerin de 34
Pseudo-Māshā'allāh *9*, 7 n.5, 34, *29*, 50, 55 n.191
Ptolemy 11 n.13, 156, 229;
  Analemma 162;
  Planisphaerium 10 & n.10, 12, 18

Qimh family 19
quadrans novus see under astrolabe quadrants
quadrants *see* astrolabe quadrants

Quadratum Nauticum 132–6, *117, 118*, 140
Qur'ān: 'throne' verse from 86
Qustà b. Lūqā 186

Ragimbold 17
Raymond of Marseilles 19, 33
Reeves, E.A. 166
Regiomontanus 41, 42, 157, 158, 232
Regnartius, Valerianus *142, 143*
Reichenau 17
Reisch, Gregor 50
Remi of Trier 29 n.95
René of Anjou 34, 40, 41
Richard of Wallingford 37, 157
Ripoll: astrolabe treatise from 16, *15*
Ritter, Franciscus 55
Robert of Ketton (Robert of Chester) 18, 19, 52
Rojas, Juan de 49, 143, *140*, 164–5;
  astrolabe named after 162, *142, 143*;
  –, modern imitation of 182–3, *156–7*
Rudolph of Bruges 18, 19
Rudolph of Liège 17
Ryther, Augustin 48–9

Sacrobosco 30, 34 n.120
Sā'dī: *Gulistan*, quoted 89
Ṣafī Qulī Beg 90
Ṣā'id al-Andalusī 152, 154
saphea Azarchelis *135, 136*, 155, 156–8, *137*, 163, 165–6;
  Maghribi 174–9, *147–53*;
  Syro-Egyptian 168–73, *145–6*;
  see also astrolabes, universal
Savasorda, Abraham 19
Schissler, Christoph (the Elder) 192
Schöner, Andreas 55
Schöner, Johann 43, 158
Scot, Michael 2 n.3, 32 & n.109
Selden, John 34 n.120
Sebokt, Severus 13
Shāhinshāh-nama, miniature from *18*
ash-Shajjar *see* 'Alī b. Khalaf b. Aḥmar
ash-Shakkaz *see* 'Alī b. Khalaf b. Aḥmar
Sharaf ad-Dīn al-Muẓaffar b. Muḥammad b. al-Muẓaffar aṭ-Ṭūsī 184
as-Sijzī 28, 83
Simler, Josias *136*, 158, 163, *141*
Sorbonne 30
Spencer, Thomas 195 n.18, 200
Stalburch, Jan van *34*
Stempel, Gerard 158
Stephanus the Philosopher 13
Stibonius, Andreas (Stöberl) 158
Stöffler, Johann *35*, 50–4, 160, *142, 143*
Stone, Edmund 210
aṣ-Ṣūfī 28, 63, 83
sundials, portable 10–1, *13*, 12
Sutton, Henry 50, 208–10, 225;
  quadrant named after 210;
  quadrant made by 226–8, *183–4*
Sylvester II, Pope *see* Gerbert
Synesius of Cyrene 10 n.7, 12 & n.17

Tanner, Robert 54
Taqī ad-Dīn, observatory of *18*
Theon of Alexandria 12, 13
Thompson, John 222
Thynne, William 54
Triens 192, *see also* horizontal instruments
Tunsted, Simon 157

'Umar Khayyām 92 n.242

Vitruvius 10, 44, 162, 163, 164
della Volpaia family 45, 49
Volmar, Johannes 158
Vopel, Caspar 50

Walcher of Lorraine 17, 18 n. 48
al-Wāsitī, Yahyā b. Muḥmud *19*
water-clocks *see* clepsydrae
Werner, Johann 158, 192
Whitwell, Charles 49
William the Englishman 156
William of Moerbeke 162, 229

Yahyā b. Maḥmud al-Wāsitī *see* al-Wāsitī, Yahyā b. Maḥmud
Yaḥyā-Ma'mūn b. Ismā'īl 152, 154, 156
Yehuda bar Mosé 156

az-Zarquellu 152, 153, 154, 155, 156, 158, 172
Zeelst, Adrian 48, 158
Ziegler, Jacob 158

Anthony J. Turner

After reading History at Oxford, Anthony Turner concentrated on research in the fields of early scientific instruments, clocks and watches, and the social history of science. He has worked in several museums in Great Britain and is the author of a number of books and articles, including *The Clockwork of the Heavens* (1973); *Scientific Instruments* (1975, with Harriet Wynter); *Science and Music in 18th Century Bath* (1977). He is currently working on a general history of instruments before *c.* 1780. When not engaged in historical research, Anthony Turner organizes exhibitions and is a partner in an antiquarian-book business based in Greenwich (England). He has lived in France since 1979.

Seth G. Atwood

Founder and Director of The Time Museum, Seth G. Atwood has been a resident of Rockford, Illinois, all his life. He graduated Phi Beta Kappa, with a B.A. in Economics from Stanford University in 1938. He was awarded the M.B.A. from Harvard University Graduate School of Business in 1941 and served with the U.S. Navy from 1942 through 1946. A businessman involved principally in manufacturing and banking, Seth Atwood is the owner of the Clock Tower Inn and The Time Museum. He is a Fellow of the National Association of Watch and Clock Collectors, member of U.S. and English Antiquarian Horological Societies, and member of The International Society for the Study of Time.

Bruce Chandler

The general editor of the Catalogue of the Collection, Bruce Chandler, is Professor of Mathematics at the City University of New York. He received a Ph.D. from New York University in 1962 and has written and edited articles and books in the fields of horology, Group Theory, and the history of mathematics. He is editor of *Horological Dialogues* and founding editor (with Harold Edwards) of *The Mathematical Intelligencer*.

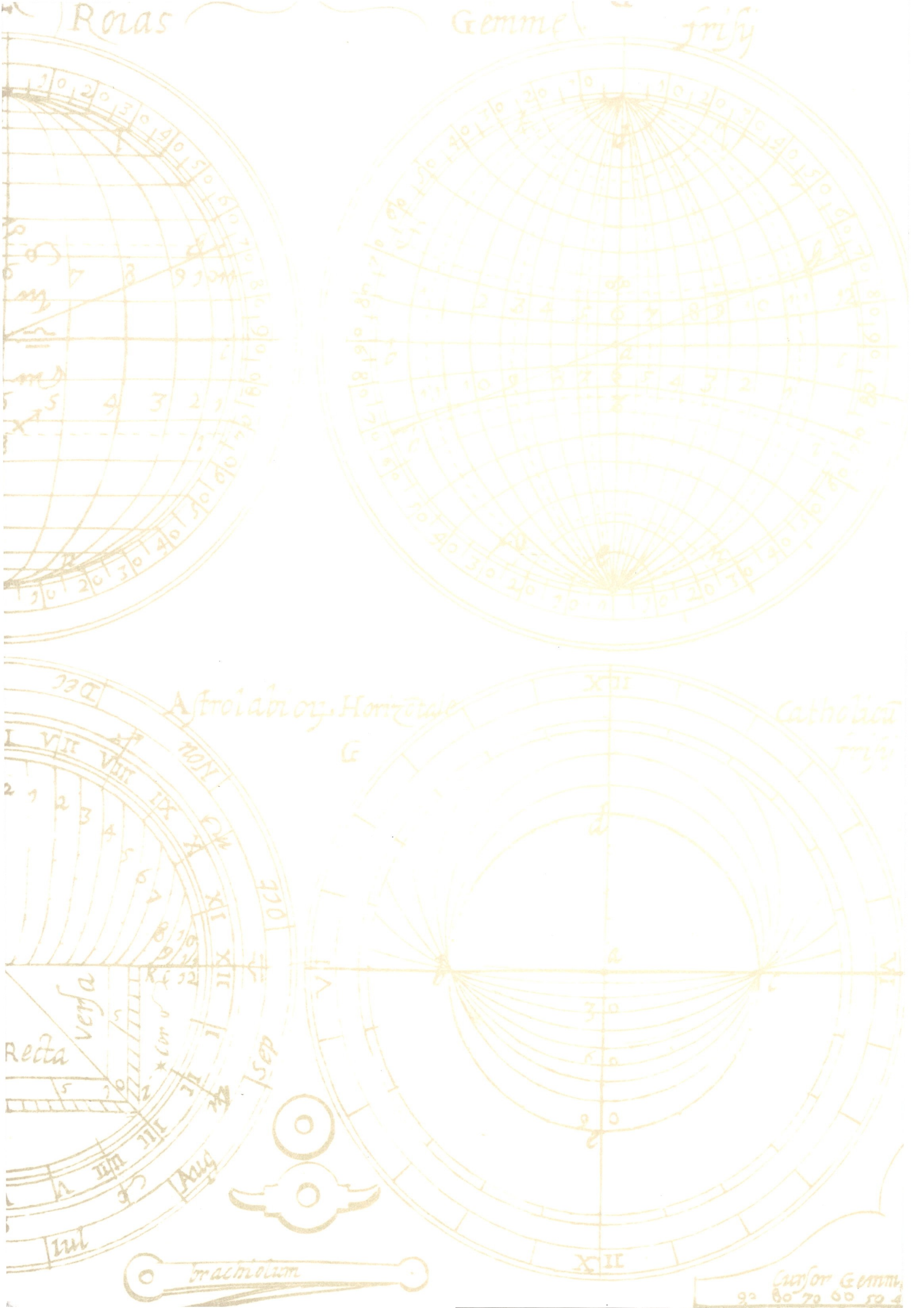
Rotas
Gemme
Recta
versa